Gel Permeation Chromatography

Gel Permeation Chromatography

Edited by *KLAUS H. ALTGELT*

CHEVRON RESEARCH COMPANY
RICHMOND, CALIFORNIA

and

LEON SEGAL

SOUTHERN REGIONAL RESEARCH LABORATORY
U.S. DEPARTMENT OF AGRICULTURE
NEW ORLEANS, LOUISIANA

1971

MARCEL DEKKER, INC., New York

CHEMISTRY

Preface

Published herein are proceedings of the American Chemical Society Symposium on Gel Permeation Chromatography. These papers are collected from issues of *Separation Science*, where they first appeared in 1970–71. The Symposium constituted part of the 159th ACS National Meeting in Houston, February 22–27, 1970, and was sponsored jointly by the Divisions of Analytical Chemistry, Petroleum Chemistry, and Cellulose, Wood and Fiber Chemistry. The organizers, who introduce the Symposium on page v, assembled an outstanding collection of authors with unusually wide interests. Their summed contributions mark an important advance in the field of GPC and, more broadly, in the entire field of macromolecular separations.

J. CALVIN GIDDINGS
Executive Editor

Introduction

Gel permeation chromatography has grown at a stunning rate both in technical sophistication and in the scope of application since J. C. Moore made it practical around 1964. GPC has now gained a key position in polymer chemistry. It is advancing rapidly to find its place among the important chromatographic methods in new fields outside polymer science where its unique ability to separate by molecular size can also be put to profitable use, for example in cellulose chemistry and technology.

In attempting to bring together recent progress in GPC, a four-day symposium was organized jointly by the Divisions of Petroleum Chemistry, Analytical Chemistry, and of Cellulose, Wood, and Fiber Chemistry of the American Chemical Society at the 159th National Meeting of the Society held in Houston, Texas, in February 1970. The papers presented demonstrated once again the manifold applications GPC has found in the polymer field, and they show further how it supplements existing methods or opens up new approaches in petroleum and cellulose chemistry. Indeed, part of this symposium was held with the purpose of exposing the petroleum chemist in particular to this new, powerful tool and to indicate to him some of the results that have been obtained with its use.

In setting up the symposium it appeared desirable to provide an introduction that would allow a chemist who had never worked with GPC to understand the basics and then to enable him to follow the advanced presentations. Therefore, the symposium was divided into four parts:

1. A simple, very basic introduction for the novice.
2. A lucid, yet sophisticated review of the latest theories and evaluation methods.
3. New developments in technique.
4. Applications in polymer, cellulose, and petroleum chemistry.

Hopefully, this approach will serve to familiarize many chemists with GPC and help them to conduct their research or process control even more efficiently than before.

Klaus H. Altgelt
Leon Segal

Contributors

H. E. ADAMS, Central Research Laboratories, The Firestone Tire & Rubber Company, Akron, Ohio 44317

ALAN D. ADLER, New England Institute, Ridgefield, Connecticut 06877, and Department of Internal Medicine and Howard Hughes Laboratories for Hematologic Research, University of Miami, School of Medicine, P.O. Box 875—Biscayne Annex, Miami, Florida 33152

E. W. ALBAUGH, Gulf Research & Development Company, Pittsburgh, Pennsylvania 15230

W. J. ALEXANDER, Eastern Research Division, ITT Rayonier, Inc., Whippany, New Jersey 07981

DAVID F. ALLIET, Materials Analyses Area, Xerox Corporation, Rochester, New York 14603

KLAUS H. ALTGELT, Chevron Research Company, Richmond, California 94802

M. R. AMBLER, Chemical Materials Development, The Goodyear Tire and Rubber Company, Akron, Ohio

EDWARD M. BARRALL II, IBM Research Laboratory, San Jose, California 95114

K. C. BERGER, Institut für Physikalische Chemie der Universität Mainz, Mainz, West Germany

FRED W. BILLMEYER, JR., Department of Chemistry, Rensselaer Polytechnic Institute, Troy, New York 12181

F. A. BLOUIN, Southern Regional Research Laboratory, New Orleans, Louisiana 70119

KARL J. BOMBAUGH, Waters Associates, Inc., Framingham, Massachusetts 01701

J. H. CAIN, Chevron Research Company, Richmond, California 94802

BRUCE F. CAMERON, Department of Internal Medicine and the Howard Hughes Laboratories of Hematologic Research, University of Miami School of Medicine, P.O. Box 875—Biscayne Annex, Miami, Florida 33152

EDWARD F. CASASSA, Mellon Institute and Department of Chemistry, Carnegie-Mellon University, Pittsburgh, Pennsylvania 15213

R. H. CASPER, Institut für Physikalische Chemie der Universität Mainz, Mainz, West Germany

JAMES H. CLARK, Monsanto Company, St. Louis, Missouri

HANS COLL, Shell Development Company, Emeryville, California 94608

A. R. COOPER, Chevron Research Company, Richmond, California 94802

B. E. DAVIS, Gulf Research & Development Company, Pittsburgh, Pennsylvania 15230

F. E. DICKSON, Gulf Research & Development Company, Pittsburgh, Pennsylvania 15230

J. N. DONE, BP Research Centre, The British Petroleum Company Limited, Sunbury-on-Thames, Middlesex, England

BARRY F. DOWDEN, Research Division, IBM Corporation, San Jose, California 95114

E. E. DROTT, Monsanto Company, Texas City, Texas 77590

J. H. DUERKSEN, Chevron Research Company, Richmond, California 94802

JOHN DYER, Research and Development, American Viscose Division, FMC Corporation, Marcus Hook, Pennsylvania 19061

VERONIKA GREENFIELD, New England Institute, Ridgefield, Connecticut 06877, and Department of Internal Medicine and Howard Hughes Laboratories for Hematologic Research, University of

Miami, School of Medicine, P.O. Box 875—Biscayne Annex, Miami, Florida 33152

ANNIE R. GREGGES, Research Division, IBM Corporation, San Jose, California 95114

W. E. HAINES, Laramie Petroleum Research Center, Bureau of Mines, U.S. Department of the Interior, Laramie, Wyoming 82070

DALE J. HARMON, The B. F. Goodrich Research Center, Brecksville, Ohio 44141

E. HIRSCH, Chevron Research Company, Richmond, California 94802

TERUO T. HORIKAWA, Research Division, IBM Corporation, San Jose, California 95114

JULIAN F. JOHNSON, Department of Chemistry and Institute of Materials Science, University of Connecticut, Storrs, Connecticut 06268

RICHARD N. KELLEY, Departments of Materials and Chemistry, Rensselaer Polytechnic Institute, Troy, New York 12181

D. R. LATHAM, Laramie Petroleum Research Center, Bureau of Mines, U.S. Department of the Interior, Laramie, Wyoming 82070

ROBERT F. LEVANGIE, Waters Associates, Inc., Framingham, Massachusetts 01701

JAMES N. LITTLE, Waters Associates, Inc., Framingham, Massachusetts 01701

C. P. MALONE, Engineering Physics Laboratory, E. I. Du Pont de Nemours & Co., Inc., Wilmington, Delaware 19898

L. F. MARTIN, Southern Regional Research Laboratory, New Orleans, Louisiana 70119

R. D. MATE, Chemical Materials Development, The Goodyear Tire and Rubber Company, Akron, Ohio

R. A. MENDELSON, Monsanto Company, Texas City, Texas 77590

G. MEYERHOFF, Institut für Physikalische Chemie, Universität Mainz, Mainz, West Germany

J. C. MOORE, Basic Research Department, Texas Division, The Dow Chemical Company, Freeport, Texas 77541

T. E. MULLER, Eastern Research Division, ITT Rayonier, Inc., Whippany, New Jersey 07981

NOBUYUKI NAKAJIMA, Plastics Division, Allied Chemical Corporation, Morris Township, New Jersey 07960

S. W. NICKSIC, Chevron Oil Field Research Company, La Habra, California 90631

H. H. OELERT,* Laramie Petroleum Research Center, Bureau of Mines, U.S. Department of the Interior, Laramie, Wyoming 82070

JEANNE M. PACCO, Materials Analyses Area, Xerox Corporation, Rochester, New York 14603

WILLIAM J. PAUPLIS, Waters Associates, Inc., Framingham, Massachusetts 01701

L. H. PHIFER, Research and Development, American Viscose Division, FMC Corporation, Marcus Hook, Pennsylvania 19061

THEODORE PROVDER,† Monsanto Company, St. Louis, Missouri 63166

W. K. REID, BP Research Centre, The British Petroleum Company Limited, Sunbury-on-Thames, Middlesex, England

EDWARD M. ROSEN, Monsanto Company, St. Louis, Missouri 63166

S. P. ROWLAND, Southern Regional Research Laboratory, New Orleans, Louisiana 70119

JAMES R. RUNYON, The Dow Chemical Co., Midland, Michigan 48640

G. V. SCHULZ, Institut für Physikalische Chemie der Universität Mainz, Mainz, West Germany

LARRY SKLAR, New England Institute, Ridgefield, Connecticut 06877, and Department of Internal Medicine and Howard Hughes Laboratories for Hematologic Research, University of Miami, School of Medicine, P.O. Box 875—Biscayne Annex, Miami, Florida 33152

C. A. STOUT, Chevron Oil Field Research Company, La Habra, California 90631

Permanent address: University of Clausthal, Clausthal-Zellerfeld, Germany.
† *Present address:* Dwight P. Joyce Research Center, Glidden-Durkee Division. SCM Corporation, 16551 Sprague Road, Strongsville, Ohio 44136.

H. L. SUCHAN, Engineering Physics Laboratory, E. I. Du Pont de Nemours & Co., Inc., Wilmington, Delaware 19898

P. C. TALARICO, Gulf Research & Development Company, Pittsburgh, Pennsylvania 15230

L. H. TUNG, Physical Research Laboratory, The Dow Chemical Company, Midland, Michigan 48640

JAMES L. WATERS, Waters Associates, Inc., Framingham, Massachusetts 01701

J. H. WEBER, Laramie Petroleum Research Center, Bureau of Mines, U.S. Department of the Interior, Laramie, Wyoming 82070

R. A. WIRKKALA, Gulf Research & Development Company, Pittsburgh, Pennsylvania 15230

JAMES C. WOODBREY, Monsanto Company, St. Louis, Missouri 63166

W. W. YAU, Engineering Physics Laboratory, E. I. Du Pont de Nemours & Co., Inc., Wilmington, Delaware 19898

Contents

IV. Applications of GPC to Problems in Polymer and Petroleum Chemistry

Gel Permeation Chromatography

FUNDAMENTAL INTRODUCTION TO GEL PERMEATION CHROMATOGRAPHY

The Sizes of Polymer Molecules and the GPC Separation*

FRED W. BILLMEYER, JR.

DEPARTMENT OF CHEMISTRY
RENSSELAER POLYTECHNIC INSTITUTE
TROY, NEW YORK 12181

and K. H. ALTGELT

CHEVRON RESEARCH COMPANY
RICHMOND, CALIFORNIA 94802

Summary

This introductory review explains in simplest terms the separation mechanism in GPC and the concept of size as its discriminant. Sample molecules permeate the gel to different degrees depending on their size and are kept out of the solvent stream in the interstices in correspondingly different time ratios. For rigid molecules the size is determined either by the volume or by the most prominent linear dimension. A better approximation seems to be Giddings's "mean external length."

With polymers the decisive size parameter is the hydrodynamic volume. Its calculation from molecular weight must take into account the coiling of the polymer, its flexibility, and its interaction with the solvent. Another important consideration is the statistical nature of polymer properties which results in average values for molecular weight and size. Chain statistics yield polymer sizes that are compatible with pore dimensions of appropriate gels.

Gel permeation chromatography (GPC) is a method to separate molecules by size. Basically, any soluble molecules can be separated by GPC, small ones of less than 100 molecular weight (MW) as well as large ones of several millions MW.

---

*Presented at the ACS Symposium on Gel Permeation Chromatography, sponsored by the Division of Petroleum Chemistry at the 159th National Meeting of the American Chemical Society, Houston, Texas, February, 1970.

3

The separation is usually carried out on columns that are tightly packed with a gel or some other porous material and completely filled with solvent. The same solvent is used to dissolve the sample before introducing it into the column and also for elution. Small sample molecules can diffuse into the pores of the gel, large ones are excluded, others of intermediate size can penetrate some of the larger pores. The molecules are constantly diffusing back and forth between the pores and the interstices. Solvent pumped through the column flows only in the interstices, sweeping along all sample molecules present there. The molecules in the pores stay behind until they diffuse back out. The large molecules which are always or mostly excluded from the pores are, therefore, eluted first; the small ones which are mostly inside the pores come out last.

A species is eluted at a volume exactly equal to the volume available to it in the column. For large completely excluded molecules, the elution volume V_e is equal to the interstitial volume V_0; for small molecules which can completely penetrate all pores of the gel it is equal to the total liquid volume of the column, i.e., equal to the sum of V_0 and the internal (pore) volume V_i. For molecules of intermediate size, the elution volume is:

$$V_e = V_0 + K_d V_i \tag{1}$$

where K_d, the partition coefficient, is equal to the ratio of accessible pore volume to total pore volume:

$$K_d = \frac{V_{i,\text{acc}}}{V_i} \tag{2}$$

Figure 1 illustrates the statement made in Eq. (1). The center shows a schematic of a gel column with the interstitial volume as the core and the internal volume toward the walls. The small column on the right has a volume equal to the interstitial column volume; the large column on the left has the size of the whole gel column. Visualize both model columns, left and right, filled with solvent and an immiscible lighter sample put on top of both. If we drain both columns, the samples will come out at exactly the column volumes, i.e., at V_0 in the case of the small one and at $V_0 + V_i$ in the case of the large one. If a real sample can only penetrate part of the internal volume, its imaginary column volume in the sense of Fig. 1 is $V_0 + V_{i,\text{acc}}$, or $V_0 + K_d V_i$ as stated in Eq. (1). This equation then holds for all cases including those of complete exclusion or complete permeation with

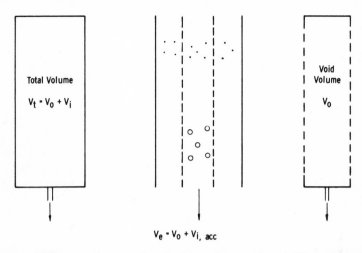

FIG. 1. Elution volume and accessible column volume. Illustration of the equation $V_e = V_0 + V_{i,acc}$ as explained in the text.

$K_d = 0$ and $K_d = 1$, respectively. GPC may thus be considered a special case of partition chromatography where partition occurs between like solvents but different locations, viz., between the spaces outside and inside the gel particles (mobile and stationary phases).

It is generally assumed that under ordinary GPC run conditions, a molecule can diffuse in and out of the gel pores several times before the sample zone has passed by a gel particle. The system is then considered in equilibrium. Other papers given in this Symposium discuss equilibrium conditions and deviations in greater detail. Here it may suffice to point out that complete or incomplete attainment of equilibrium affects primarily the peak width and only slightly the elution volume (1, 2).

Various workers have tried to predict V_e in terms of molecular size parameters. All of the rather diverse models gave good agreement with experimental results, even in cases where the models were obviously unsuitable (3). The reason for such apparent agreement was a rather insensitive square or cube root relation between elution volume and molecular size on one hand and an insufficient range of molecular sizes in the experiments on the other hand.

Today the emphasis is not so much on a detailed theoretical model of GPC as on either a fundamental understanding of the separation process or on a universal calibration method. On both aspects we will hear more in later papers in this Symposium. For those who have not worked with GPC yet, some basic considerations on calibration in GPC may be of value.

Linear relation was found empirically between the elution volume of a species and the logarithm of its molecular weight. The upper and the lower ends of these calibration curves bend to 90° slopes at the limits of resolution. The total curve is S-shaped. The broader the pore size distribution of the gel or the more types of gels used in a GPC system, the longer is the linear part of the curve. Gels with very narrow pore ranges give S-shaped curves with little or no linearity. Better than plotting the logarithm of molecular weight is plotting the logarithm of molecular size versus elution time. In this way different shapes, flexibilities, and degrees of swelling do not affect the calibration curves. Since molecular size is the primary selective parameter in GPC, such a plot yields a universal curve which seems to hold for all kinds of molecules (4, 5).

What exactly is molecular size in our context? Is it the length, the cross section, or the volume of a molecule? In the case of spherical molecules, any of these parameters could be used equally well. For rodlike molecules, either the length or the volume is suitable; the same would hold for prolate ellipsoids. For oblate ellipsoids it would be cross section or volume. It is always the most prominent dimension or the volume that determines size for GPC. Giddings et al. (6) proposed a "mean external length" $\bar{L}$ which is a mean length of projection of the molecule along an infinite number of axes. For rigid molecules, this length $\bar{L}$ seems to correlate better than other size parameters with V_e (6).

In contrast to rigid molecules, in the case of polymers we must distinguish between the size of a molecule, i.e., the amount of space it takes up in solution, and its mass. We shall have to develop the relationships, if any, between these two independent quantities.

With the aid of a model, we can demonstrate the long-chain nature of a typical polymer, the basic flexibility of the molecular chain, and its randomly coiling nature when it is surrounded by small solvent molecules. Let us consider for the moment a single polymer molecule with a fixed chain length, say 2000 carbon-carbon bond repeat units, each 1.54 Å long. We can calculate the size of this chain if we make enough simplifying assumptions.

If we assume that there are absolutely no restrictions on the positions of successive atoms of the chain except that they be 1.54 Å apart, the calculation is the very old one developed by Lord Rayleigh around

1870 and known as the random flight calculation. The answer has to be a statistical one, for the chain can take up a vast number of arrangements or conformations at different times, and it is appropriate only to ask about its *average* size. If we take, as the particular average quantity to calculate, the root-mean-square distance between the ends of the chain, $\sqrt{r^2}$, the answer for the random-flight calculation is $\sqrt{r_f^2} = l\sqrt{n}$. Putting $n = 2000$ and $l = 1.54$ Å, we obtain an end-to-end distance of about 69 Å.

This calculation disregards short-range interactions restricting the arrangement of successive atoms in the chain. We know that in real polymer chains the carbon-carbon bond angle must be preserved and that even in a very simple polymer such as polyethylene there are restrictions to free rotation about the carbon-carbon bonds, including those which rule out conformations putting near-neighbor atoms (say, the first and fifth in a sequence) on top of one another. In recent years, polymer chemists have been able to calculate the results of these restrictions with good accuracy; the contributions of Flory and his colleagues (7) have been outstanding in this way. He has shown, for example, that the root-mean-square end-to-end distance for polyethylene is increased by a factor of about 2.6 by all these restrictions. For our model chain, the resulting value, known as the unperturbed dimension, is $\sqrt{r_0^2} = 178$ Å.

Flory's calculation is in good agreement with the experimental value for the unperturbed dimensions of polyethylene, as computed from their hydrodynamic volume at the theta temperature. In good solvents, the chains are further expanded, of course, because of favorable polymer-solvent interactions and long-range interactions. We will come back to this later.

At this stage, we can draw the following conclusions:

1. For linear chains, the relation between size and the square root of the number of atoms (and this means the square root of molecular weight) is preserved, though the proportionality constant depends on polymer type, solvent, and temperature, and very slightly on molecular weight. This implies that for a given linear system there is a unique relation between size and mass, justifying the assumption that almost all GPC workers take for granted.

2. Even though we can talk only about the average behavior of a polymer chain, this is enough to explain its behavior in the GPC experiment. We can think of the average either as an instantaneous one

over many polymer chains of the same length or as a time average over the many conformations taken up by a single chain. During the time of a GPC experiment, a given molecule will take up so many conformations that its behavior is very well approximated by assuming it has its average size all the time.

3. The randomly coiling nature of the polymer chain also implies something about its shape. It can be shown that, as long as the chain has more than about ten segments, the Gaussian statistics required by the random-flight calculation hold. They require that the average shape for all conformations of the freely coiling chain, without any outside force fields or restraints, is spherical. Only if restraints are introduced do deviations from spherical symmetry occur. Thus, the average shape of all conformations in which both ends of the chain are at fixed locations is a prolate ellipsoid of revolution; the average when the center segment is fixed also is described as "bean-shaped" and so on. There is no reason to expect that in the GPC experiment the chain will be constrained to assume only these special conformations so that its shape can be considered spherical for our purposes.

To get some feel for the relation between the sizes of polymer molecules and GPC gel structures, we can make the following calculations. Consider a monodisperse polystyrene dissolved in tetrahydrofuran at room temperature for the usual GPC experiment. For any given molecular weight, we can calculate the random-flight end-to-end distance of this chain by the usual methods. Next, taking Flory's result (7) that the effect of short-range interactions in polystyrene is to expand the chain by a factor of about 3.2, we can calculate the unperturbed dimensions of the molecule.

The easiest way to get the effect of long-range interactions on the size of the chain is to compare its viscosity in the good solvent and in a θ-solvent. We can do this using Benoit's relationship (8) between viscosity and molecular weight in THF and that of Cantow (9) for a θ-solvent. The ratio of the viscosities gives the cube of the factor by which the chain dimensions are expanded in THF over their unperturbed dimensions. For MW = 100,000 the expansion factor is about 1.24. (Both this number and the factor due to short-range interactions are slightly molecular-weight dependent, but we have neglected this.)

We can further ask, what is the amount of space effectively occupied by a polymer with a certain end-to-end distance? Roughly speaking, it is that of a sphere whose diameter is about five times the end-to-end distance; all the segments of the polymer chain ought to be inside a sphere of this diameter about 95% of the time.

Putting all this together, we estimate the effective diameter of a polystyrene molecule in THF whose molecular weight is 100,000 to be just under 1000 Å; and that of a molecule with MW = 1,000,000 to be about 3000 Å. In Figs. 2 and 3 "random coils" of these sizes are sketched very crudely onto electron photomicrographs (10) of Styragel.

Figure 2 shows the sketch at MW = 100,000 in relation to a 10^4 Å Styragel. It seems to fit comfortably in the larger openings of the gel structure, which is reassuring since the exclusion limit for this material is about 400,000 molecular weight.

In Fig. 3, both this sketch and that corresponding to MW = 1,000,000

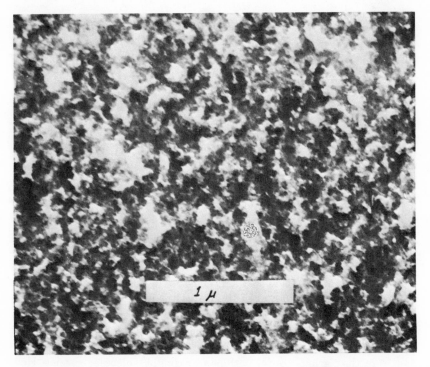

FIG. 2. Sketch of the effective size of a polystyrene molecule in THF at MW = 100,000, on an electron photomicrograph of 10^4 Å Styragel.

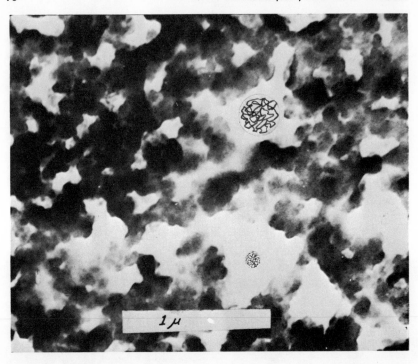

FIG. 3. Sketch of the effective sizes of polystyrene molecules in THF at MW = 100,000 and MW = 1,000,000 on an electron photomicrograph of a 10^6 (old designation) Styragel.

are shown on a 10^6 Styragel (old designation; now designated 10^5) where the exclusion limit is about MW = 4,000,000. Again, things seem to look about reasonable.

Now let us turn to the more usual type of polymer sample where we are always faced with a distribution of molecular weights and corresponding chain lengths. Again assuming that a fixed relation exists between over-all chain length and end-to-end distance, it follows that a distribution of end-to-end distances exists as a result of the various molecular species present in the sample. It would seem that two distributions exist simultaneously, the one just mentioned and that considered previously, the distribution of end-to-end distances resulting from the various conformations of a single molecular species. How can GPC separate the effects of these two?

The answer lies in the averaging nature of the many repeated steps of permeation in the GPC process. During the experiment the polymer chains assume many conformations so that the average behavior of any single species corresponds to a fixed size parameter, such as the root-mean-square end-to-end distance, uniquely related to total chain length or molecular weight. The different molecular species display average behavior characteristic of their position in the distribution of such species present, and the separation occurs on the basis of this latter distribution.

What is the result? In the usual experiment, it is a plot of some measure of the amount of material existing in the column as a function of elution volume. One of the major efforts in the development of GPC has been to provide methods allowing the correlation of elution volume with molecular size or molecular weight. Details of these studies will be reported in later papers in this Symposium; here we shall make only one point:

The GPC experiment alone does not provide any information on either average molecular weights or molecular weight distribution. It is solely a separation technique. All the rest of the information must come from the calibration step. Ultimately, this requires the use of absolute methods for determining the average molecular weights of polymers. These methods have been reviewed in the literature from time to time (11) and will not be discussed further here.

Acknowledgments

One of us (FWB) would like to acknowledge support of his research on GPC by the Texas Division, Dow Chemical Co., and Waters Associates, Inc. The research is carried out in Rensselaer's Materials Research Center, a facility supported by the National Aeronautics and Space Administration.

REFERENCES

1. T. C. Laurent and E. P. Laurent, *J. Chromatogr.*, **16**, 89 (1962).
2. H. Vink, *J. Chromatogr.*, **25**, 71 (1966).
3. K. H. Altgelt, *Advances in Chromatography*, Vol. 7 (J. C. Giddings and R. Keller, eds.), Dekker, New York, 1968.
4. Z. Grubisic, P. Rempp, and H. Benoit, *J. Polym. Sci., Part B,* **5**, 753 (1967).
5. J. Cazes and D. R. Gaskill, *Separ. Sci.,* **2**, 421 (1967).
6. J. C. Giddings, E. Kucera, C. P. Russell, and M. N. Myers, *J. Phys. Chem.,* **72**, 4397 (1968).
7. P. J. Flory, *Statistical Mechanics of Chain Molecules*, Wiley, New York, 1969.

8. H. Benoit, Z. Grubisic, P. Rempp, D. Decker, and J. G. Zilliox, *J. Chim. Phys.* **63**, 1507 (1966).

9. H. J. Cantow, *Makromol. Chem.*, **30**, 169 (1959).

10. Electron micrographs kindly supplied by J. C. Moore.

11. F. W. Billmeyer, Jr., *J. Polym. Sci., Part C*, **8**, 161 (1965); *Polym. Eng. Sci.*, **6**, 359 (1966); *Appl. Polym. Symp.*, **10**, 1 (1969).

Gel Permeation Chromatography Column Packings—Types and Uses*

DALE J. HARMON

THE B. F. GOODRICH RESEARCH CENTER
BRECKSVILLE, OHIO 44141

Summary

Proper selection of the column packing material is an essential part of a successful gel permeation analysis. Generally the desirable properties of a packing are good chemical, thermal, and mechanical stabilities combined with good resolution and low resistance to liquid flow. Pore size distribution, particle size distribution, polar characteristics, as well as other physical parameters play a role in how well a packing will perform. This paper describes the different types of column packings which are commercially available, as well as briefly mentioning several packings used on an experimental scale. Physical characteristics of the various packings are compared. Suggested uses along with some limitations are given.

A variety of porous packing materials are available to the chromatographer, and it is the task of the individual to choose the one best suited to his needs. Generally, the desirable properties of a packing are good chemical, thermal, and mechanical stability combined with good resolution and low resistance to liquid flow. A combination of the above properties would allow one to use a column in different solvents, at different temperatures, and at various flow rates with little or no loss of resolution.

Most commonly used for synthetic polymer analysis are the cross-linked polystyrenes. These came about through the work of Moore

* Presented at the ACS Symposium on Gel Permeation Chromatography, sponsored by the Division of Petroleum Chemistry at the 159th National Meeting of the American Chemical Society, Houston, Texas, February, 1970.

13

(*1*) who demonstrated that by controlling the nature and amount of diluent present, cross-linked polystyrene beads of controlled pore sizes could be obtained. Table 1 illustrates the different recipes for obtaining certain pore sizes. These gels are sold under the name Styragel and are available only from Waters Associates, Inc., Framingham, Massachusetts. The gels can be used with most organic solvents; however, they cannot be used without specific precautions with acetone,

TABLE 1

Permeabilities of Gels with Various Diluents, all Made from 30% Styrene, 10% Divinylbenzene, 60% Diluent,[a] According to Moore (*1*)

Diluents, parts/100 parts of gel	Excluded molecular weight (M_e)
60 Toluene	7×10^3
30 Toluene, 30 diethylbenzene	1.5×10^4
60 Diethylbenzene	1.2×10^4
45 Toluene, 15 n-dodecane	1×10^5
30 Toluene, 30 n-dodecane	3×10^5
15 Toluene, 45 n-dodecane	2×10^6
10 Toluene, 50 n-dodecane	2×10^3
40 Diethylbenzene, 20 isoamyl alcohol	3.6×10^3
20 Diethylbenzene, 40 isoamyl alcohol	8×10^6
13.3 Diethylbenzene, 46.7 isoamyl alcohol	10^{10}
60 Isoamyl alcohol	Extremely high

[a] "Styrene" is a mixture of styrene and ethylvinylbenzene. "Divinylbenzene" is a mixture of about 54% divinylbenzene, 42% vinyl ethylbenzene, and 4% diethylbenzene.

alcohols, most acids, and water. The gels as sold have a particle range usually of 20–60 μ and are available in exclusion ranges of from 40–10^7 Å expanded chain length, i.e., from about 200 to 50,000,000 molecular weight. The upper exclusion limit is determined by the largest pores present. Molecules larger than this limit will be eluted at the interstitial volume, V_0, of the column. Styragel is a rigid gel as compared to soft gels such as reported by Heitz (*2*). The difference is in the degree of cross-linking. The soft gels will swell appreciably in many solvents and are limited to lower operating pressures than the rigid gels. Figure 1 shows the separation of a polystyrene made on soft gel. Rather remarkable separations can be achieved at low flow rates. The upper temperature limit of the Styragel packing is about 140°F.

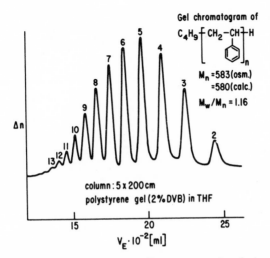

Gel chromatogram of

$$C_4H_9 \left[CH_2-CH \right]_n H$$

M_n =583(osm.)
     =580(calc.)
M_w/M_n = 1.16

Δn

column: 5 x 200 cm
polystyrene gel (2% DVB) in THF

$V_E \cdot 10^{-2}$ [ml]

FIG. 1. Separation of polystyrene oligomers on soft polystyrene gel.

A series of soft cross-linked polystyrene gels is marketed by Bio-Rad Laboratories under the trade name Bio-Beads S. They have an operating range of 100 to 14,000 in molecular weight based on polystyrene dissolved in tetrahydrofuran. Their physical characteristics are given in Table 2.

Other organic polymers have been used. One which is commercially available is EM-Gel-OR, a cross-linked polyvinyl acetate. These gels, developed by Heitz (*3*) and manufactured by E. Merck, are produced by copolymerization of vinyl acetate and butanediol-1, 4-divinyl ether or divinyl esters of dicarboxylic acids. Up to an exclusion limit

TABLE 2

Physical Characteristics of Bio-Beads S

Type	Mesh size dry	Exclusion limit	Separation range, mol wt	Swollen bed volume (ml/g)[a]
S-X1	28–74 μ	14,000	600–14,000	9.8
S-X2	28–79 μ	2,700	100–2,700	6.2
S-X3	28–74 μ	2,000	Up to 2,000	5.1
S-X9	28–74 μ	1,400	Up to 1,400	4.2
S-X8	28–74 μ	1,000	Up to 1,000	3.9
SM-1	20–50 mesh	14,000	600–14,000	3.1
SM-2	20–50 mesh	14,000	600–14,000	2.9

[a] In benzene.

TABLE 3

Physical Characteristics of EM-Gel-OR

Gel type	Exclusion limit	Swelling factor
OR 750	ca. 750	ca. 2
OR 1,000	ca. 1,500	ca. 3
OR 5,000	ca. 5×10^3	ca. 4.5
OR 20,000	ca. 2×10^4	ca. 4.0
OR 100,000	ca. 1×10^5	ca. 4.3
OR 1,000,000	ca. 1×10^6	ca. 4.1

of 5000 molecular weight, the gels are homogeneously cross-linked. Higher limits are produced by polymerization in the presence of an inert diluent. The gels are lipophilic and could be used for adsorption chromatography. Table 3 shows the gels which have been prepared. The exclusion limits and swelling factors are for polystyrene in tetrahydrofuran. Only the lower four exclusion ranges are commercially available at the present. These gels are primary for use in polar organic solvents such as acetone, methanol, ethanol, and tetrahydrofuran. The upper limit of thermal stability is 100°C. The swelling porosity enables a separation of low molecular weight compounds even with gels of high exclusion limits. Figure 2 shows the separation of polystyrenes and oligophenylenes covering the molecular weight range, of 92 to 8.3×10^5. The separation was done using EM-Gel-OR 1,000,000, an open glass column of 1 m $\times$ 1.5 cm, and THF as the solvent. Even though the gels swell, they are rigid, and columns can be operated with or without pressure.

Of increasing interest and popularity are the various glass or silica packings which are available on the market. Best known of these are the spherical porous silica beads of de Vries (4) manufactured by

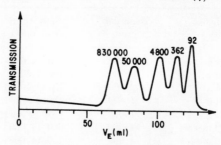

FIG. 2. Separations on Merckogel OR 1.000,000 0.020–0.055 mm, open glass column, 1 m length, 1.5 cm diameter. Solvent: THF.

TABLE 4

Physical Characteristics of Porasil[a]

Type	Surface area (m²/g)	Average pore diameter (Å)	Excluded mol wt
A	480	100	6×10^4
B	200	100–200	2.5×10^5
C	50	200–400	4×10^5
D	25	400–800	1×10^6
E	4	800–1500	1.5×10^6
F	1.5	>1500	$>2 \times 10^6$

[a] Particle size ranges available: 36–75 μ, 75–125 μ.

Pechiney-Saint-Govain under the name Spherosil and sold in this country by Waters Associates under the name Porasil. Table 4 shows the types available and the upper pore size for each type. Two particle sizes are available, and the size of 75–125 μ is recommended for per-

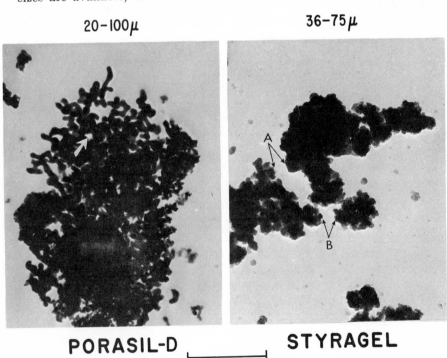

FIG. 3. Electron photomicrographs of Porasil-D and Styragel.

meation chromatography. Figure 3 is an electron photomicrograph of
Porasil and Styragel. The Porasil appears to be made of overlapping
saddlelike structures which have a smooth surface. These structures
fused together form a porous particle. The size of the pore marked by
the arrow is about 550 Å, which agrees with the designated pore size.
The Styragel appears to be made up of very fine, almost round
particles. These particles are fused together to form agglomerates with
rough textured surfaces. The agglomerates, in turn, are joined to form
the macro particles which appear to the eye as very small spheres.
The manner in which the agglomerates go together leads to the formation of voids. The entrances to the two voids pointed out by the
arrows are about 1500 Å. Since the Porasil-D has an upper exclusion
limit of about 1×10^6 in molecular weight and the Styragel of about
3×10^6, this void size is reasonable. This structural difference may
explain why Yau (5) observed that separation on porous glass appears
to be controlled by steric exclusion while separation on Styragel appears to be influenced by lateral diffusion as well as steric exclusion.

Other glass packings are the Corning Porous Glass packings reported
in the literature by Haller (6). The pores are all within ±15% of the
reported value. These are porous glass granules with a packing density
of 0.3 to 0.7 g/cc. Table 5 gives the characteristics of this material.
Due to the small range of pore size, good resolution is obtained over
a relatively narrow range as shown in Fig. 4. Best results are obtained
using more than one column in series. Another commercially available
porous glass with a wider pore size distribution than the glasses made

TABLE 5

Physical Characteristics of Corning Glass Packings[a]

Pore diameter (Å)	Exclusion limits,[b] mol wt	Operating range[b]
75	28,000	300–28,000
125	48,000	650–48,000
175	68,000	1,050–68,000
240	95,000	1,150–95,000
370	150,000	5,000–150,000
700	300,000	15,000–300,000
1,250	550,000	40,000–550,000
2,000	1,200,000	120,000–1,200,000

[a] Particle sizes available: 36–75 μ, 75–125 μ, 125–175 μ.
[b] Based on dextrans in distilled water.

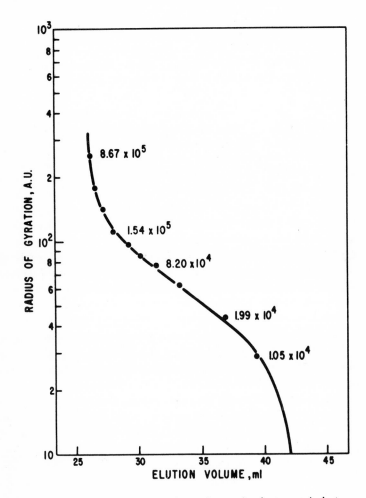

FIG. 4. Radius of gyration vs elution volume of polystyrene in butanone-isopropanol. Small size porous glass.

by Haller's process is Bio Glass sold by Bio-Rad Laboratories, Richmond, California. The physical characteristics of the Bio Glass packings are given in Table 6.

Another glass-type packing which only recently became available is EM-Gel-Si. These packings are irregularly shaped, crushed silica particles. They are manufactured by E. Merck and like the EM-Gel-OR are marketed in the United States by Waters Associates Inc. These

TABLE 6
Physical Characteristics of Bio Glass Packings[a]

Type	Exclusion limit	Separation range, mol wt	Average pore diameter (Å)
200	30,000	3,000–30,000	200
500	100,000	10,000–100,000	500
1,000	500,000	50,000–500,000	1000
1,500	2,000,000	400,000–2,000,000	1500
2,500	9,000,000	800,000–9,000,000	2500

[a] Particle sizes available: 147–297 μ, 147–187 μ, 125–147 μ, 74–147 μ, 44–74 μ, <44 μ.

materials exhibit high-surface area and have active sites typical of silicalike materials. Table 7 lists the physical characteristics of the materials which are available.

All of the glass packings mentioned have some polar sites present and care must be used to prevent tailing. In some cases a liquid modifier such as a silylating agent is required. Cooper and Johnson (7) have described the effect of treatment with hexamethyldisilazane on porous glass packings. Recently Bombaugh et al. (8) described the GPC analysis of polyvinyl alcohol using Porasil columns with water as the carrier solvent. A temperature of 65°C and a flow rate of 1 ml/min was employed. Figure 5 illustrates the chromatograms obtained for three different polymerization systems. Polyvinyl alcohol is normally permanently retained on Porasil. The separation was made possible by development of a deactivated Porasil in which the polar sites are blocked. Columns in use up to 6 months have shown no increase in adsorption.

For molecules which are soluble in water or electrolyte solutions, one can use Sephadex packings. Sephadex is dextran crosslinked to give a three-dimensional network of polysaccharide chains. Because of Sephadex's high content of hydroxyl groups, the beads swell consider-

TABLE 7
Physical Characteristics of EM-Gel-Si

Gel type	Exclusion limit	Surface (m²/g)	Pore size (Å)	Pore volume (ml/g)
150	ca. 50,000	120–170	120–220	0.81
500	ca. 400,000	35–65	300–700	0.78
1,000	ca. 1 × 10⁶	10–20	700–1300	0.70

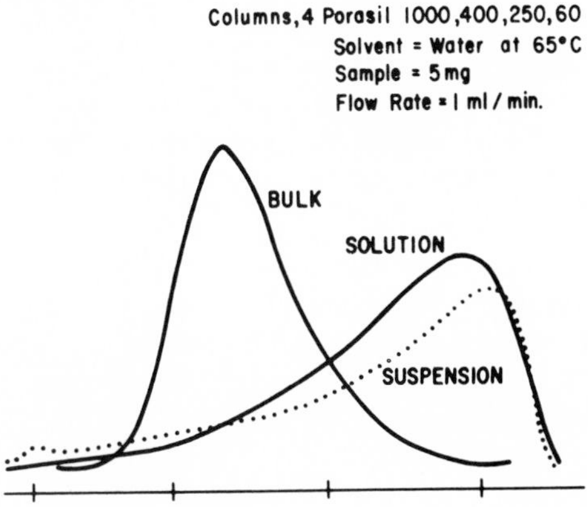
Columns, 4 Porasil 1000,400,250,60
Solvent = Water at 65°C
Sample = 5 mg
Flow Rate = 1 ml / min.
BULK
SOLUTION
SUSPENSION

FIG. 5. GPC chromatograms of polyvinyl alcohol from various polymerization methods.

ably in water and electrolyte solutions. Table 8 lists the grades available and some of their characteristics. Sepharose is an agarose gel which complements Sephadex by extending the separation range to include viruses and high molecular weight proteins, polysaccharides, and nucleic acids.

By alkylation of most of the hydroxyl groups of Sephadex G-25, a material which swells in organic solvents is obtained. This is Sephadex

TABLE 8

Physical Characteristics of Sephadex

Type	Particle size (μ)	Excluded mol wt[a]	Bed volume (ml/g)
Sephadex G-10	40–120	7×10^2	2
G-15	40–120	1.5×10^3	3
G-25[b]	50–150	5×10^3	5
G-50[b]	50–150	1×10^4	10
G-75[c]	40–120	5×10^4	15
G-100[c]	40–120	1×10^5	20
G-150[c]	40–120	1.5×10^5	30
G-200[c]	40–120	2.0×10^5	40
Sepharose 6B	40–210	1×10^6	—
4B	40–190	5×10^6	—
2B	60–250	2×10^7	—
Sephadex LH-20	25–100	4×10^3	—

[a] Based on dextrans.

[b] Available in fine, 20–80 μ; coarse, 100–300 μ.

[b] Available in superfine, 10–40 μ.

LH-20 which can be used with polar organic solvents as well as water and has the same separation range as G-25.

Other column packings for aqueous systems are the Bio-Gel P and Bio-Gel A packings of Bio-Rad Laboratories. The Bio-Gel P is obtained by polymerizing varying concentrations of acrylamide and methylene bisacrylamide. Each is readily hydrated for use in a wide variety of aqueous media. Four particle size ranges are available: 50–100 mesh for large columns where speed is important, 100–200 mesh for general purpose work, 200–400 mesh for high resolution chromatography, and —400 mesh for thin layer chromatography. Table 9 lists the physical characteristics of Bio-Gel P.

TABLE 9

Physical Characteristics of Bio-Gel P[a]

Type	Exclusion limit	Separation range, mol wt	Hydrated bed volume (ml/g)	Water regain (g/g)
P2	1,800	200–1,800	3.8	1.5
P4	4,000	800–4,000	5.8	2.4
P6	6,000	1,000–6,000	8.8	3.7
P10	20,000	1,500–20,000	12.4	4.5
P30	40,000	2,500–40,000	14.8	5.7
P60	60,000	3,000–60,000	19.0	7.2
P100	100,000	5,000–100,000	19.0	7.5
P150	150,000	15,000–150,000	24.0	9.2
P200	200,000	30,000–200,000	34.0	14.7
P300	400,000	60,000–400,000	40.0	18.0

[a] Particle sizes available: 147–297 μ, 74–147 μ, 28–74 μ (P2 through P10 only), <28 μ.

Bio-Gel A complements Bio-Gel P in that it extends the upper operating range from 400,000 to 1.5×10^8 molecular weight. Bio-Gel A is agarose, the nonionic constituent of agar. The use of agarose eliminates the undesirable side effects associated with agar: ion exchange and adsorption, electroendosmosis, and specific chemical interactions. These agarose gels have a useful temperature range of 0–30°C. The physical characteristics are shown in Table 10.

To summarize, it would seem that acceptable packing materials are now available for almost any polymer solvent system one may desire to work with. While the glass packings offer greater stability and ease of packing, the cross-linked polymers give the least amount of peak spreading or the highest plate counts. Problems of polarity

TABLE 10

Physical Characteristics of Bio-Gel A[a]

Type	Exclusion limit	Separation range, mol wt	Approximate % agarose in gel
A-0.5 m	5×10^5	10^4 to 5×10^5	10
A-1.5 m	1.5×10^6	10^4 to 1.5×10^6	8
A-5 m	5×10^6	10^4 to 5×10^6	6
A-15 m	1.5×10^7	4×10^4 to 1.5×10^7	4
A-50 m	5×10^7	10^5 to 5×10^7	2
A-150 m	1.4×10^8	10^6 to 1.5×10^8	1

[a] Particle sizes available: 147–297 μ, 74–147 μ, 28–74 μ.

can be encountered with both glass and polymer packings, and one must be alert for signs of adsorption when working with unfamiliar systems. Lightly cross-linked gels may be of interest where low flow rates and low pressures can be used. Extraordinary separations have been achieved with such gels.

Acknowledgments

The author wishes to acknowledge the courtesy of the following editors, publishers and authors for permission to reproduce the designated material: Huethig and Wepf Verlag, publisher of *Die Makromolekulare Chemie*, Fig. 1; Dr. W. Heitz, University of Mainz, Fig. 2; J. C. Moore, Dow Chemical Company, Fig. 4; and American Chemical Society, publisher of *Analytical Chemistry*, Fig. 5.

REFERENCES

1. J. C. Moore, *J. Polym. Sci., Part A*, **2**, 835 (1964).
2. W. Heitz and W. Kern, *Angew. Makromol. Chem.*, **1**, 150 (1967).
3. W. Heitz, *Makromol. Chem.*, **127**, 113 (1969).
4. A. J. de Vries et al., *Anal. Chem.*, **39**, 935 (1967).
5. W. W. Yau, *J. Polym. Sci., Part A-2*, **7**, 483 (1969).
6. W. J. Haller, *Chem. Phys.*, **42**, 686 (1965).
7. A. R. Cooper and J. J. Johnson, *J. Appl. Polym. Sci.*, **13**, 1487 (1969).
8. K. J. Bombaugh and R. F. Levengie, *Anal. Chem.*, **41**, 1337 (1969).

Chromatographic Instrumentation and Detection of Gel Permeation Effluents*

EDWARD M. BARRALL II and JULIAN F. JOHNSON†

IBM RESEARCH LABORATORY
SAN JOSE, CALIFORNIA 95114

Summary

The basic instrumentation required for gel permeation chromatography are described. The various choices for sample injectors, pumping systems, detectors, and recording techniques are discussed and critically evaluated. Column choice, construction, etc., are omitted since these items are very well covered by other Symposium papers.

INTRODUCTION

The instrumentation for gel permeation chromatography (GPC) is relatively simple in concept but somewhat involved in execution. All functional instruments contain the following systems: (a) sample injection, (b) pumping, (c) chromatographic column, (d) sample detection, (e) eluent volume detection, and (f) data recording. External to the apparatus flow system are the sample preparation apparatus and column temperature regulation control. A detailed discussion of each of the above items with the exception of the chromatographic column will be the subject matter of this paper. It is not intended that this discussion be an exhaustive survey of the very large literature in this field. However, examples will be cited to illustrate particular points when necessary for clarity.

*Presented at the ACS Symposium on Gel Permeation Chromatography, sponsored by the Division of Petroleum Chemistry at the 159th National Meeting of the American Chemical Society, Houston, Texas. February, 1970.

† Present address: Department of Chemistry and Institute of Materials Science, University of Connecticut, Storrs, Connecticut 06268.

25

SAMPLE INJECTION

Some means must be provided for the introduction of the sample in solution into the chromatograph. This problem is complicated by three factors: (a) relatively high hydrostatic pressure in the GPC system before the columns and after the pump, (b) flow properties of two liquids of different densities, and (c) danger of introducing gas bubbles. The objective of the instrumentation for sample addition is to introduce a known amount of sample onto the column with a minimum loss of resolution.

Systems for GPC sample addition fall into two main types: multi-port valves and hypodermic injection systems. Figure 1 shows an injection system based on a multiport valve described by Bombaugh, Dark, and King (1). The sample loop is filled with a polymer solution of the desired concentration while solvent is flowing through the column. Interchangeable loops of different volumes are available. To inject the sample the valve is turned 90°, causing solvent flow to be directed through the sample loop. Valving of the sample injection system has been a problem in the past. The most common difficulty reported is solvation of the valve gasketing by the GPC solvent. Leak-

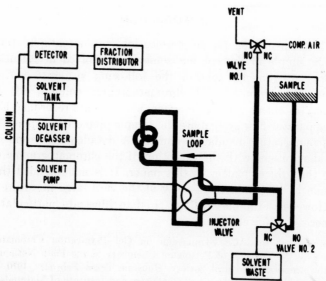

FIG. 1. Loop injection system. Sample filling loop is shown in bold lines (1).

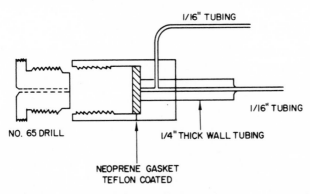

FIG. 2. Injection assembly (*5*).

ing valves and plugged lines are the general result. It is now possible to obtain reasonably long lasting sliding plate and O ring valves with insoluble Teflon surfaces. This type of sample injection system has the ability to introduce reproducible amounts of sample and is readily automated. Improvements in valve packing and construction have made possible the introduction of automatic injection systems. A commercially available system, Waters Associates, has six sample loops and will automatically inject six previously loaded samples at any selected fixed interval.

In general, for a given amount of polymer, reducing the volume of the sample loop reduces the peak broadening, see for example the work of Billmeyer and Kelley (*2*). However, since this necessarily causes an increase in the concentration of the polymer in solution, there is a point where the increase in resolution due to decreasing solution injected is counteracted by the irregular shape of injection of a very viscous sample. This "viscous fingering" effect has been discussed by Little et al. (*3*). The balance between sample size and concentration for optimum resolution is a complex function depending at least on flow rate and molecular weight and probably on other factors (*4*).

A typical hypodermic injection system developed by Gamble, McCracken, and Wade is shown in Fig. 2 (*5*). The polymer solution is added from the syringe which is inserted through the Neoprene gasket positioned by the guide hole. The solution is added directly into the flowing solvent at the head of the column. Sample volumes are less reproducible than with the multiport valve but the apparatus is compact, inexpensive, and easy to use. Jentoft and Gouw (*6*) have described modifications of the septum-hypodermic sample system to allow injection onto columns operating at pressures up to 1000 lb/in.²

PUMPING SYSTEM

The pumping system is required to produce a constant but readily variable flow rate typically in the range of 0.1 to 5 ml/min, at back pressures up to 250 lb/in.[2] These limits are, of course, a rather arbitrary selection, for higher and lower flow rates have been used and higher back pressures are occasionally encountered. Pressure fluctuations must be minimized as these are frequently a source of noise in the detector. The more common pumping systems are based on commercially available reciprocating or pulsing positive displacement pumps with an adjustable delivery volume. These pumps are used in conjunction with a damping system to reduce pressure pulses. A particularly effective small pulsation damper has been described by Ross and Castro and is shown in Fig. 3 (7).

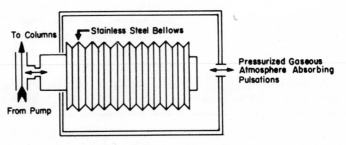

FIG. 3. Pulsation damper (7).

Jentoft and Gouw (8) and others (9) have reported on pulseless pump designs. Figures 4a and 4b illustrate typical designs. In the system shown in Fig. 4b constant volume flow is obtained using commercially available constant volume pumps (Zenith Products Co., West Newton, Mass., No. 1/2B-4391 with Teflon U-seals). The necessary low flow rates are obtained by operating two pumps differentially. That is, one pump delivers a volume V-1 as the source for the second pump which delivers a slightly smaller volume V-2. The excess volume, V-1 minus V-2, from a T-connection between the two pumps is the flow delivered to the chromatograph. Since these are gear-type pumps, the pumped fluid must have a considerable viscosity to achieve constant volume flow. High viscosity for constant volume requires an intermediate pumping fluid. Mineral oil of about 200 cSt was used to displace mercury which subsequently displaces the chromatographic solvent. The pump delivery rate is very constant against variable pressure heads, easily variable by the adjustment of system parameters and gear ratios, and essentially pulse free. The pump has one small disadvantage, the solvent reservoir must remain sealed during pump operation.

A particularly simple and quite satisfactory pump consists of a pneumatic pump (Waters Associates, Framingham, Mass.) discharging through a capillary restrictor into the gel permeation column (10). The pump is simply a polyethylene wash bottle in a metal container to which air pressure can be applied. While not a continuous pump, the flow is constant for a volume of about 300 ml which is sufficient for most analytical GPC measurements.

SAMPLE DETECTORS

It is possible to characterize fractions collected by gel permeation in the classical manner, that is by collection, evaporation of stabilizer-free solvent, weighing, and determination of molecular weight. However, one of the major advantages of GPC is that continuous detectors may be used to monitor and record the polymer concentration in the eluent.

Differential refractometers to detect polymers in organic solvents were first introduced by Moore (11) and are the most widely used detectors. A differential detector is required because of the low concentrations involved. The usual sample of polymer contains 2–10 mg of polymer which elutes from the column in 25–75 ml of solvent. With the refractometer detector the solvent-polymer system is chosen to give a maximum difference in the refractive index. Quantitatively, the detector response depends on the relationship that Δn is proportional to Δc where Δc is the concentration of the polymer. In turn this implies that the refractive index of the polymer is independent of molecular weight. While this is generally true for homologous series with molecular weights above a few thousand, some error may be introduced at lower molecular weights unless a correction factor is introduced to account for the small dependence on molecular weight (12).

Differential refractometry is not satisfactory for monitoring the fractionation of a mixture unless response factors are known for each component. There is no *ab initio* rule for determining if. a small peak in a mixture represents a trace component with a large refractive index difference from the solvent or a major component with a refractive index very similar to the solvent. Also, low liquid holdup in the cell is necessary to prevent excessive loss of resolution. For example, when the volume of the detector cell was decreased from 0.070

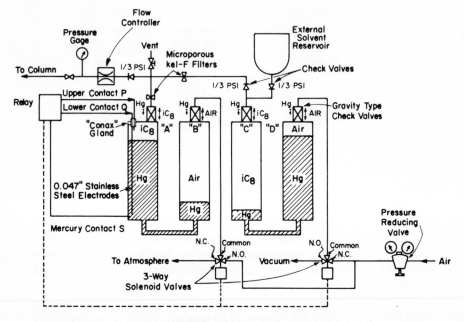

FIG. 4a. Pulseless high pressure pump (8).

to 0.010 ml, Billmeyer and Kelley observed a threefold gain in sensitivity and a reduction of one-half in the zone broadening due to the cell (2). Johnson, Campanile, and Lefebre (12a) reported on the design of a low holdup reflection type refractometer with a limit of detection for Δn of 1×10^{-6} and a noise and drift for 1 hr periods of only 5×10^{-7}.

Since the refractometer is a differential detector, a reference, flowing stream of solvent must be supplied to one side of the refractometer cell. The reference stream is usually provided by the eluent of a relatively short section of packed column. Since the differential refractometer is measuring differences between the sixth and seventh decimal of refractive index during a chromatogram, it is very sensitive to minor solvent and flow variations (surges). Care must be used to balance the flow rate and pressure drop across the reference and working columns if a smooth base line is to be obtained.

In some cases infrared spectrometers may also be used as detectors with certain advantages over the refractometer type. An example is given by Ross and Castro (7), who found that very satisfactory

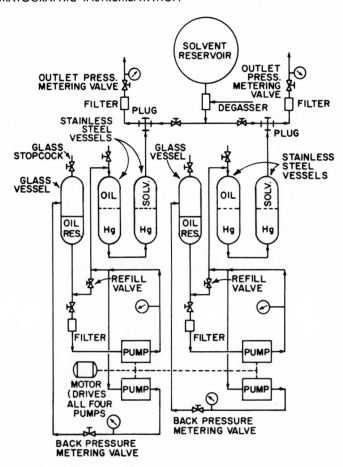

FIG. 4b. Dual constant volume pumping unit (9).

results could be obtained by monitoring the 3.4 μ band for polyethylene molecular weight distributions using perchloroethylene at 110°C as a solvent. Sensitivity was stated to be better and noise due to temperature fluctuations less than with a differential refractometer at high temperatures. It was necessary to add a zero suppressor circuit so that the 80–100% transmission range appeared as full scale on the recorder. In this region concentration was found to be a linear function of optical density, see Fig. 5.

Rodriguez, Kulakowski, and Clark (13) have utilized infrared detection to determine relative concentrations of functional groups as

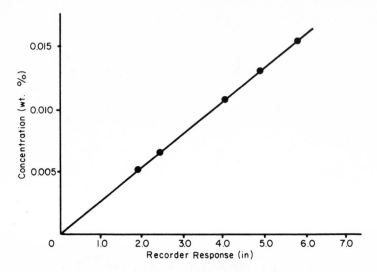

FIG. 5. Response of recorder with polyethylene concentration at $3.4\,\mu$
($2940\ \mathrm{cm^{-1}}$) (7).

well as total polymer concentration. The importance of such measurements in copolymer analysis is apparent. For example, as shown in Fig. 6, Terry and Rodriguez determined methyl methacrylate from the $1731\ \mathrm{cm^{-1}}$ band and styrene from the $698\ \mathrm{cm^{-1}}$ band (14). Phenyl

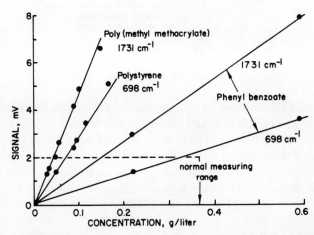

FIG. 6. Calibration curves for infrared detector. Solvent: trichloro-ethylene (14).

benzoate, which elutes separately from the polymer, was measured
at both frequencies to provide an internal standard. In addition, the
flow may be stopped when a peak appears and a complete spectrum
recorded. This is particularly useful for low molecular weight compounds and in identification of additives in polymers. The spectrometer need not be a rapid scan type as essentially no column resolution
is lost when flow is interrupted for periods of several hours (15).

The search for more sensitive detectors to permit smaller sample
sizes and higher resolution led to the development by several research
groups of a detector based on flash evaporation of the eluent. This
deposits the nonvolatile solute on a moving chain or cord which carries it into a pyrolyzer after which the amount of pyrolysis products
is determined by a very sensitive detector. In the above case this was
the well-known flame ionization detector used in gas chromatography
(16, 17). A very sensitive detector of this type developed by Johnson,
Seibert, and Stross (18) is shown in Fig. 7. While the construction and
operation is much more complex than previous detectors, the sensitivity
is much higher. As Fig. 8 shows, 0.1 μg amounts are detectable. The

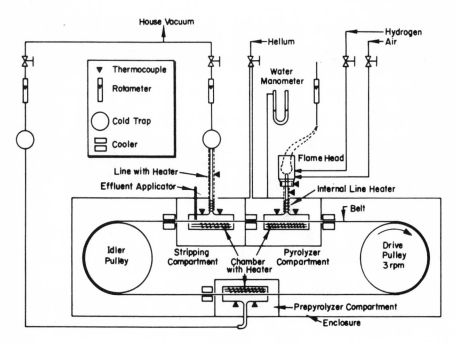

FIG. 7. Schematic of belt detector (18).

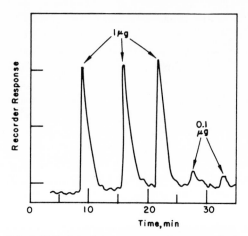

FIG. 8. Belt detector response, polystyrene (*18*).

nickel chromium alloy belt moves at a speed of about 2.5 cm/sec. A temperature gradient is used to remove the solvent in the stripper section. The gradient is usually somewhat below the solvent boiling point at the entrance to 100°C above the boiling point at the exit. The optimum temperature for the pyrolysis chamber depends on the specific polymer, ranging from 450–550°C. A flame ionization detector and electrometer comprise the detector.

Polyether polyols have been separated by GPC and the concen-

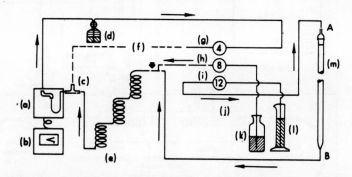

FIG. 9. Flow diagram of gel permeation chromatograph using a recording colorimeter detector (*19*): (a) colorimeter, (b) recorder, (c) separator, (d) displacement bottle, (e) mixing coil, (f) waste, (g) Tygon tube, (h) Tygon tube, (i) Acidflex, (j) proportioning pump, (k) cobalt thiocyanate reagent, (l) EDC reservoir, (m) column.

tration determined from the cobalt thiocyanate complex using a recording colorimeter by Kondo, Hori, and Hattori (19). The apparatus shown in Fig. 9 was sensitive with a linear response versus concentration. Its applicability is limited to systems where a suitable reagent exists.

For solvent-polymer systems where they are applicable, conductivity detectors (20) and ultraviolet spectrometers (21, 22) are sensitive, inexpensive instruments.

ELUENT VOLUME DETECTORS

To record the volume of eluent through the column as a function of time, a siphon that discharges when full and at each discharge sends a signal to the recorder where a blip is produced is most frequently used. This is a convenient arrangement, but is subject to error. Yau, Suchan, and Malone (23) in studies of the volume delivered by the siphon as a function of flow rate identified two sources of error. One was due to evaporation from the siphon and the other due to eluent continuing to flow into the siphon while it is discharging. Similar results were reported by Little et al. (3). Various methods of saturating the air in the siphon chamber with solvent vapor have been used to minimize evaporation. Cooper has found (10) that a siphon designed by Gray (24) gives more reproducible results. It would seem that developing a more accurate method for measuring the eluent volume would be a rewarding research area.

COMMERCIAL INSTRUMENTS

Recently, the number of instrument manufacturers interested in the liquid–solid and gel permeation market has increased. Older manufacturers have diversified their lines and now offer a large number of instruments in various price ranges for specific applications. Instruments satisfactory for GPC are currently being produced by Waters Associates, 16 Fountain St., Framingham, Mass.; du Pont Instrument Products Division, Wilmington, Delaware; Varian-Aerograph, Walnut Creek, Calif.; and Nester-Faust, 2401 Ogletown Rd., Newark, Del. Most of these companies also sell components as well as complete chromatographs. The cost of many commercial instruments is below the cost of custom construction in industry. Academic research workers can probably construct chromatographs more cheaply from components in highly specialized situations. However, the instrumentation scene is rapidly changing and should be reviewed in detail prior to construction or purchase of GPC equipment.

DATA RECORDING

Conventionally, the signal from the polymer concentration detector and the eluent volume detector are recorded on standard potentiometer recorders. Other more sophisticated systems may be used. For example, one commercially available system (Waters Associates, Framingham, Mass.) uses a digital curve translator to monitor the signal from three gel permeation chromatographs simultaneously and provides both a printed and punched paper tape record of the output. Gregges, Dowden, Barrall, and Horikawa (25) have reported a system for direct reading of GPC data into the analog/digital interface of an IBM 1800 computer. The system was designed to work with the Waters Associates automatic sample injection system or manual injection. The final read out is both graphical and punched cards. The cards are in a form suitable for most of the currently employed data reduction programs.

Acknowledgments

The authors are pleased to acknowledge the courtesy of the editors and publishers of the following journals for permission to reproduce the figures cited: *Analytical Chemistry*, Figs. 4a, 7, and 8; *Bunseki Kagaku*, Fig. 9; *Journal of Polymer Science*, Figs. 1, 3, 5, and 6; *Journal of Chromatography*, Fig. 4b; and Waters Associates, Inc., Fig. 2.

REFERENCES

1. K. J. Bombaugh, W. A. Dark. and R. N. King, *J. Polym. Sci., Part C*, **21**, 131 (1968).
2. F. W. Billmeyer, Jr., and R. N. Kelley, *J. Chromatogr.*, **34**, 322 (1968).
3. J. N. Little, J. L. Waters, K. J. Bombaugh, and W. J. Pauplis, *Amer. Chem. Soc. Div. Polym. Chem. Preprints*, **10**, 326 (1969).
4. J. N. Little, J. L. Waters, K. J. Bombaugh, and W. J. Pauplis, paper presented at Division of Polymer Chemistry, 159th National Meeting of the American Chemical Society, Houston, Texas, February, 1970.
5. L. W. Gamble, E. A. McCracken, and J. T. Wade, *Preprints of Third International Seminar on Gel Permeation Chromatography*, Geneva, May 1966.
6. R. E. Jentoft, and T. H. Gouw, *Anal. Chem.*, **40**, 923 (1968).
7. J. H. Ross and M. E. Castro, *J. Polym. Sci., Part C*, **21**, 143 (1968).

8. R. E. Jentoft and T. H. Gouw, *Anal. Chem.*, **38**, 949 (1966).
9. J. F. Johnson and M. J. R. Cantow, *J. Chromatogr.*, **28**, 128 (1967).
10. A. R. Cooper, A. R. Bruzzone, and J. F. Johnson, *J. Polym. Sci., Part A-2*, In press.
11. J. C. Moore, *J. Polym. Sci., Part A*, **2**, 835 (1964).
12. E. M. Barrall, II, M. J. R. Cantow, and J. F. Johnson, *J. Appl. Polym. Sci.*, **12**, 1373 (1968).
12a. H. W. Johnson, V. A. Campanile, and H. A. Lefebre, *Anal. Chem.*, **39**, 33 (1967).
13. F. Rodriguez, R. A. Kulakowski, and O. K. Clark, *Ind. Eng. Chem., Prod. Res. Develop.*, **5**, 121 (1966).
14. S. L. Terry and F. Rodriguez, *J. Polym. Sci. Part C*, **21**, 191 (1968).
15. A. R. Cooper, A. R. Bruzzone, and J. F. Johnson, *J. Appl. Polym. Sci.*, In press.
16. A. Karmen, *Anal. Chem.*, **38**, 286 (1966).
17. A. T. James, J. R. Ravenhill, and R. P. W. Scott, *Chem. Ind.* (London), **1964**, 746.
18. H. W. Johnson, Jr., E. E. Seibert, and F. H. Stross, *Anal. Chem.*, **40**, 403 (1968).
19. K. Kondo, M. Hori, and M. Hattori, *Bunseki Kagaku*, **16**, 414 (1967).
20. D. Saunders and R. L. Pecsok, *Anal. Chem.*, **40**, 44 (1968).
21. J. S. Fawcett and C. J. O. R. Morris, *Separ. Sci.*, **1**, 9 (1966).
22. W. Haller, Nature, **206**, 693 (1965).
23. W. W. Yau, H. L. Suchan, and C. P. Malone, *J. Polym. Sci., Part A-2*, **6**, 1349 (1968).
24. D. O. Gray, *J. Chromatogr.*, **37**, 320 (1968).
25. A. R. Gregges, B. F. Dowden, E. M. Barrall, and T. T. Horikawa, ACS Symposium on Gel Permeation Chromatography, Division of Cellulose, Wood and Fiber Chemistry, Houston, Texas, February, 1970.

Peak Resolution and Separation Power In Gel Permeation Chromatography

DALE J. HARMON

THE B. F. GOODRICH RESEARCH CENTER
BRECKSVILLE, OHIO 44141

Summary

Plate count alone cannot predict the separation ability of permeation columns. Separation is a function of the size range of the molecules in the sample, the size range of the pores in the packing, and how well a column is packed. Ideally one should select columns which have a pore size range exactly covering the size distribution of the sample and which do not spread the peaks. In practice, one usually settles for columns which at least cover the size range of the sample and have some pores of every size present. The lower the slope of the calibration plot (log, A, M or [n]M vs V_R), the better the resolution. The more nearly the plot approaches a straight line the less the distortion and the simpler the treatment of the data.

Since the objective of GPC is to separate species by their molecular size in solution, it is desirable to have some means of describing how well a column will do this job. Not only is the information of value to the practitioner to guide him in selecting the proper column or columns to successfully carry out a specific analysis, but such information is necessary for comparison of data between laboratories. Any published data should also contain descriptions of the columns used and of their resolution capability.

The separation efficiency of columns has usually been compared by measuring the number of theoretical plates using equations originally developed for liquid-liquid chromatography (1)

$$N = 16 \left(\frac{V_R}{W} \right)^2 \tag{1}$$

39

where V_R equals retention volume and W equals peak base width. As pointed out by Cazes (2) "The theoretical plate concept is borrowed from that area of chemical engineering involving fractional distillation. A theoretical plate, in the case of distillation, refers to a discreet distillation stage constituting a simple distillation in which complete equilibrium is established between the liquid and vapor phases. In the case of gel permeation chromatography where the two phases are in constant motion, i.e., the solvent in the interstitial volume and the solvent within the gel pores, equilibrium is probably never achieved. The true significance of the theoretical plate is lost. It must be realized, moreover, that the calculated theoretical plate in a chromatographic column represents smaller separating ability than the theoretical plate in a distillation column by a factor of twenty-five to fifty." Quite often plate count, which is the number of theoretical plates per foot,

$$PC = \frac{N}{L} = \left(\frac{16}{L}\right)\left(\frac{V_R}{W}\right)^2 \qquad (2)$$

with $L =$ total column length in feet, or its reciprocal, the "height equivalent to a theoretical plate"

$$H = \frac{1}{PC} = \left(\frac{L}{16}\right)\left(\frac{W}{V_R}\right)^2 \qquad (3)$$

are used instead of theoretical plates since these terms take into consideration the total column length.

As pointed out by Giddings (3), gel permeation is unique in that there is a well-defined limit to peak capacity. The reason is that the peaks are confined to a definite retention volume range determined by the interstitial volume at one end (V_0) and the maximum solvent volume (both in and out of the pores) at the other end (V). The largest molecules with no penetration will elute at the first limit and the smallest molecules at the second. In most other forms of chromatography, while there is a similar lower limit to retention volume, the upper limit is indefinite and may be 10 or 100 times larger than the lower limit.

Assuming a fixed number of plates, N, equal for each solute and a minimum base line separation, Giddings (3) developed the following highly simplified expression for peak capacity of a permeation column.

$$n \cong 1 + 0.2N^{1/2} \qquad (4)$$

where n is the number of peaks resolvable. Table 1 compares the

TABLE 1

Comparative Peak Capacity of Gel Filtration and Other Columns
for Given Numbers of Theoretical Plates[a]

Theoretical plates (N)	Peak capacities (n)		
	Gel chromatography	Gas chromatography	Liquid chromatography
100	3	11	7
400	5	21	13
1,000	7	33	20
2,500	11	51	31
10,000	21	101	61

[a] From Ref. *3*.

number of peaks resolvable at various theoretical plate levels, using
Eq. (4), for gas, liquid, and gel permeation chromatography. Gel
columns generally have around 1000 theoretical plates and therefore
may be expected to separate a maximum of about seven peaks.

Plate counts determined by Eq. (2) are more a measure of how well
a column is packed, that is, how much peak spreading it will cause,
than how well it will resolve. This is clearly illustrated in Fig. 1. Two
GPC traces for a polystyrene are shown. The traces were obtained
under identical conditions except in one case (A) four Porasil columns

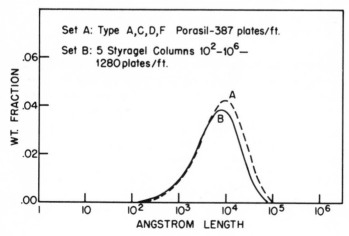

FIG. 1. Chromatograms of polystyrene. Conditions identical except for
columns.

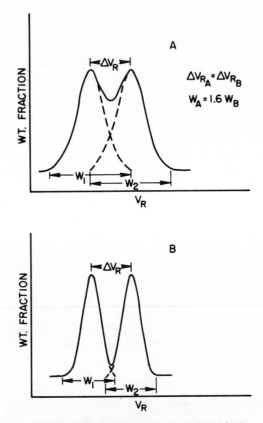

FIG. 2. Effect of band spreading on peak resolution.

were employed and in the other (B) five Styragel columns. The plate counts were 387 and 1280/ft, respectively, yet the two curves show nearly equal resolution. No corrections were applied to the data.

The effectiveness of the GPC separation is determined by the distance separating the centers of the peaks of two species and the width of the two peaks. This is illustrated in Fig. 2. The conventional definition of resolution (R) in chromatography is

$$R = \frac{2(V_{R_2} - V_{R_1})}{W_1 + W_2} \tag{5}$$

where V_{R_1} and V_{R_2} are the retention volumes of species 1 and 2, respectively, measured from injection to the peak maximum and W_1 and W_2 are the base widths measured between two tangents drawn on the

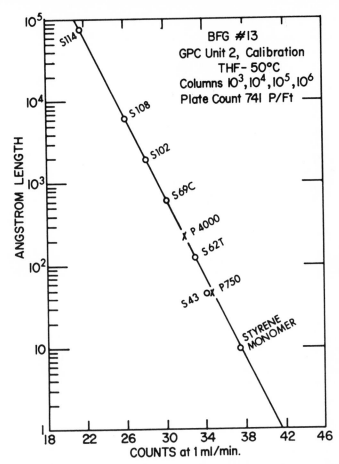

FIG. 3. Typical calibration curve for Styragel columns.

points of inflection of the curve and extended to the base line (4).
For complete separation R must be equal to or greater than 1.

For high molecular weight, polydispersed materials resolution decreases with increasing molecular weight. This is readily apparent if one looks at a calibration curve such as shown in Fig. 3 where Angstrom size or molecular weight is plotted against elution volume for a 20-ft Styragel column system. The reason resolution decreases with increasing molecular weight is that fewer pores are available to the large molecules than are available to the small molecules. A realistic approach, then, is to calculate some sort of resolution index

TABLE 2

Resolution Indices for GPC Columns

Col- umn	Pore size (A°)	Plates/ ft.	Solvent	Flow rate (ml/ min)	Resolution index	
					2×10^5	8.6×10^5
A_1	10^6	450	THF	1	0.097	0.10
A_2	10^6	990	THF	1	0.21	0.24
B	7×10^5	1370	THF	1	0.062	0.11
C	4×10^5	243	THF	1	0.15	0.23
D	3×10^4	103	THF	1	0.19	0.17
E	10^4	800	THF	1	0.28	0.03
Combination of columns					Measured Calc.	Measured Calc.
$A_2 + C$			THF	1	0.32 0.29	0.38 0.36
$A_2 + C + E$			THF	1	0.43 0.41	0.38 0.35
$A_1 + C + D + E$			THF	1	0.39 0.42	0.36 0.37
$A_1 + C + D + E$			THF	1	0.39	0.40

[a] Data taken from Ref. 6.

that takes into consideration molecular weight and base width. Bly (5) defines a specific resolution (R_s) based on the assumption that the two polymer species used to determine R_s have the same molecular weight distribution and that

$$V_e = A - B \log M \tag{6}$$

which usually holds over a fairly wide range of molecular weight. Bly's equation is

$$R_s = \frac{2(V_2 - V_1)}{(W_1 + W_2)(\log \bar{M}_{w_1} - \log \bar{M}_{w_2})} \tag{7}$$

where V and W are as before and $\bar{M}_{w_1}$ and $\bar{M}_{w_2}$ are the molecular weights of the two species used to determine R_s.

Feldman and Smith (6) have proposed a similar sort of resolution

$$RI = \left(\frac{M_1}{M_2}\right)^{W_{12}/(P_1 - P_2)} \tag{8}$$

index where W_{12} is the average width of the two species and P is the peak position or retention volume. If one assumes that the relative displacements of two polymer peaks are additive in a series combination of columns and that the square of the peak width for the series is equal to the sum of the squares of the individual peak widths, i.e., $W^2 = \Sigma_i W_i^2$, the resolution index (RI) of a series combination of columns is given by

$$\frac{1}{\log (RI)} = \Sigma_i \left[\frac{1}{\log (RK)_i} \right] \left[\frac{W_i}{(\Sigma_j W_j^2)^{1/2}} \right] \qquad (9)$$

Table 2 illustrates how plate count may be misleading as an indicator of separation ability and how RI values calculated using Eq. (9) agree with measured values.

Acknowledgments

The author wishes to acknowledge the courtesy of the following editors and publishers for permission to reproduce the designated material: The American Chemical Society, publishers of *Analytical Chemistry*, Table 1; Interscience Publishers, a division of John Wiley and Sons, Inc., publishers of *Journal of Polymer Science*, Table 2.

REFERENCES

1. A. J. P. Martin and R. L. M. Synge, *Biochem. J.*, **35**, 1358 (1941).

2. J. J. Cazes, *J. Chem. Educ.*, **43**, A625 (1966).

3. J. C. Giddings, *Anal. Chem.*, **39**, 1027 (1967).

4. J. C. Giddings, *Dynamics of Chromatography*, Dekker, New York. 1965.

5. D. D. Bly, *J. Polym. Sci., Part C*, **21**, 13 (1968).

6. W. V. Smith and G. A. Feldman, *J. Polym. Sci., Part A-2*, **7**, 163 (1969)

A Review of Peak Broadening in Gel Chromatography

RICHARD N. KELLEY* and FRED W. BILLMEYER, JR.

DEPARTMENTS OF MATERIALS AND CHEMISTRY
RENSSELAER POLYTECHNIC INSTITUTE
TROY, NEW YORK 12181

Summary

Current theories of peak broadening in gel permeation chromatography are discussed in detail. Factors contributing to dispersion in liquid chromatographic systems are reviewed with regard to their significance in GPC. Published data on column efficiency obtained with various columns and packings (both GPC and gel filtration) are compared and interpreted using existing theories.

INTRODUCTION

When a monodisperse substance is passed through a chromatographic column, it is eluted with a distribution of retention times. Everyone is familiar with such broadening phenomena, yet the detailed mechanisms responsible are so complex and varied in nature that they are not well understood. Knowledge of the variables affecting broadening in chromatographic and packed-bed processes is essential for the optimization of process conditions, as well as for the design of improved separation systems. Several theories (1–12) have been developed predicting broadening behavior based upon idealized macroscopic models. These theories do not imply, however, that the same type of idealized dispersion mechanisms occur on the microscopic scale during the passage of individual molecules through a packed column. Rather, they are mathematical models of idealized physical processes that allow us to predict broadening behavior.

* Present address: Eastman Kodak Company, Roll Coating Division, Kodak Park, Rochester, N.Y. 14650.

A chromatographic separation process can be considered to occur in a series of hypothetical steps, within each of which equilibrium is achieved between solute concentrations in the mobile and stationary phases. Each step is termed a plate and is considered to correspond to a specific height of the column. Although the actual separation does not occur in this manner, the concept of height equivalent to a theoretical plate (H) has proved to be extremely useful for characterizing column efficiency.

The number of theoretical plates, N, can be easily calculated from the chromatogram of a monodisperse substance as

$$N = (T_r/\sigma)^2 \tag{1}$$

where T_r is the mean retention time and σ is the standard deviation. The number of theoretical plates, N, is mathematically defined in terms of the chromatogram (concentration–time distribution), since the retention time, T_r, is the first moment of the distribution and the variance, σ^2, is the second moment of the distribution taken about the mean. In practice, both T_r and σ^2 should be corrected for the finite time and width associated with the sample injection. Once N has been determined, the plate height H may be obtained from

$$H = L/N \tag{2}$$

where L is the column length.

Because of the mathematical nature of the plate height definitions, most of the theoretical expressions developed to predict broadening have utilized this concept. In general, plate height H is predicted in terms of operating variables such as the average carrier (solvent) interstitial velocity (U), the solute molecular diffusivity D_m, and column geometry factors such as packing particle size, column radius, and column length. When correlating plate height data, it is often convenient to define a reduced plate height h and a reduced velocity v according to the following relationships:

$$h = H/d_p \tag{3}$$
$$v = Ud_p/D_m \tag{4}$$

where d_p is the effective particle diameter. It should be noted that the reduced velocity is analogous to the product of the Reynolds number times the Schmidt number and is also a Peclet number based on the molecular diffusivity. The utility of these relationships will become apparent later in the paper.

Phenomena such as eddy diffusion, molecular diffusion, velocity-profile effects (or nonequilibrium effects) in the mobile phase, dead-volume effects, sorptive effects, viscosity effects (such as viscous fingering), and dispersion due to diffusion into and out of the pores (the permeation process) may be encountered to varying degrees in separation processes such as gel permeation chromatography (GPC) or gel filtration (9). Broadening can be further separated into mobile-phase dispersion occurring in the absence of permeation or adsorption, and broadening associated with the separation mass-transfer process itself. It is important to understand the dispersion contributions occurring in the mobile phase (in the absence of mass transfer) before attempting to interpret the overall dispersion present in actual chromatographic systems.

There are often objections to the use of plate height since it depends upon so many variables. In another approach, Hendrickson (13) postulated that peak spreading obeys an equation of the form

$$\bar{W}_b^2 = \bar{W}_m^2 + \bar{W}_a^2 + \bar{W}_i^2 + \bar{W}_d^2 + \bar{W}_s^2 \tag{5}$$

where $\bar{W}_b$ is the observed width (at the base) of the chromatogram, and the other terms represent, in order, the contributions to that width from the molecular-size distribution of the test sample, the apparatus, spreading in the interstitial volume within the column, diffusional spreading due to holdup of molecules within the pores of the gel, and sorption.

Excellent reviews of GPC and gel filtration have been presented by Altgelt and Moore (14), Altgelt (15), and Determan (16), but no comprehensive review correlating broadening theory and experimental data has appeared. It is the purpose of this paper to discuss current theories of peak broadening with respect to published data on column efficiency obtained with various columns and packings. Mobile-phase dispersion effects will be reviewed to provide a background for understanding broadening behavior in gel chromatography. Factors contributing to peak broadening in liquid chromatographic systems will be reviewed with regard to their significance in gel chromatography.

THEORY

In general, plate height H may be related by the equation:

$$H = 2D/U + \text{(a mass transfer contribution)} \tag{6}$$

where D is an overall dispersion number. Utilizing this equation, van

Deemter (1) assumed that the dispersion number was composed of a longitudinal molecular-diffusion term and an eddy-diffusion term, and that the mass-transfer equation was a linear function of carrier velocity:

$$D = \underbrace{\phi D_m}_{\substack{\text{molecular} \\ \text{diffusion}}} + \underbrace{\lambda U d_p}_{\substack{\text{eddy} \\ \text{diffusion}}} \tag{7}$$

where ϕ is a tortuosity factor ($\phi = 2/3$) and λ is an eddy-diffusion proportionality factor ($\lambda \approx 1/11$). This resulted in an overall equation for plate height of the general form

$$H = A + (B/U) + CU \tag{8}$$

where the first term accounts for eddy diffusion, the second term for molecular diffusion, and the third term for mass-transfer resistances in the stationary and mobile phases. In the absence of mass-transfer processes, and with appropriate substitutions, Eq. (8) becomes

$$H = 2\lambda d_p + 2\phi D_m/U \tag{9}$$

which applies to mobile-phase dispersion effects.

For some chromatographic systems, however, van Deemter's approach did not correlate well with experimental results. Perhaps the most important reason for lack of correlation was the difficulty in accounting adequately for the complex velocity and flow profiles existing within a packed column. The concept of eddy diffusion is highly idealized and assumes that particles remain fixed in stream lines, that stream lines split when they impinge directly on a particle, and that perfect mixing occurs between particles when stream lines combine. In addition, this approach did not account adequately for variations in aspect ratio (δ = column diameter/particle diameter) and variations in packing uniformity. Further objections arose to the form of the mass-transfer contribution to plate height. It was to explain these deviations between theory and experiments that Giddings (2–4) developed his "coupling theory." In this theory, the eddy diffusivity is coupled in a nonadditive manner with the mobile-phase resistance to nonequilibrium mass transfer. In general form, the coupling equation theory of Giddings may be written

$$H = (B/U) + C_s U + \sum_i \frac{1}{(1/A_i) + (1/C_{mi}U)} \tag{10}$$

where U is the carrier velocity, B takes into account molecular diffusion, C_s accounts for mass-transfer effects in the stationary phase, A_i accounts for eddy diffusion, and C_{mi} takes into account mass-transfer effects in the mobile phase. For describing mobile-phase dispersion in the absence of mass transfer between a stationary and a mobile phase, Giddings' coupling theory predicts

$$H = \frac{2\gamma D_m}{U} + \sum_i \frac{1}{(1/2\lambda_i d_p) + (D_m/\omega_i d_p^2 U)} \tag{11}$$

where γ, λ_i, and ω_i are geometrical constants. In terms of the reduced parameters h and v, Eq. (11) becomes

$$h = 2\gamma/v + \sum_i \frac{1}{(1/2\lambda_i) + (1/\omega_i v)} \tag{12}$$

Giddings and Mallik (7) applied Eq. (10) specifically to the theory of zone broadening in gel filtration (permeation) chromatography, defining the parameters A_i, B, C_s, and C_{mi} in the context of the GPC process. Because of the obstructions to diffusion posed by the gel network, the diffusion coefficient for the solute in the mobile phase (D_m) differs from that in the stationary phase (D_s). An obstruction factor for the gel was defined as $r = D_s/D_m$ and was estimated to be approximately equal to 2/3. A reduced velocity (R) was defined as the ratio of the zone velocity to the mobile-phase velocity. The final form of the equation proposed for plate height is:

$$H = \frac{4}{3}\frac{D_m}{RU} + \frac{1}{20} R(1-R) \frac{d_p^2 U}{D_m} + \sum_i \frac{1}{(1/2\lambda_i d_p) + (D_m/\omega_i d_p^2 U)} \tag{13}$$

where λ_i and ω_i are geometrical factors of order unity.

For gel chromatography, the variation of reduced plate height with reduced velocity according to Eq. (13) is shown in Fig. 1. The curve is concave downward at low flow velocities, becoming linear at higher reduced velocities. The magnitude of the plate height predicted is quite dependent upon the evaluation of the coupling term.

Another approach (7–9) which helps to explain the origin of coupling has been to extend van Deemter's theory by incorporating an additional term to account for velocity-profile effects caused by a nonuniform velocity over the column cross section. A nonuniform velocity profile causes a spread in retention times whose magnitude is deter-

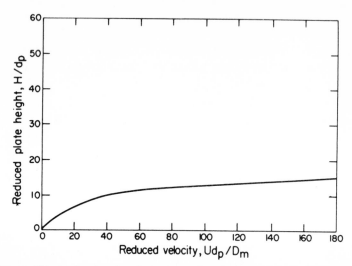

FIG. 1. Effect of reduced velocity on reduced plate height according to the theory of Giddings and Mallik (7).

mined primarily by radial (transverse) diffusion. Johnson (12) applied this approach to gaseous systems and stated that a major portion of the observed broadening in liquid systems results from velocity-profile effects. Sie and Rijnders (6) found this approach valuable for describing band broadening in packed chromatographic columns for gaseous systems, and pointed out that it may also be useful in liquid chromatography. We have previously indicated (8, 9, 17) the value of utilizing a velocity-profile approach for describing dispersion effects occurring in the mobile phase and of its incorporation into an overall model for describing broadening in GPC and gel filtration (8, 18).

Most chromatographic column-packing materials have a relatively wide particle-size range. During column packing, the particles can segregate, producing some cross sections having a small-particle-diameter packing and others having large-diameter packing. Such particle-size segregation, coupled with variations in packing density, presents variable resistance to the flowing fluid and leads to a nonuniform velocity profile. Because the column void fraction is greater near the wall, the flowing fluid encounters less resistance there, and the average velocity near the wall is correspondingly greater than in the center of the bed. Several investigators have reported this behavior (12, 19–22).

The overall longitudinal-dispersion number D is therefore assumed to be the sum of contributions from molecular diffusion, eddy diffusion, and velocity-profile effects operating in the column:

$$D = \underbrace{\phi D_m}_{\substack{\text{molecular} \\ \text{diffusion}}} + \underbrace{\lambda U d_p}_{\substack{\text{eddy} \\ \text{diffusion}}} + \underbrace{\hbar R_c^2 U^2 / \bar{D}_r}_{\substack{\text{velocity-profile} \\ \text{effects}}} \tag{14}$$

where $\hbar$ is a velocity-profile constant, R_c is the column radius, and $\bar{D}_r$ is an average radial diffusivity. This form of the expression for the overall dispersion number was developed by Taylor (23) and Aris (24).

In cases where a mass-transfer process such as adsorption or permeation is absent or negligible, the plate-height equation becomes:

$$H = (2\phi D_m / U) + 2\lambda d_p + (2\hbar R_c^2 U / \bar{D}_r) \tag{15}$$

The utility of this equation for describing mobile-phase dispersion in actual liquid chromatographic systems depends greatly on the evaluation of the average radial diffusivity $\bar{D}_r$ and the appropriate velocity-profile constant $\hbar$ for the experimental column used. Previous work (6, 9, 25–27) has shown that the radial diffusivity is determined by radial gradients existing within the column which come about from both molecular-diffusion and eddy-diffusion processes. These in turn depend upon the variation of velocity across the column:

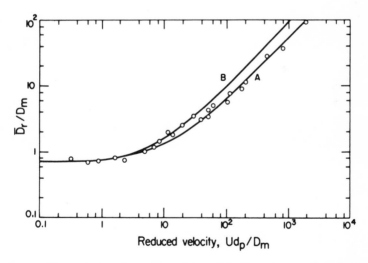

FIG. 2. Theoretical and experimental radial dispersion coefficients for randomly packed beds of uniform-size beads. Curve A was calculated using the equation $\bar{D}_r / D_m = 2/3 + 0.091\ U d_p / D_m$. Curve B was calculated using the equation $\bar{D}_r / D_m = 0.7 + 0.055\ U d_p / D_m$. Experimental data were taken from Grane and Gardner (28) and Blackwell (29). This figure was, in part, redrawn from Perkins and Johnston (25).

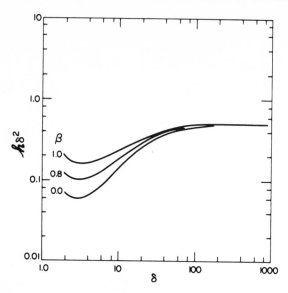

FIG. 3. Effect of aspect ratio (δ) on the dispersion parameter ($\hbar\delta^2$) for packed beds of uniform-size beads. Figure redrawn from Johnson (*12*). β = a dispersion intensity ratio = $\lambda U d_p / (\lambda U d_p + \phi D_m)$.

$$D_r(\rho) = \phi D_m + \lambda d_p U(\rho) \qquad (16)$$

where ρ is a dimensionless radial position variable: $\rho = r/R_c$, where r is the actual radial position variable. From Eq. (16), the average radial diffusivity is:

$$\bar{D}_r = \phi D_m + \lambda d_p U \qquad (17)$$

Figure 2 shows that Eq. (17) gives excellent agreement with experimental data from the literature, and therefore the average radial diffusivity may be predicted accurately.

Perhaps the biggest drawback for utilizing the velocity-profile model in the past has been the difficulty in evaluating the velocity-profile constant $\hbar$. Johnson (*12*) has rigorously evaluated $\hbar$ for packed-bed systems as a function of aspect ratio and β (which is a measure of the relative importance of eddy diffusion versus the combination of eddy and molecular diffusion). As a result of Johnson's work, it is now possible to determine graphically with the aid of Fig. 3 (*12*) the velocity-profile constant for the particular column system of interest (*9, 12*). The calculations with the velocity-profile model then become relatively simple.

Incorporating these concepts, the plate height equation describing *mobile-phase dispersion* with a velocity-profile model becomes:

$$H = 2[(\phi D_m/U) + \lambda d_p] + 2\hbar R_c^2/[(\phi D_m/U) + \lambda d_p] \qquad (18)$$

It is obvious that the denominator of the second term of Eq. (18) represents a "coupling" of the eddy and molecular diffusivities, similar to that of Giddings. From the form of Eq. (18) it appears that for a given particle diameter the contribution to plate height from the velocity-profile term increases with the square of the column radius. This is not so, since the last term of Eq. (18) may be rewritten

$$2\hbar R_c^2/[(\phi D_m/U) + \lambda d_p] = \hbar\delta^2 d_p^2/2[(\phi D_m/U) + \lambda d_p] \qquad (19$$

At constant d_p this contribution to plate height varies linearly with the dispersion parameter $\hbar\delta^2$ which is nearly constant at aspect ratios above 40, the region of practical importance.

Kelley and Billmeyer (*18*) recently proposed that the effect of increased broadening in GPC and gel filtration due to the permeation of solute molecules into and out of the pores may be diffusion controlled. To account for such broadening, a term which is linear with carrier velocity is added to Eq. (18) to provide an overall equation for plate height in GPC:

$$H = 2[(\phi D_m/U) + \lambda d_p] + 2\hbar R_c^2/[(\phi D_m/U) + \lambda d_p] + CU \qquad (20)$$

A schematic representation of the relative importance of the terms in this equation is given in Fig. 4.

Huber (*10, 11*) has recently developed a theory to describe broadening in liquid chromatographic systems. The overall expression for plate height is expressed as a sum of individual contributions which include

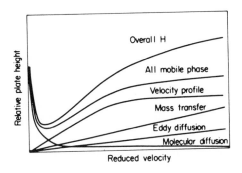

FIG. 4. Schematic representation of the relative contributions to plate height with a velocity-profile model.

mixing arising from diffusion in the mobile phase (H_{md}), mixing due to convection in the mobile phase (H_{mc}), and contributions from mass-transfer effects in the mobile (H_{em}) and in the stationary phases (H_{es}). These effects were assumed to be independent and additive quantities (as has been assumed in previous theories) such that

$$H = H_{md} + H_{mc} + H_{em} + H_{es} \qquad (21)$$

where

$$H_{md} = (2\epsilon_m/\tau_m)(D_{ia}/U) \qquad (22)$$

$$H_{mc} = 2\lambda d_p/\{1 + \lambda_2(D_{ia}/Ud_p)^{1/2}\} \qquad (23)$$

$$H_{em} = (1/5.7)\left(\frac{\epsilon_a - \epsilon_m + K_i\epsilon_\beta}{\epsilon_a + K_i\epsilon_\beta}\right)^2 \frac{\epsilon_m^{1/2}d_p^{3/2}\nu_\alpha^{1/6}U_{ia}^{2/3}}{(1 - \epsilon_m)D_{ia}^{1/2}} \qquad (24)$$

$$H_{es} = (1/30)\frac{\epsilon_a - \epsilon_m + K_i\epsilon_\beta}{(\epsilon_a + K_i\epsilon_\beta)^2}\frac{\epsilon_m(1 - \epsilon_m)\tau_s}{\epsilon_s}\frac{d_p^2 U}{D_{is}} \qquad (25)$$

ϵ_m = the interstitial volume fraction

ϵ_s = the internal pore volume fraction occupied by stationary fluid

ϵ_a = volume fraction occupied by the moving phase

ϵ_β = volume fraction occupied by the stationary phase

λ_1 and λ_2 = geometric constants

τ_m, τ_s = tortuosity factors

K_i = distribution coefficient equal to the ratio of the equilibrium concentrations in the stationary and mobile phases

ν_α = the kinematic viscosity of the mobile phase

D_{ia} = the diffusion coefficient in the mobile phase

D_{is} = the diffusion coefficient in the stationary phase

Huber's approach takes into account the specific volume fractions in which the broadening processes are occurring. The overall plate height curve as a function of fluid velocity is concave downward with decreasing flow rate until molecular diffusion effects begin to dominate where a sharp rise in H is observed. The effect of H_{md} is less important at high flow rates. Contributions due to H_{em} and H_{mc} are both concave downward with decreasing flow rate, with the larger contribution coming from H_{mc}. The term H_{es} is linear with flow rate. The overall plate height behavior is qualitatively very similar to that of Giddings (7) and Kelley and Billmeyer (9, 18).

Broadening occurring in the mobile phase results by definition from all dispersion processes that affect the residence time distribution but in which the solute molecules are not removed from the mobile phase. Adsorption and permeation are mass transfer processes which involve a stationary phase in which the solute molecules become temporarily entrapped during their passage through the column. Major sources of mobile phase dispersion included in the theoretical models just discussed are molecular diffusion, eddy diffusion, and nonequilibrium effects associated with velocity variations due to packing geometry. Dispersion associated with stagnant regions, boundary-layer phenomena, natural convection effects resulting from density gradients, viscous fingering, and other concentration effects have not been incorporated into the mathematical dispersion models. All of these effects could be present to varying degrees in an actual column system and would lead to disagreement between theory and data. It is probable that stagnant regions or "dead volume" areas do exist in real systems. Transfer into and out of such dead volume is diffusion controlled and would give a linear contribution to plate height. At flow rates normally encountered in gel chromatography, boundary layers developed around the particles should be very thin and would not appreciably affect the observed dispersion. Grashoff numbers (measures of natural convection) are estimated to be well below 1.0 for gel chromatography at normal flow rates, indicating that natural convection should not be important. The low Grashoff numbers result from the generally small particle sizes employed, together with the very low solute concentrations.

MOBILE-PHASE DISPERSION IN NONPOROUS SYSTEMS

Many excellent reviews of mobile-phase dispersion phenomena in gaseous (4, 12, 30, 31) and liquid (32–35) systems have been published. There is, however, a need for further research into the basic nature of dispersion occurring under conditions such as those encountered in the liquid chromatography of macromolecules. The following discussion deals only with dispersion data applicable to gel chromatography.

1. Extra-Column Effects

Before accurate dispersion measurements can be made with packed columns, the extra-column broadening associated with the sample injection and detection system must be evaluated and minimized. To evaluate these contributions to peak broadening, samples are usually injected through the sample loop directly into the detector with no column in the system. Billmeyer and Kelley (17) noted considerable tailing (asymmetry) and other anomalies in the injection-detection pattern of samples of macromolecular solutes utilizing the Waters Model 100 Gel Permeation Chromatograph with a standard refracto-meter cell (volume 70 μl) and a 1/16-in. null glass. Similar effects have been noted by other investigators (10, 36–38).

Huber and Hulsman (10) utilized a modified flow arrangement with a Waters type R4 differential refractometer containing a micro re-fractometer cell having a volume of 10 μl to reduce mixing in the ap-paratus outside the column during the liquid chromatography of small molecules. Billmeyer and Kelley (17) minimized extra-column broaden-ing and associated flow anomalies by changing to a micro refracto-meter cell, minimizing the amount of tubing in the system (and elim-inating the heat exchanger), using short injection times, and reducing the solute concentration. By making the above modifications, the extra-column contributions to broadening were made practically negligible compared to the broadening occurring during passage through a single GPC column.

Carmichael (39) has formally evaluated the extra-column and column contributions to the chromatogram. Through knowledge of the individual extra-column and column effects associated with mono-disperse substances, the column contribution to the elution curve for an arbitrary molecular size distribution can be determined. The in-dividual column contributions can be mathematically evaluated for monodisperse substances by using a stochastic model for gel chroma-tography (40–42) and assuming a Gaussian distribution (43) for the chromatogram. Extra-column effects are separately evaluated for the system without the column.

2. Diffusivity and Flow Rate Effects

Axial dispersion of molecules during flow through packed beds, such as chromatographic columns, is known to be greatly dependent upon the amount of radial mixing. The importance of radial processes necessitates further discussion of the mechanisms involved. In Fig. 2, radial dispersion was shown to be due to the combination of molecular diffusion and eddy diffusion. Fickian molecular diffusion due to a con-centration gradient, occurring in both the radial and axial directions, is a well-known concept. Molecular diffusion becomes extremely im-portant at very low flow rates. Eddy, or as it is sometimes referred to, convective diffusion, is lateral transport caused by "stream splitting." A stream line impinging upon a particle cannot go through,

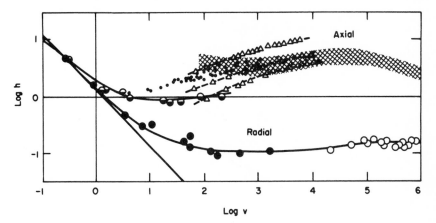

FIG. 5. Variation of axial and radial reduced plate height as a function of reduced velocity. Symbols: ○, Bernard and Wilhelm (*45*); ●, Blackwell (*29*); ◑, Brigham, Reed, and Dew (*46*); ◒, Kelley and Billmeyer (*9, 47*), data for 120–140 mesh glass bead column with solutes of varying diffusivity; △, Knox (*48*). Cross-hatched area represents the range of data of Ebach and White (*49*), a total of 50 data points.

so it splits, with part of the original stream line going one direction and the other part in the opposite direction. By this idealized process the original stream line is displaced laterally by one-half the particle diameter. Eddy diffusion predominates at high flow rates. Sie and Rijnders (*6*) have explained the mechanism of eddy diffusion in detail and have pointed out that the magnitudes of the eddy and molecular diffusivities are often nearly equivalent in liquid chromatography. Hiby (*44*) has experimentally demonstrated stream splitting by following stream lines of dye in an idealized two-dimensional bed of spheres.

Axial and radial dispersion data of several investigators (*9, 29, 45–49*) are summarized in Fig. 5. Radial plate heights are considerably lower than those observed from axial dispersion except at low reduced velocities where molecular diffusion dominates. (A range of reduced velocity between 10 and 10^4 is commonly encountered in the gel chromathography of macromolecules.) Axial dispersion results (*9, 47*), obtained from measurements utilizing high molecular weight solutes and nonporous glass bead GPC columns show good agreement with literature data.

The effect of molecular diffusivity on the resulting band broadening as a function of flow rate (expressed in terms of Reynolds number) is illustrated in Fig. 6. The diffusivities cover a wide range, from approxi-

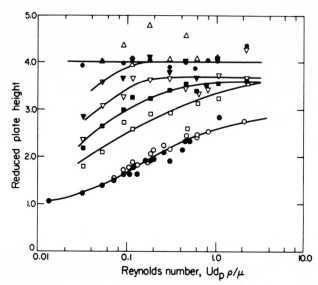

FIG. 6. Reduced plate height vs Reynolds number for 120–140 mesh non-porous glass bead column. ●, Hexane; ○, cyclohexane; □, n-$C_{36}H_{74}$; ■, 2,000 PS; ▽, 3,600 PS; ▼, 10,300 PS; ⊕, 19,800 PS; △, 97,200 PS. Toluene solvent at room temperature. Data of Kelley and Billmeyer *(9, 47)*.

mately 2×10^{-5} cm²/sec for hexane to 5×10^{-7} cm²/sec for the 97,200 molecular weight polystyrene standard. Plate height curves for low molecular weight materials are concave downward with decreasing flow rate, while dispersion is relatively unaffected by flow rate for high molecular weight substances such as 97,200 polystyrene. Similar concave downward curvature has been noted by Huber *(11)* for small solute molecules with inert diatomaceous earth packings. Qualitatively, this dependence of broadening on diffusivity is predicted theoretically by the models of Giddings *(7)* and Huber *(10, 11)*, and by the velocity-profile model *(8, 9)*. It is not consistent, however, with van Deemter's original theory *(1)*.

Horne, Knox, and McLaren *(34)* stated that the major factor producing dispersion between the flow region where molecular diffusion dominates and that where turbulence dominates is the slow rate at which transcolumn equilibrium is achieved. Kelley and Billmeyer *(9)* found that the major source of mobile phase dispersion was the velocity-profile term which includes the radial diffusivity in the denominator.

Dispersion caused by lack of equilibrium or velocity-profile effects usually increases with decreasing radial diffusion. This term becomes especially important with high molecular weight solutes having very low molecular diffusivities.

3. Aspect Ratio and Particle Size Distribution Effects

The aspect ratio, or the column diameter divided by the particle diameter, is very important in determining the magnitude of wall, velocity-profile, or transcolumn nonequilibrium effects. Experimentally, aspect ratio dependence may be investigated either by changing the particle diameter at constant column diameter or by changing the column diameter for a given packing. Both procedures have limitations. Changing the particle diameter may change the particle size distribution (when nonuniform diameter materials are employed) by introducing an additional variable. Changing the column radius may lead to additional dispersion associated with entrance and exit effects. Column packing uniformity is generally difficult to repeat experimentally since it is very dependent upon the packing techniques employed.

Particle size variations in a given packing material are known to cause segregation during packing which leads to increased dispersion in gaseous systems. Such dependence of dispersion on the breadth of the particle-size distribution has not been extensively investigated in liquid chromatography until recently.

Kelley and Billmeyer (9) found that decreasing the particle size leads to improved column efficiency, as shown in the plate height curves in Fig. 7. Smaller particle sizes both reduce the eddy-diffusion contribution to broadening and lead to reduced wall effects associated with the larger aspect ratios. Such behavior is confirmed by experimental results of Huber (11). At very high aspect ratios (greater than 100), theory predicts that the dispersion should be relatively insensitive to further increases in aspect ratio. Most gel chromatography systems employ aspect ratios over 100, as in GPC with standard column diameters of 0.775 cm and particle diameters of less than 40 μ.

Knox and Parcher (27) examined the effects of aspect ratio on mobile-phase dispersion in liquid chromatography. Most of the data were obtained at low aspect ratios to determine dispersion variations caused by wall effects and associated transcolumn nonequilibrium. Giddings' nonequilibrium theory (2–4) was extended to include the effects of aspect ratio and varying velocity profile. At high aspect

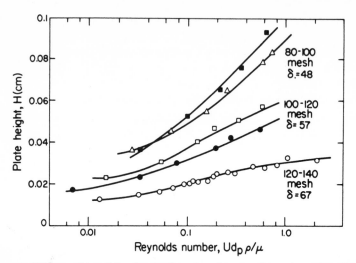

FIG. 7. Effects of particle size and segregation on dispersion of cyclo-hexane. Conditions same as in Fig. 6. ○, 120–140 mesh; □, 100–120 mesh; △, 80–100 mesh; ■, 1/1 by volume 80–100 mesh/100–120 mesh; ●, 1/1/1 by volume 80–100 mesh/100–120 mesh/120–140 mesh.

ratios (about 10), the trans-column contribution to plate height was found to become independent of column diameter. A sharp increase in plate height was observed in the range $6 < \delta < 8$.

Horne, Knox, and MaLaren (*34*) examined the importance of particle-size distribution on plate height and found only slight increases in broadening as the distribution was increased. Kelley and Billmeyer (*9*) reexamined the importance of the particle-size distribution utilizing higher aspect ratios, as shown in Fig. 7, and found that the breadth of the distribution did not alter the resulting dispersion significantly (with calculations carried out at the average particle size for the distribution). These results indicate that the breadth of the particle-size distribution is relatively unimportant for mobile phase dispersion in liquid systems. This is not true with gaseous systems. Further research is necessary to explain such differences between gaseous and liquid systems.

4. Discussion

The broadening models proposed by Giddings (*2–4*) and Huber (*10, 11*), and the velocity-profile model adapted to liquid systems by Billmeyer and Kelley (*8, 9*) seem to predict the observed trends qualitatively for a wide range of variables. These theories may be utilized to assess the relative importance of operational variables on mobile phase dispersion in gel chromatography. Before their relative merits for predicting dispersion quantitatively can be determined, more extensive experimental data are required. It is clear, however, that the experimental results do not correlate with van Deemter's Eq. (1). It is also apparent that the major source of dispersion is often trans-column nonequilibrium effects associated with the column walls and the velocity profile across the bed.

BROADENING IN POROUS SYSTEMS

The broadening effect caused by a mass-transfer process such as adsorption or permeation can often be accounted for by an additional term in the plate-height equation. In the case of gas chromatography, it has been established (1) that the variation of plate height due to the sorption-desorption process is linear with gas velocity. Such a linear and additive term arises, in part, from a linear adsorption isotherm.

Kelley and Billmeyer (18) studied peak broadening in gel chromatography with nonporous and porous column packings using identical operation conditions. The difference between the results, that is, the contribution to plate height due to permeation into and out of the pores, was a linear function of flow velocity. Such a linear dependence suggests that the broadening associated with permeation may be diffusion controlled.

Variations in plate height were found to be dependent upon the particular gel chromatography packing material employed (18). For a porous glass (Porasil) packing, dispersion arising from permeation was the major contribution to peak broadening, whereas for a poly-styrene-divinylbenzene) (Styragel) packing, dispersion due to mobile-phase effects was predominant. In order to explain these results qualitatively, it is necessary to study the differences in pore structure of the two materials.

For convenience, broadening in gel chromatography will be discussed separately according to the porous packing employed. The packing material most commonly used in analytical GPC is cross-linked poly(styrene-divinylbenzene) manufactured and marketed by Waters Associates under the trade name Styragel. Cross-linked porous dextran gels (Sephadex, Pharmacia Fine Chemicals Inc.) are widely used in biological applications for separations described as gel filtration. The properties of these gels have been described by Altgelt and Moore (*14*). In addition, recently-developed porous glass beads are finding many applications in gel chromatography, especially where chemical inertness and structural stability are prime considerations in selecting the column packing.

Three different types of porous glass beads have been developed for gel chromatography. LePage and de Vries (*38, 50–52*) reported the use of porous silica beads which are manufactured by Pechiney-Saint-Gobain (*53*) and currently distributed by Waters Associates under the trade name Porasil. Haller (*54–56*) developed a different process for producing porous silica beads now made by the Corning Glass Co., with Waters Associates as distributors. Another type of glass beads with the trade name Bio-Glas, having controlled pore size, has been developed by Bio-Rad Laboratories of Richmond, California (*57*). They also produce porous polyacrylamide gels, agarose gels, and poly(styrene-divinylbenzene) beads for gel chromatography. The structure of Bio-Glas has been described by Barrall and Cain (*58*) and its properties as a packing material for GPC have been discussed by Cantow and Johnson (*59, 60*). See also Cooper et al. (*60a*).

1. Gel Filtration

Flodin (*61*) examined the effect of flow rate and particle size on plate height in gel chromatography with Sephadex G-25 gel. Data indicate that as the flow rate was increased, plate height also increased from 0.39 mm at 0.053 ml/cm^2-min to 5.49 mm at 1.01 ml/cm^2-min with uridylic acid as the solute. The increase probably resulted primarily from the increasing importance of nonequilibrium effects at higher flow rates. A decrease in plate height from 1.05 mm at 0.074 ml/cm^2-min to 0.51 mm at 0.51 ml/cm^2-min was noted for hydrochloric acid solute with increased flow rate. The higher plate height at the lower flow rate resulted from axial molecular diffusion due to the higher diffusion coefficient for hydrochloric acid solute. Higher plate heights were associated with higher particle sizes.

Giddings and Mallik (*7*) tabulated plate height results from gel filtration data of Porath (*62, 63*) with Sephadex gel under varying operating conditions. Reduced plate heights were generally above 10 at reduced velocities lower than 50. The high values of plate height and scatter in the data may have resulted from extra-column effects, microscopic packing variations, and particle-size effects. It was concluded that the contribution to plate height from the stationary phase term was relatively small, since stationary-phase diffusion was not rate-controlling.

Edwards (64) applied a reaction kinetic model for fixed-bed processes to gel-chromatography data of Flodin (61). By calculating the distribution coefficients and overall mass-transfer coefficient from the various resistances, the height of a reaction unit together with the number of reaction units could be obtained. Relatively close agreement between the predicted (855) and the observed (653) number of theoretical plates was obtained with sodium chloride solute. However, predicted values for hemoglobin (1630 plates) gave poor agreement with observation (625 plates). This lack of agreement was thought to result from a viscous effect, and the fact that the equations were developed for low molecular weight compounds and may have required modification to account for the larger axial dispersion coefficients of macromolecules.

2. Poly (styrene–divinylbenzene) Gels

Plate-height data of several investigators (8, 13, 65, 66) utilizing Styragel columns are summarized in Fig. 8. Plate heights increase with increasing solvent flow rate. Generally, the curves are concave downward with decreasing flow rate.

Giddings and Mallik (7) reviewed the concave downward plate-height data of Smith and Kollmansberger (65), concluding that they are in qualitative agreement with the coupling theory expressed in Eq. (10) and shown in Fig. 1. Such a concave-downward relationship is predicted by the theories of Giddings (7), Huber (10, 11), and Billmeyer and co-workers (8, 18), but not by that of van Deemter et al. (1).

Heitz (67–69) investigated the variation of plate height with reduced velocity for poly (styrene-divinylbenzene) gels of varying cross-linking density. Separation efficiency was lowered with decreasing cross-linking density. At low reduced velocities (<2), plate height increases due to axial molecular diffusion were observed. Deviations from broadening theory were found to be caused by an alteration of the diffusion coefficient in the stationary phase with respect to that in the mobile phase.

Recently, Little et al. (70) examined the broadening behavior of Styragel columns for fast GPC at flow rates near 10 ml/min. Over the flow rate range from 0.1 to 12.5 ml/min, elution volume was found to be independent of flow rate for both small and large molecules. The flow rate dependence of peak width was found to be significantly lower than predicted by the van Deemter equation. Peak width was found

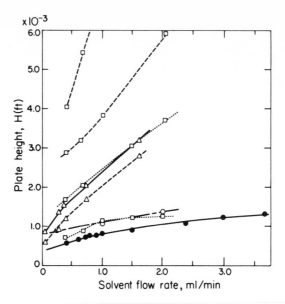

FIG. 8. Comparison of plate height data obtained on Styragel columns. ●, Billmeyer et al. (8); △, Smith and Kollmansberger (65); □, Hendrickson (13); ○, Duerksen and Hamielec (66).

generally to be independent of solute concentration (over a range from 0.05 to 0.5% for small molecules), suggesting that "viscous fingering" effects were minimized by the increased mixing (shear) associated with high velocities. Peak symmetry also increased with increasing flow rate.

A marked resemblance exists between the concave-downward plate height curves obtained in gel chromatography with poly(styrene-divinylbenzene) gels and the plate-height relationships observed for mobile phase dispersion using nonporous beads. Giddings and Mallik (7) stated that the stationary-phase term in gel permeation should not make a significant contribution to plate height until the reduced velocity becomes considerably greater than 100. High relative velocities can therefore be employed to give more rapid separations. These conclusions are supported by the data of Kelley and Billmeyer (18) for Styragel systems.

3. Porous Glass Systems

Reduced plate height data obtained by LePage et al. (38) and by Kelley and Billmeyer (18, 47) for Porasil E columns are compared

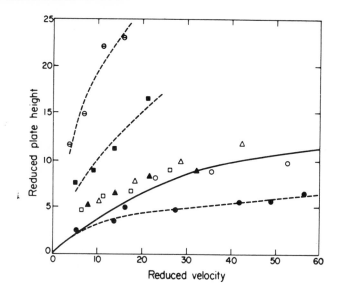

FIG. 9. Comparison of reduced plate height data as a function of reduced velocity for Porasil E. Effect of particle size. Data of LePage et al. (*38*) : ⊖, 60–80 μ; ■, 80–100 μ; □, 100–125 μ; ▲, 125–150 μ; △, 160–200 μ; ○, 200–250 μ. Data of Kelley and Billmeyer (*18, 47*): ●, 100–125 μ.

in Fig. 9. Reduced plate height curves, as a function of reduced velocity, are concave downward and show good agreement between investigators.

When the plate height data of Kelley and Billmeyer (*18, 47*) are plotted as a function of the log of the Reynolds number, a concave-downward relationship is not obtained. Instead, the resulting curve is concave upward showing very rapid increases in plate height with increasing Reynolds number. This serves to illustrate the point that the same data may convey outwardly two entirely different messages depending upon the plotting technique.

The data of LePage et al. (*38*) in Fig. 9 show that the reduced plate height (at constant reduced velocity) decreases as the particle size increases. Reduced broadening at higher particle sizes is difficult to explain in light of the theories of peak broadening which are supported by the data of Flodin (*61*) for gel filtration, together with the mobile phase data of Knox (*27, 34*) and Kelley and Billmeyer (*7*). Variation of pore geometry with particle size, differences in packing techniques, combined with the possibility of agglomeration of the finer particles during packing may account for the discrepancy (*70a*).

Dispersion is generally greater for Porasil columns than for Styragel columns at equivalent reduced velocities (18) and increases more rapidly with increasing flow rate. Reasons for the increased dispersion have been attributed to internal pore structure differences between Porasil and Styragel (18). Essentially, if a molecule can enter a pore in a Porasil bead, it has a greater probability (compared to the case in Styragel) of continuing to diffuse toward the center of the bead before leaving to return to the mobile phase. Such diffusion within the stationary phase leads to increased broadening. We have stated (18) that while mobile-phase dispersion effects are predominant in Styragel columns, the stationary-phase effects (which were found to be linear with flow rate) are predominant in the Porasil systems studied. Pore depth generally increases with particle size in porous glass systems. Broadening occurring in the stationary phase may be minimized by reducing the particle diameter. Reduced particle diameters also lead to shorter interparticle distances and therefore more rapid lateral mixing.

Moore and Arrington (71) investigated the separation mechanism in GPC with porous glass beads furnished by W. Haller of the National Bureau of Standards. Plate height was concave downward with flow rate. Column efficiencies greater than 1000 plates were attained with 122 cm columns.

Sliemers et al. (72) examined factors affecting the efficiency of GPC columns packed with Porasil beads. Column efficiency (as evaluated by plate height) increased with increasing aspect ratio, i.e., efficiency was improved by using smaller size beads. Literature data were compared to the broadening theory of Giddings and Mallik (7).

4. Discussion

Data (18) indicate that mobile phase and stationary phase dispersion processes are independent of each other and contribute additive terms to the resulting plate height in gel chromatography. When the overall plate height is large compared to that due to the mobile phase, it is possible to obtain the stationary phase contribution by subtracting the mobile phase dispersion. Kelley and Billmeyer (18) have noted that the plate height term (stationary phase) arising from the permeation into and out of porous glass particles was a linear function of solvent velocity. The linear relationship suggested that broadening due to permeation is controlled by a diffusion mechanism. A diffusion mechanism for retention is consistent with various theories proposed previously, including those of Yau and Malone (*73*), Casassa (*74*, *75*), and Hermans (*76*).

Generally, the plate height equations proposed by Giddings (*7*), Billmeyer (*8*, *18*), and by Huber (*10*, *11*) show good qualitative agreement with experimental data. Additional results will be required before their relative merits for quantitatively predicting broadening behavior can be assessed accurately. Van Deemter's equation (*1*) does not apply to most broadening results obtained in gel chromatography.

Column efficiency differences related to differences in pore structure between Styragel and Porasil packings suggest that pore geometry is an important factor to be optimized for improved gel filtration systems. A deep interconnected pore structure within the particle should lead to reduced efficiency.

The slowness of achieving radial mixing due to the low molecular diffusivities of macromolecules can be compensated for, in part, by reducing the interparticle distances by using smaller diameter particles while maintaining a relatively high aspect ratio. Smaller particles, however, are more difficult to pack uniformly and agglomeration effects may significantly increase mobile phase dispersion (*70a*). Since molecular diffusivity varies directly with temperature, significant increases cannot be obtained with increased operating temperature.

Little et al. (*7*) have shown that efficient separations are possible at high flow rates. Clearly, for process control applications, future developments will be made for high-speed gel chromatography utilizing small particles, high flow rates, and shorter column lengths, combined with optimization of the pore structure.

Acknowledgments

We wish to acknowledge support of research on GPC by the Texas Division, Dow Chemical Co., and Waters Associates, Inc. The research is carried out in Rensselaer's Materials Research Center, a facility supported by the National Aeronautics and Space Administration.

REFERENCES

1. J. J. van Deemter, F. Zuiderweg, and A. Klinkenberg, *Chem. Eng. Sci.*, **5**, 271 (1956).

2. J. C. Giddings, *Anal. Chem.*, **34**, 1186 (1962).

3. J. C. Giddings, *ibid.*, **35**, 439 (1963); **35**, 1338 (1963).

4. J. C. Giddings, *Dynamics of Chromatography, Part 1*, Dekker, New York, 1965.

5. H. F. Walton, in *Chromatography* (E. Heftmann, ed.), Reinhold, New York, 1961, p. 299.
6. S. T. Sie and G. W. A. Rijnders, *Anal. Chim. Acta,* **38**, 3 (1967).
7. J. C. Giddings and K. L. Mallik, *Anal. Chem.,* **38**, 997 (1966).
8. F. W. Billmeyer, Jr., G. W. Johnson, and R. N. Kelley, *J. Chromatogr.,* **34**, 316 (1968).
9. R. N. Kelley and F. W. Billmeyer, Jr., *Anal. Chem.,* **41**, 874 (1969).
10. J. F. K. Huber and J. A. R. J. Hulsman, *Anal. Chim. Acta,* **38**, 305 (1967).
11. J. F. K. Huber, *J. Chromatogr. Sci.,* **7**, 85 (1969).
12. G. W. Johnson, Ph.D. Thesis, Rensselaer Polytechnic Institute, Troy, N.Y., February 1967.
13. J. G. Hendrickson, *J. Polym. Sci., Part A-2,* **6**, 1903 (1968); "Basic Gel Permeation Chromatography Studies VI. Peak Spreading Causes and Evaluation," presented at the 4th International Seminar on Gel Permeation Chromatography, Miami Beach, Florida, May 22–24, 1967.
14. K. H. Altgelt and J. C. Moore, "Gel Permeation Chromatography," in *Polymer Fractionation* (M. Cantow, ed.), Academic, New York, 1967, pp. 123–179.
15. K. H. Altgelt, "Theory and Mechanics of GPC," in *Advances in Chromatography,* Vol. 7 (J. C. Giddings and R. A. Keller, eds.), Dekker, New York, 1968.
16. H. Determan, *Gel Chromatography: Gel Filtration, Gel Permeation, Molecular Sieves,* Springer, New York, 1968.
17. F. W. Billmeyer, Jr., and R. N. Kelley, *J. Chromatogr.,* **34**, 322 (1968).
18. R. N. Kelley and F. W. Billmeyer, Jr., *Anal. Chem.,* **42**, 399 (1970).
19. R. Rhoades, Ph.D. Thesis, Rensselaer Polytechnic Institute, Troy, N.Y., 1963.
20. C. E. Schwartz and J. M. Smith, *Ind. Eng. Chem.,* **45**, 1209 (1953).
21. M. Morales, C. W. Spinn, and J. M. Smith, *ibid.,* **43**, 225 (1951).
22. E. J. Cairns and J. M. Prausnitz, *ibid.,* **51**, 1441 (1959).
23. G. I. Taylor, *Proc. Roy. Soc.,* **A219**, 186 (1953); **A223**, 446 (1954); **A225**, 473 (1954).
24. R. Aris, *ibid.,* **A235**, 67 (1956).
25. T. K. Perkins and O. C. Johnston, *Soc. Petrol. Eng. J.,* **3**, 70 (1963).
26. V. P. Dorweiler and R. W. Fahien, *Amer. Inst. Chem. Eng. J.,* **5**, 139 (1959).
27. J. H. Knox and J. F. Parcher, *Anal. Chem.,* **41**, 1599 (1969).
28. F. W. Grane and G. H. F. Gardner, *J. Chem. Eng. Data,* **6**, 283 (1961).
29. R. J. Blackwell, *Soc. Petrol. Engrs. J.,* **2**, 1 (1962).
30. K. B. Bischoff, *Ind. Eng. Chem.,* **58**, 18 (1966).
31. O. Levenspiel and K. B. Bischoff, *Adv. Chem. Eng.,* **4**, 95 (1963).
32. S. F. Miller and C. J. King, *University of California Radiation Laboratory Report 11951,* May 1965.
33. A. Hennico, G. Jacques, and T. Vermeulen, *University of California Radiation Laboratory Report 10696,* March 18, 1963.
34. D. S. Horne, J. H. Knox, and L. McLaren, *Separ. Sci.,* **1**(5), 531 (1966).
35. M. C. Hawley, Ph.D. Thesis, Michigan State University, East Lansing, Mich., 1964.
36. H. W. Osterhoudt and L. N. Ray, Jr., *J. Polym. Sci., Part A-2,* **5**, 569 (1967).

37. P. E. Pierce, The Glidden Company, Cleveland, Ohio, Private communication, Sept. 1967.
38. M. LePage, R. Beau, and A. J. De Vries, *Polymer Preprints*, **8**(2), 1211 (1967); *J. Polym. Sci., Part C*, **21**, 119 (1968).
39. J. B. Carmichael, *Makromol. Chem.*, **122**, 291 (1969).
40. J. B. Carmichael, *J. Chem. Phys.*, **49**, 5161 (1968).
41. J. B. Carmichael, *Macromolecules*, **1**, 526 (1968).
42. J. B. Carmichael, *J. Polym. Sci., Part A-2*, **6**, 572 (1968).
43. H. Vink, *Makromol. Chem.*, **116**, 241 (1968).
44. J. W. Hiby, in *Proceedings of the Symposium on the Interaction Between Fluids and Particles* (P. A. Rottenburg and N. T. Shepherd, eds.), Institution of Chemical Engineers, London, June 20–22, 1962, pp. 312–325.
45. R. A. Bernard and R. H. Wilhelm, *Chem. Eng. Progr.*, **46**, 233 (1950).
46. W. E. Brigham, P. W. Reed, and J. N. Dew, *Soc. Petrol. Eng. J.*, **1**, 1 (1961).
47. R. N. Kelley, Ph.D. Thesis, Rensselaer Polytechnic Institute, Troy, N.Y., June 1969.
48. J. H. Knox, *Anal. Chem.*, **38**, 253 (1966).
49. E. A. Ebach and R. R. White, *Amer. Inst. Chem. Eng. J.*, **4**, 161 (1958).
50. M. LePage and A. J. de Vries, "Evaluation of a New, Versatile Support for G. P. C. Columns: Porous Silica Beads," presented at the 3rd International Seminar on Gel Permeation Chromatography, Geneva, Switzerland, May 1966.
51. A. J. de Vries, M. LePage, R. Beau, and C. L. Guillemin, *Anal. Chem.*, **39**, 935 (1967).
52. C. L. Guillemin, M. LePage, R. Beau, and A. J. de Vries, *ibid.*, **39**, 940 (1967).
53. Pechiney-Saint-Gobain, French Patents 1,473,240; 1,475,929 (1967).
54. W. Haller, *Nature*, **206**, 693 (1965).
55. W. Haller, *J. Chem. Phys.*, **42**, 686 (1965).
56. W. Haller, *J. Chromatogr.*, **32**, 676 (1968).
57. *Price List T*, Bio-Rad Laboratories, Richmond, California, July 1, 1968.
58. E. M. Barrall and J. H. Cain, *J. Polym. Sci., Part C*, **21**, 253 (1967).
59. M. J. R. Cantow and J. F. Johnson, *J. Polym. Sci., Part A-1*, **5**, 2835 (1967).
60. M. J. R. Cantow and J. F. Johnson, *J. Appl. Polym. Sci.*, **11**, 1851 (1967).
60a. A. R. Cooper, J. H. Cain, E. M. Barrall, II, and J. F. Johnson, ACS Div. of Petroleum Chemistry, *Preprints*, **15**, A-95 (1970); Separ. Sci., in press.
61. P. Flodin, *J. Chromatogr.*, **5**, 103 (1961).
62. J. Porath, in *Advances in Protein Chemistry*, Vol. 17 (C. B. Anfinsen, M. L. Anson, K. Bailey, and J. T. Edsall, eds.), Academic, New York, 1962, p. 209.
63. J. Porath and P. Flodin, in *Protides of the Biological Fluids* (H. Peeters, ed.), 10th ed., American Elsevier, New York, 1963, p. 290.
64. V. H. Edwards, Unpublished data.
65. W. B. Smith and A. Kollmansberger, *J. Phys. Chem.*, **69**, 4157 (1965).
66. J. H. Duerksen and A. E. Hamielec, *J. Appl. Polym. Sci.*, **12**, 2225 (1968).
67. W. Heitz and W. Kern, *Angew. Makromol. Chem.*, **1**, 150 (1967).
68. J. Coupek and W. Heitz, *Makromol. Chem.*, **112**, 286 (1968).
69. W. Heitz and J. Coupek, "Column Efficiency in GPC," presented at the 5th International Seminar on Gel Permeation Chromatography, London, England, May 19–22, 1968.

70. J. N. Little, J. L. Waters, K. J. Bombaugh, and W. J. Pauplis, *J. Polym. Sci.,* *Part A-2,* **7,** 1775 (1969).
70a. K. H. Altgelt, Chevron Research Company, Richmond, California, Private communication, Feb. 1970.
71. J. C. Moore and M. C. Arrington, "The Separation Mechanism of Gel Permeation Chromatography Experiments With Porous Glass Column Packing Materials," presented at the 3rd International Seminar on Gel Permeation Chromatography, Geneva, Switzerland, May 1966.
72. F. A. Sliemers, K. A. Boni, D. E. Nemzer, and G. P. Nance, "An Examination of Factors Affecting the Efficiency of GPC Columns Packed With Porous Silica Beads," presented at the 6th International Seminar on Gel Permeation Chromatography, Miami Beach, Florida, October 7–9, 1968.
73. W. W. Yau and C. P. Malone, *J. Polym. Sci., Part B,* **5,** 663 (1967).
74. E. F. Casassa, *ibid.,* **5,** 773 (1967).
75. E. F. Casassa and Y. Tagami, *Macromolecules,* **1,** 526 (1969).
76. J. J. Hermans, *J. Polym. Sci., Part A-2,* **6,** 1217 (1968).

Mathematical Methods of Correcting Instrumental Spreading in GPC*

L. H. TUNG

PHYSICAL RESEARCH LABORATORY
THE DOW CHEMICAL COMPANY
MIDLAND, MICHIGAN 48640

Summary

Various mathematical approaches to correct instrumental spreading in GPC are summarized. The basic equation describing the spreading correction is identical to that used in X-ray diffraction for correcting its instrumental spreading. In GPC, however, artificial oscillation is easily induced in the solution of the basic equation. This difficulty is partially overcome by data smoothing procedures.

INTRODUCTION

Like any other type of chromatography, the GPC chromatogram of a monomeric compound appears as a curve of finite width as shown in Fig. 1. The position of the peak of the curve depends on the molecular weight of the compound; the area under the curve is proportional to the amount of the compound in the total sample; and the width of the curve depends on various band spreading mechanisms in the GPC instrument, both within and without the columns. For a polydispersed sample such as those normally encountered in high polymers, the chromatogram is a composite of the curves of all its components. The total area under the curve is still proportional to the amount of the entire sample but the height of the curve does not reflect the relative abundance of the components at the corresponding elution volumes, as

---

* Presented at the ACS Symposium on Gel Permeation Chromatography, sponsored by the Division of Petroleum Chemistry at the 159th National Meeting of the American Chemical Society, Houston, Texas, February, 1970.

73

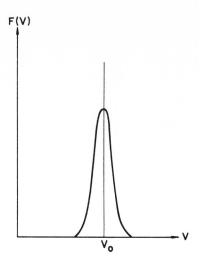

FIG. 1. Chromatogram of a monodisperse sample.

it depends also on the abundance of the neighboring components. At the ends of the chromatograms there are curve portions representing components which do not even exist in the sample. For accurate molecular weight distribution analysis, this overlapping and diffused pat-

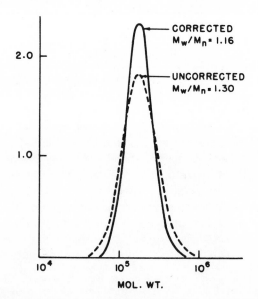

FIG. 2. Comparison of an instrumental spreading corrected distribution with an uncorrected distribution.

tern of the chromatogram must be corrected. Figure 2 shows the difference between a corrected and uncorrected chromatogram.

DERIVATION OF THE CORRECTION EQUATION

Let us assume that for the moment the instrumental spreading function is Gaussian. Thus, the chromatogram of a monomeric compound has the shape of a Gaussian curve. Let $f(v)$ denote the chromatogram as a function of the elution volume (or count) v. For a monomeric compound then

$$f(v) = A(h/\sqrt{\pi}) \exp [-h^2(v - v_0)^2] \tag{1}$$

where v_0 is the elution volume at the peak, h is a parameter related to the width of the Gaussian curve, and A is the total area under the curve. For a multicomponent system

$$f(v) = \sum_i A_i(h_i/\sqrt{\pi}) \exp [-h_i^2(v - v_{0i})^2] \tag{2}$$

The area A_i under the Gaussian curve is proportional to the amount of that component in the sample. Thus, when the number of components in the sample becomes very large we may replace A_i with a continuous function $w(y)$ that has a value at elution volume y proportional to the amount of the component with its peak at y. Equation (2) becomes now

$$f(v) = \int w(y)(h/\sqrt{\pi}) \exp [-h^2(v - y)^2] \, dy \tag{3}$$

The function $w(y)$ is the chromatogram free from the effect of instrumental spreading and therefore the unknown to be solved. If we let $g(v\text{-}y)$ denote the instrumental spreading function in general then

$$f(v) = \int w(y)g(v - y) \, dy \tag{4}$$

Equation (4) has the form of a convolution integral equation and is the same equation that describes the instrumental spreading correction in X-ray diffraction.

METHODS OF SOLUTION

There are apparently three different approaches used by GPC investigators in solving the above integral equation.

1. Solution by Minimization

Aside from the method of steepest descent in the function space used by Chang and Huang (1), other methods of minimization as reported by Hess and Kratz (*2*), by Smith (*3*), by Pickett et al. (*4*), and by the author (*5*) all involved the approximation of eq. (4) by a set of linear algebraic equations in the following form.

$$f(v_j) = \sum_{i}^{n} w(y_i)g_j(v_j - y_i)(\Delta y)_i \tag{5}$$

Equation (5) is for the jth equation. For each point on the chromatogram an equation like Eq. (5) can be written. The unknown w function is now represented by n unknown points $w(y_i)$ spaced in suitable intervals $(\Delta y)_i$ apart. The products $g_j(v_j\text{-}y_i)(\Delta y)_i$ are known and they are the coefficients for the unknowns $w(y_i)$. The unknowns can be solved by methods of minimization if we read from the chromatogram a total number of points larger than n.

Solution by way of linear algebraic equations has the flexibility of using any form for the g-function. The g-function can be made to vary with elution volume v. These methods, however, generally require large computer storage spaces and often long computation time. The computation for the method of Chang and Huang (*1*) was reported to be fast.

2. Solution by Fourier Transform

This is the approach used by Stokes (*6*) for the case of X-ray diffraction. Pierce and Armonas (*7*) have published an attractive simplification of this approach for GPC. The author (*8*) has also adopted Stoke's method to GPC problems.

The Fourier transforms for the three functions involved in Eq. (4) are:

$$F(k) = (1/\sqrt{2}\pi) \int_{-\infty}^{\infty} f(v)e^{ikv}\, dv \tag{6}$$

$$G(k) = (1/\sqrt{2}\pi) \int_{-\infty}^{\infty} g(v)e^{ikv}\, dv \tag{7}$$

$$W(k) = (1/\sqrt{2}\pi) \int_{-\infty}^{\infty} w(v)e^{ikv}\, dv \tag{8}$$

The limits of integration in Eq. (4) can be extended to $+\infty$ and $-\infty$ even though both $f(v)$ and $w(y)$ have values of zero beyond the ends of the chromatogram. Then according to the Faltung theorem

$$W(k) = (1/\sqrt{2}\pi)[F(k)/G(k)] \tag{9}$$

Since $F(k)$ and $G(k)$ can be computed from the given functions,

$W(k)$ is now known. By the following inverse transform we may obtain the corrected chromatogram $w(v)$.

$$w(v) = (1/\sqrt{2\pi}) \int_{-\infty}^{\infty} W(k)e^{-ivk}\, dk \qquad (10)$$

The equations involved in this approach imply that a constant g-function with respect to v is required. But this inflexibility can be circumvented by treating the chromatogram one section at a time using the proper g-function for each section.

3. Solution through Polynomial Representation of the Chromatogram

Three published methods used this approach, one by Aldhouse and Stanford (9) and two by the author (5, 8).

In general, the functions $f(v)$ and $w(y)$ for the chromatograms can be represented by polynomials. If the product of $w(y)$ and $g(v\text{-}y)$ is integrable, then by a comparison of the coefficients of $f(v)$ with those of the polynomial after integration, the coefficients for $w(y)$ may be solved. A convenient polynomial to use is

$$f(v) = \exp\left[-q^2(v - v_o)^2\right] \sum_{i=0}^{n} U_i(v - v_o)^i \qquad (11)$$

where q, v_0, and U_i are adjustable parameters and coefficients. Because of the exponential factor, the right-hand side of Eq. (11) ap-

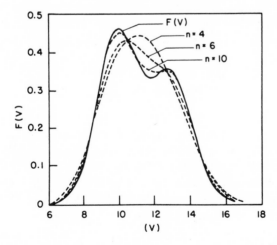

FIG. 3. Fitting of a two-peak distribution by polynomials.

proaches zero when v approaches $+\infty$ and $-\infty$. The fit of Eq. (11) to a chromatogram is not difficult. Figure 3 shows the fit to a complex chromatogram. When n in Eq. (11) is 16, the curve calculated is indistinguishable from the given chromatogram.

In the polynomial approach the form of the g-function is more restrictive. The Gaussian function works well. Some asymmetrical g-function may also be used but it must be integrable when combined with $w(y)$.

Our current correction method uses a fourth-degree polynomial to correlate nine points on the chromatogram at a time. The g-function used is Gaussian. The procedure is repeated with every point on the chromatogram as the center point of the nine-point fit. In this way the h parameter in the Gaussian g-function may be varied with the elution volume. The calculation on a computer for this scheme is extremely fast and uses a relatively small amount of storage spaces.

The methods discussed so far all require high speed digital computers to execute the calculation. Frank, Ward, and Williams (10), however, have described a simple method, the calculation for which may be managed by a desk calculator. From the chromatogram they separated one or several Gaussian curves, depending on the number of peaks in the chromatogram. The residual smooth function was left uncorrected. The Gaussian peaks were narrowed by subtracting from them the instrumental spreading which was assumed to be also Gaussian. A similar approach has been used in our laboratory and proved to be very useful for the chromatograms of extremely narrow distribution samples. For these narrow distribution chromatograms the more complex method could be difficult to use because of the problem of oscillation.

THE PROBLEM OF OSCILLATION

Duerksen and Hamielec (11) have made a comparison of some of the above-mentioned methods. In all methods examined by them some degree of oscillation induced by the computation was suspected. Figure 4 shows a chromatogram $f(v)$ calculated from a known $w(y)$ function when the instrumental spreading g is relatively broad with respect to w. It can be seen that a small variation in the slope of $f(v)$ will bring about a considerable larger variation in the slope of $w(y)$. This sensitivity of $w(y)$ varies with the breadth of the g-function or the extent of correction. In the limiting case where there is no correction or $g = 1$,

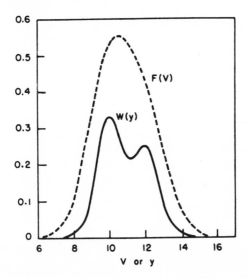

FIG. 4. Relation between corrected chromatogram $W(y)$ and uncorrected chromatogram $F(v)$ when the instrumental correction is large.

$w(y)$ becomes identical to $f(v)$. In any other cases variation of the slope for $f(v)$ is always less than that for $w(y)$. If in experiments $f(v)$ can be determined to a high degree of precision and if the g-function used describes the spreading characteristics extremely accurately, then the solution for $w(y)$ can be obtained with a high degree of confidence regardless of the method used. In reality neither of the conditions can be fulfilled and as a result the uncertainties in $f(v)$ and in $g(v\text{-}y)$ are easily transformed into oscillations in $w(y)$. This problem is more severe when the correction is large or when the sample contains very narrow peaks. It is also more pronounced at the ends of a chromatogram where the $f(v)$ function is even less precisely known. To minimize the fluctuations in the raw data, mathematical correlations are used to smooth out $f(v)$ before calculation. In fact, in many of the above-mentioned methods such a data-smoothing procedure is implicitly or explicitly carried out in the computer program for the method. Whether one method of solution is better than another depends often more on this smoothing procedure than the mathematics involved. If smoothing is too drastically done, then some of the true features of $w(y)$ may be lost; if not enough is accomplished, then oscillations may show up in the solution. It is not unusual that a smoothing procedure is found to be sufficient for one chromatogram but totally inadequate for another. Such a fact is perhaps the reason why in so short a time so many solutions were proposed for this one problem. No one is apparently completely satisfied with the correction method which he has on hand.

REFERENCES

1. K. S. Chang and R. Y. M. Huang, *J. Appl. Polym. Sci.*, **13**, 1459 (1969).
2. M. Hess and R. F. Kratz, *J. Polym. Sci., Part A-2*, **4**, 731 (1966).
3. W. N. Smith, *J. Appl. Polym. Sci.*, **11**, 639 (1967).
4. H. E. Pickett, M. J. R. Cantow, and J. F. Johnson, *J. Polym. Sci., Part C*, **21**, 67 (1968).
5. L. H. Tung, *J. Appl. Polym. Sci.*, **10**, 375 (1966).
6. A. R. Stokes, *Proc. Phys. Soc.*, **61**, 382 (1948).
7. P. E. Pierce and J. E. Armonas, *J. Polym. Sci., Part C*, **21**, 23 (1968).
8. L. H. Tung, *J. Appl. Polym. Sci.*, **13**, 775 (1969).
9. S. T. E. Aldhouse and D. M. Stanford, Paper presented at the 5th International GPC Seminar, London, May, 1968.
10. F. C. Frank, I. M. Ward, and T. Williams, *J. Polym. Sci., Part A-2*, **6**, 1357 (1968).
11. J. H. Duerksen and A. E. Hamielec, *J. Polym. Sci., Part C*, **21**, 83 (1968).

Comparison of Different Techniques of Correcting for Band Broadening in GPC

J. H. DUERKSEN

CHEVRON RESEARCH COMPANY
RICHMOND, CALIFORNIA 94802

Summary

A qualitative description is presented for two types of gel permeation chromatography (GPC) band broadening. One is symmetrical or Gaussian band broadening; the other is unsymmetrical or skewed band broadening. The effects of band broadening on chromatogram interpretation are discussed.

Available methods of correcting the GPC molecular weight distribution (MWD) for the effects of symmetrical and unsymmetrical band broadening are discussed and compared. For symmetrical band broadening, Tung's original polynomial expansion method is generally adequate. Tung's newer methods look promising for both symmetrical and unsymmetrical band broadening, but they require further evaluation. For unsymmetrical band broadening, the method of Balke and Hamielec looks most promising for unimodal MWD's, but it requires further evaluation with more complex MWD's. Except for the latter method, the corrected MWD's for all methods had inconsistent oscillations when resolution was low or skewing was significant. Since these oscillations are probably caused by noise in the chromatogram or inaccuracies in reading chromatogram heights, they could be minimized by improving chromatogram accuracy and by using correction techniques that include adequate data smoothing.

1. INTRODUCTION

This paper compares techniques for interpretation of GPC chromatograms for linear homopolymers. The techniques compared here are those which attempt to correct for the effects of band broadening (also called zone broadening, peak broadening, instrument spreading, im-

81

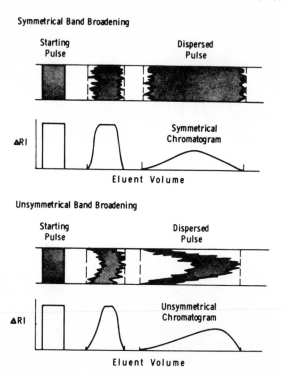

FIG. 1. Effect of axial dispersion and velocity profile on GPC band broadening.

perfect resolution, and axial dispersion). This correction is necessary if absolute molecular weight distributions (MWD's) are desired.

For linear homopolymers there appear to be at least two corrections that must be made, both involving axial dispersion. The GPC chromatogram for a monodisperse polymer may be Gaussian (symmetrical) with respect to eluent volume under certain operating conditions, or it may be highly unsymmetrical, with skewing towards higher eluent volumes and lower molecular weights. Symmetrical and unsymmetrical chromatograms are illustrated in Fig. 1. Symmetrical band broadening is caused by axial dispersion whereas unsymmetrical band broadening or skewing is usually attributed to the effect of velocity profile and radial dispersion on axial dispersion (*1*, *2*). The skewing phenomenon is particularly important for relatively viscous, high molecular weight polymer solutions, where radial dispersion is small due to the small diffusion coefficients of the polymer molecules. Under these circumstances, velocity profiles can greatly increase axial dispersion and cause unsymmetrical chromatograms.

The effect of symmetrical axial dispersion is to lower the calculated GPC number-average molecular weight, $\bar{M}_n$, and raise the calculated weight-molecular weight, $\bar{M}_w$. The effect of unsymmetrical axial dispersion is to lower both $\bar{M}_n$ and $\bar{M}_w$.

For GPC operation where the chromatograms of narrow standards are Gaussian, methods of chromatogram interpretation are well developed (3-5). However, when chromatograms of narrow standards are unsymmetrical, techniques of interpretation are not nearly so well developed (6-8). This paper compares the available methods that correct for symmetrical and unsymmetrical band broadening. These are methods by Tung (3, 9), Smith (10), Hess and Kratz (11), Pickett, Cantow, and Johnson (12), and Balke and Hamielec (8).

2. BASIS FOR THE MATHEMATICAL CORRECTION OF GPC BAND BROADENING

Before comparing the various methods of chromatogram interpretation, it would be instructive to consider the behavior of a pulse of monodisperse polymer solution as it progresses through the GPC columns and the chromatograms resulting therefrom. This will help to illustrate the basis of interpretation for polydisperse samples used by the various methods compared here.

Figure 1 illustrates the two types of undesirable GPC band broadening for an input pulse of monodisperse polymer solution. Symmetrical band broadening is caused by eddy and molecular diffusivity at the leading and trailing edges of the pulse. This type of flow has been referred to as dispersed plug flow (13). Its effect is the same on both edges of the pulse and causes symmetrical broadening and dilution of the pulse with a resulting symmetrical chromatogram.

Unsymmetrical band broadening is caused by an interaction between a nonuniform velocity profile and eddy and molecular diffusivity, as illustrated in Fig. 1. Since the pulse velocity ranges from almost zero near the wall to a maximum at the tube center, the resulting chromatogram is skewed toward higher elution volumes. This tailing toward higher elution volumes is more pronounced for higher molecular weight polymers because the larger molecules diffuse more slowly

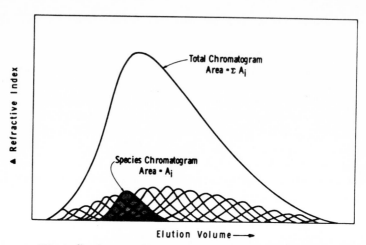

FIG. 2. Species contributions to the total chromatogram.

away from the wall toward regions of lower concentration and higher flow velocity.

For the case where we have band broadening, Fig. 2 illustrates what we might observe if we could "see" the individual molecular species that contribute to the overall chromatogram. To obtain a molecular weight distribution, we would simply measure the amount (i.e., area) of each species present and divide by the total amount. Since we cannot see or measure the amounts of individual species present, these amounts must be calculated from the overall, measured chromatogram. Since more than one species contributes to the chromatogram height at a particular elution volume, the calculation of species amounts is not straightforward. To convert a chromatogram into a molecular weight distribution, each of the techniques compared here must assume a shape for the single molecular species and calibrate for the parameters that define the single species shape. Each method also requires a calibration of molecular weight versus elution volume. From the assumed shape for the single species, the measured height at a particular elution volume can be expressed in terms of the unknown amounts of species contributing at that point. In principle, if n unknown species are present, the MWD can be calculated by reading n heights off the chromatogram to give n equations in n unknowns. The methods of chromatogram interpretation compared here differ only in the techniques they use to solve for the unknown species amounts from measured chromatogram heights.

3. METHODS BY TUNG

3.1. Development

Tung was one of the first to develop methods to correct for band broadening (3). In his early development he assumed that the chromatograms of single species in a polydisperse polymer were Gaussian (symmetrical). The Gaussian-shaped chromatogram $F(v)$ was represented by

$$F(v) = A \sqrt{h/\pi} \exp\left[-h(v - v_0)^2\right] \tag{1}$$

where v is eluent volume, v_0 is the eluent volume at the peak of the curve, A is a constant related to the area and weight of polymer injected, and h is the resolution factor $[= \frac{1}{2}(\text{variance})^2]$.

For a multicomponent polymer system with n species, the chromatogram height $F(v)$ is the sum of the height contributions of the individual species, i.e.,

$$F(v) = \sum_{i=1}^{n} A_i \sqrt{h_i/\pi} \exp\left[-h_i(v - v_{0i})^2\right] \tag{2}$$

If the number of species is large, a continuous distribution function $W(y)$ can be used to denote the abundance of components in the mixture. The chromatogram can then be represented by

$$F(v) = \int_{v_a}^{v_b} W(y) \sqrt{h/\pi} \exp\left[-h(v - y)^2\right] dy \tag{3}$$

where v_a is the initial eluent volume and v_b is the final eluent volume of the chromatogram.

Equation (3) was proposed by Tung (3) for the purpose of GPC chromatogram interpretation. It is generally referred to as his integral dispersion equation and is used to solve for species amounts from measured chromatogram heights.

In his early work (3) Tung developed two methods of solving his integral dispersion equation to obtain a chromatogram corrected for band broadening. One method used the Gaussian quadrature formula and linear programming. This method has not been extensively evaluated. Tung found that it gave satisfactory results but required excessive computation time (9). The other method used a polynomial expansion technique. It has been evaluated for a wide range of GPC operating conditions (6, 7).

The polynomial expansion method assumes that the resolution factor

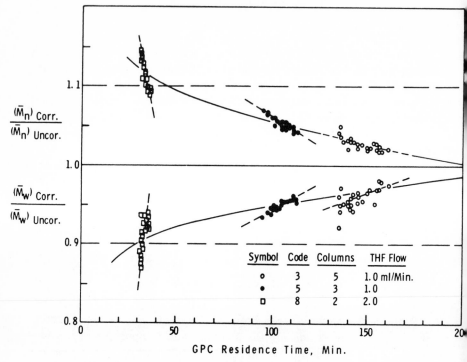

FIG. 3. Ratio of corrected to uncorrected molecular-weight averages as a function of GPC residence time and column combination for Tung's polynomial expansion method. Gaussian band broadening assumed.

h is constant over the elution volume range of a chromatogram and that the band broadening due to dispersion can be represented by a Gaussian distribution function. The method solves for the resolution corrected chromatogram $W(y)$ by using a polynomial representation for $F(v)$ and $W(y)$ and performing the integration in Eq. (3). A predetermined resolution factor is used in the solution.

3.2. Evaluation of Tung's Polynomial Expansion Method

In the evaluation of this and other methods by Duerksen and Hamielec (6, 7), polystyrene samples covering a wide range of molecular weights were analyzed over a wide range of resolutions. Different resolutions were obtained by varying GPC flow rate and column combinations. To be truly effective a correction method should give the same corrected MWD for the same sample analyzed at several widely different resolutions.

For Tung's polynomial expansion method, molecular weight averages for the same sample run at different resolutions agreed well when band broadening was Gaussian or nearly so (6). In general, this was true for low molecular weights (less than 100,000) and flow rates of 1.0 and 3.0 ml/min. At higher molecular weights or a flow rate of 2.0 ml/min, skewing was significant and agreement between molecular weight averages was poor. The averages with skewing present were lower than when no skewing was present.

The effect of assuming Gaussian band broadening and correcting for it is to raise the calculated $\bar{M}_n$ and lower the calculated $\bar{M}_w$ relative to the uncorrected values. This is illustrated in Fig. 3 for a range of GPC residence times and three different column combinations (6).

The effect of assuming Gaussian band broadening when it is actually skewed toward higher elution volume and lower molecular weight is illustrated in Fig. 4. The actual skewed single species area is represented by a Gaussian (symmetrical) area (dashed lines) having the same moments about the peak elution volume, v_0. The area between the skewed shape and Gaussian shape on both sides of v_0 is, therefore, regarded as a contribution from lower molecular weight material than is really the case. The net effect is to calculate a lower $\bar{M}_n$ and $\bar{M}_w$ than the true values and an MWD skewed toward lower molecular

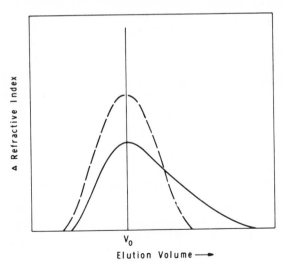

FIG. 4. Skewed vs Gaussian single species shapes.

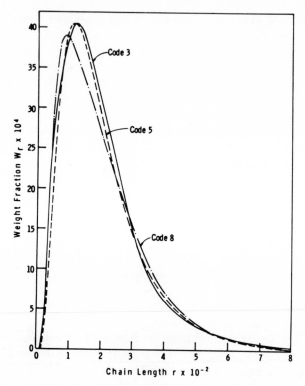

FIG. 5. Typical MWD's for a low molecular weight polystyrene corrected by Tung's polynomial expansion method.

weight. This predicted behavior agrees with the observed behavior (6, 14).

Figure 5 illustrates MWD's calculated by Tung's polynomial expansion method for a low molecular weight sample. Three column combinations were used (7), and the resolution corrections were relatively small (less than 10% on the averages). Agreement was good between Column Codes 3 and 5, which had nearly Gaussian band broadening at this molecular weight level. Column Code 8, however, had slightly skewed band broadening; this has resulted in an MWD skewed toward lower molecular weight relative to the Codes 5 and 3 MWD's.

Figure 6 illustrates MWD's calculated for a high molecular weight broad MWD sample. The resolution factors were around 1.0 counts⁻² for Code 6 and 0.5 for Codes 11 and 12, resulting in a relatively large resolution correction. Skewing for narrow standards for Code 6 was significant, and for Codes 11 and 12 it was extremely severe, with tailing extending out to the monomer elution volume. The resulting MWD's for the broad samples for Codes 11 and 12 in Fig. 6 were skewed toward lower molecular weight relative to Code 6. In addition, large inconsistent oscillations are observed at low molecular weights for Codes 11 and 12. Tung has pointed out (9) that these oscillations can be caused by chromatogram noise and by differences between the assumed and actual single species shape when the resolution correction is large. Codes 11 and 12 had significant chromatogram noise and large resolution corrections.

3.3. Other Methods for Solving Tung's Integral Dispersion Equation

Other methods for solving Tung's integral dispersion equation have recently been developed (9, 15, 16).

A method by Pierce and Armonas (15) is based on the use of Fourier transforms. Since it treats the chromatogram a point at a time, a variation in h can be handled by using a different h at each elution volume considered. The method also requires short computation time. However, from noise in the chromatogram, it is possible to generate oscillations in the corrected chromatogram in taking derivatives at a

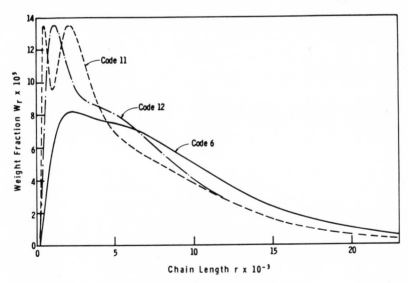

FIG. 6. Typical MWD's for a high molecular weight polystyrene corrected by Tung's polynomial expansion method.

point on the chromatogram (17). Tung also points out (9) that for large dispersion corrections the use of derivatives at a point gives inaccurate results. Aldhouse and Stanford (16) have proposed a similar point-to-point approach using a Taylor's expansion method. Since derivatives of the chromatogram are also required for their solution, the deficiencies of the method of Pierce and Armonas should still be present.

Tung has proposed two new methods of solving his integral dispersion equation (9, 18): a Fourier analysis method and a polynomial method. These compare very favorably with the above-mentioned methods.

Both of Tung's methods are simple and require little computation time. In addition, the polynomial method is less likely to generate oscillations due to the smoothing characteristics of its least squares fitting. The Fourier analysis method can use unsymmetrical functions to correct for skewed band broadening. Further testing and evaluation are required to determine how well this method can correct for unsymmetrical band broadening.

4. METHOD OF SMITH

4.1. Development

To solve for unknown species amounts from measured chromatogram heights, Smith (10) proposed an equation which is equivalent to Eq. (2) in Tung's development. In Smith's notation the chromatogram height at elution volume v_0 is

$$f(v_0) = \cdots + \frac{K_{-1}}{\sqrt{2\pi}\sigma_{-1}} \exp\left[\frac{-(v_0 - v_{-1})^2}{2\sigma_{-1}}\right] + \frac{K_0}{\sqrt{2\pi}\sigma_0}$$
$$+ \frac{K_1}{\sqrt{2\pi}\sigma_1} \exp\left[\frac{-(v_0 - v_1)^2}{2\sigma_1}\right] + \cdots \quad (4)$$

where K_j is a factor proportional to the concentration of the jth molecular species. The K_j's are comparable to Tung's A_i's in Eq. (2), and the σ's are related to Tung's h by $h = \frac{1}{2}\sigma^2$. Solution of the K_j's yields the MWD.

Smith assumed that the K_j's were proportional to the chromatogram height at v_j, i.e.,

$$K_j = k_j f(v_j) \quad (5)$$

By considering the polymer sample to consist of a finite number of species n, and by assuming that the proportionality constants k_j are the same for all species contributing to chromatogram height $f(v_j)$ at elution volume v_j, Smith was able to rewrite Eq. (4) as

$$f(v_j) = k_j \sum_{i=0}^{n} (h_i/\pi)^{1/2} f(v_i) \{\exp [-h_i(v_j - v_i)^2]\} \qquad (6)$$

A set of n equations in n unknown k_j's is obtained; one equation for each chromatogram height read. The assumption of the same k_j's for all species contributing at v_j allows the equations in k_j to be solved consecutively rather than simultaneously. This amounts to a point-to-point solution for species amounts and permits the use of a different resolution factor h for each elution volume used.

The k_j values range from zero at each end of the chromatogram to a maximum value at or near the peak elution volume. However, the assumption that the k_j's are the same for all species contributing at v_j is still quite accurate since only those species relatively close to v_j will contribute, and the variation in their k_j's is relatively small.

Smith later modified his method to eliminate the above assumption (7). The initial set of calculated k values is used to calculate the chromatogram height at each elution volume v_i using Eq. (6) and the calculated $(k_j)_i$ for each species i. If these calculated heights do not agree with the observed heights, each k is adjusted by the ratio of the observed to calculated height. This calculation is repeated until the desired agreement between observed and calculated heights is obtained.

Smith's modified method also compared the area under the chromatogram and the sum of the area contributions of the assumed species. If the ratio of the calculated to observed area was less than 1, it was necessary to assume more species (i.e., read more heights off the chromatogram). If the ratio was greater than 1, the resolution factors were assumed to be too small. They were increased according to the ratio of the areas and all calculations were repeated.

4.2. Evaluation of Smith's Method

Molecular weight averages by Smith's method using a Gaussian single species shape agreed well with averages by Tung's method (6, 14), even though a variable h was used in Smith's method. The Gaussian shape was inadequate when skewing of narrow standards was significant.

A log normal single species shape gave higher corrected molecular weight averages than did the Gaussian shape. When skewing was significant, the averages for the log normal shape also agreed better

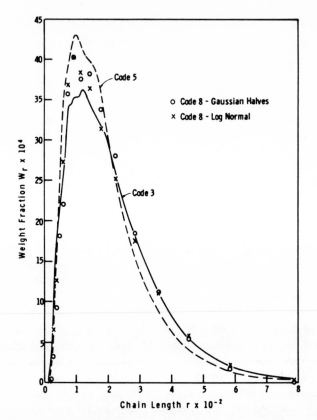

FIG. 7. Typical MWD's for a low molecular weight polystyrene corrected by Smith's method.

with averages calculated at conditions where the single species shape was Gaussian *(6,14)*. It has been shown *(14)* that the unsymmetrical log normal shape raises both the number- and weight-average molecular weights relative to the uncorrected values. This behavior is directionally correct for correction of skewed band broadening.

A single species shape made up of two Gaussian halves, each with its own *h*, did not account for skewed band broadening as well as the log normal shape did *(14)*. Using Gaussian halves, the weight-average molecular weights were significantly lower than those calculated with the log normal shape *(14)*. The Gaussian halves did not correct sufficiently for skewing.

Typical MWD's for a low molecular weight polystyrene sample

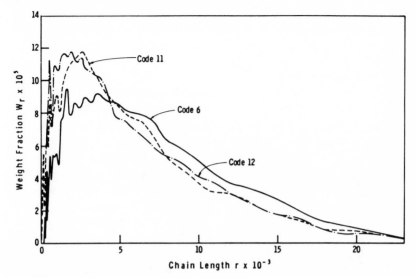

FIG. 8. Typical MWD's for a high molecular weight polystyrene corrected by Smith's method.

($\bar{M}_n = 14,000$) are shown in Fig. 7. Agreement between different column combinations is quite good except for small oscillations.

Typical MWD's for a high molecular weight sample ($\bar{M}_n = 400,000$) are shown in Fig. 8. These were calculated using a single species shape made up of unequal Gaussian halves. There are large deviations among the MWD's and large oscillations in each MWD. The oscillations may be caused by noise in the chromatogram. The deviations among the MWD's again show that the Gaussian halves did not adequately represent the effect of skewing.

5. THE METHOD OF HESS AND KRATZ

To solve for unknown species amounts from measured chromatogram heights, the method of Hess and Kratz (11) approximates the chromatogram by a set of linear algebraic equations which are solved simultaneously by matrix inversion to give species amounts. A sample of broad MWD is considered to consist of a finite number n of "pure" solutes. At least n heights are read off the chromatogram giving n equations in n unknowns, similar in form to Eq. (2). The method is based upon a dispersion model for a packed bed and requires experimental measurement of the single species dispersion coefficient E over the molecular weight range of interest.

Since the dispersion model predicts an unsymmetrical shape for the single species chromatogram (11, 13), it can be used to account for skewing. The predicted shape becomes more skewed as the dispersion coefficient E increases, ranging from almost symmetrical at very low E to very skewed at high E. The experimentally determined value of E has been observed to increase as elution volume decreases (6, 14), corresponding to increased skewing with increased molecular weight.

A very limited evaluation of this method has been made (6). For most chromatograms, the solution for species amounts was unsuccessful because the matrix of coefficients for the unknowns was ill conditioned. The successful solutions agreed reasonably well with results by Tung's polynomial expansion method (6) for the case of symmetrical band broadening. Tung has pointed out (9) that the unsymmetrical shape predicted by the dispersion model of Hess and Kratz could be used in his new Fourier analysis method to account for skewed band broadening. Further evaluation is necessary to determine the adequacy of the dispersion model in accounting for skewed band broadening.

6. THE METHOD OF PICKETT, CANTOW, AND JOHNSON

To solve for unknown species amounts from measured chromatogram heights, the method of Pickett, Cantow, and Johnson expresses the chromatogram or concentration curve as the weighted sum of the normalized concentration curves of its constituent species or fractions (12). This equation is similar in form to Eq. (2) in Tung's development. However, the method does not assume a specific shape for the chromatogram of a single species as the previously discussed methods do. Instead, it uses the observable shapes of narrow distribution polymer standards of known MWD to represent the constituent species or fractions. The reshaping principle of the method says that if the chromatogram can be represented as a weighted sum of normalized fractions, the reshaped chromatogram (i.e., the chromatogram corrected for band broadening) is represented by the same weighted sum of the reshaped fractions. The method uses a least squares technique to find the weighting factors that fit the sum of the fractions to the measured chromatogram.

Pickett, Cantow, and Johnson tested their method using two mathematically generated chromatograms (12). The method was able to resolve the chromatogram into its constituent fractions if a sufficient number of chromatogram points was used.

The method was also tested on an experimental chromatogram ob-

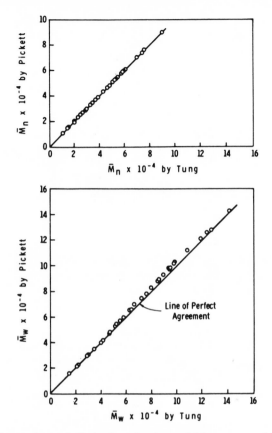

FIG. 9. Comparison of Code 5 molecular weight averages. Method of Pickett, Cantow, and Johnson vs the method of Tung.

tained from a sample consisting of equal amounts by weight of three low molecular weight polystyrene standards (12). The chromatogram was unimodal with no shoulders. The chromatograms of the three components were also added mathematically to give an expected chromatogram that was indistinguishable from the measured chromatogram, except for a slight shift in elution volume. The method resolved the synthesized chromatogram exactly into its three components. The measured chromatogram was resolved into three components, but these did not agree with the constituent components, either in proportion or elution volume. This was attributed to the slight shift in elution volume.

An evaluation of the method was also made by Duerksen and

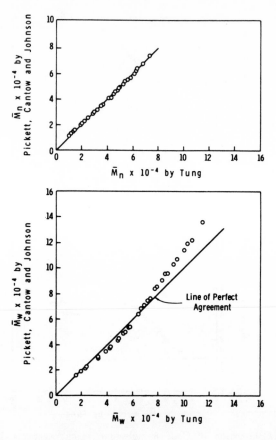

FIG. 10. Comparison of Code 8 molecular weight averages. Method of Pickett, Cantow, and Johnson vs the method of Tung.

Hamielec (*6*, *7*, *14*) who used broad distribution polystyrene samples. Code 5 standard chromatograms were nearly Gaussian; Code 8 standard chromatograms were significantly skewed. The $\bar{M}_n$ results showed reasonable agreement, but the Code 8 $\bar{M}_w$'s were generally slightly lower than the Code 5 values (*14*).

Figures 9 and 10 compare Codes 5 and 8 molecular weight averages for the method of Pickett et al. with Tung's polynomial expansion method. The Code 5 results show reasonable agreement over the entire molecular weight range. The Code 8 results show reasonable agreement in $\bar{M}_n$ but poor agreement in $\bar{M}_w$. The poor agreement in $\bar{M}_w$ is due to nonlinearities in the calibration curve and skewing effects. These

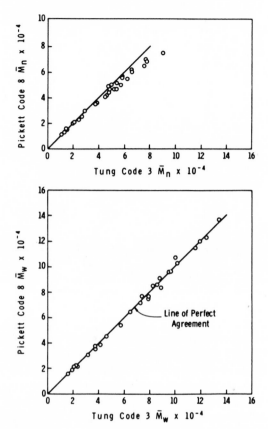

FIG. 11. Comparison of molecular weight averages. Method of Pickett, Cantow, and Johnson for Code 8 vs the method of Tung for Code 3.

effects were accounted for in the method of Pickett et al. but not in Tung's method.

If the method of Pickett et al. successfully accounts for the effect of skewing on molecular weight averages, its Code 8 averages should agree with the Code 3 averages by Tung's polynomial expansion method, since Code 3 standard chromatograms were Gaussian. These results are compared in Fig. 11. Except for the upper $\bar{M}_n$ range, reasonable agreement is indicated, certainly much better than was observed for similar comparisons with the previously discussed methods (14).

Typical MWD's for the same sample run on Codes 5 and 8 are compared in Fig. 12. Even though the molecular weight averages from Codes 5 and 8 were in good agreement, the MWD's differed signif-

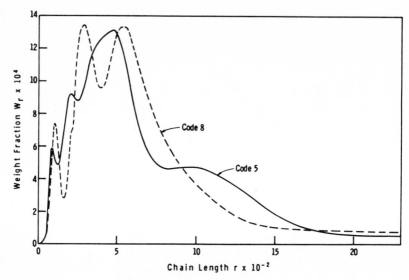

FIG. 12. Typical MWD for a polystyrene corrected by the method of Pickett, Cantow, and Johnson.

icantly in detail. Large inconsistent oscillations were observed in many of the MWD's for Codes 5 and 8, even at low molecular weights. Since the method was able to resolve synthetic chromatograms (12) without introducing oscillations, they appear to be caused by noise or inaccuracies in reading the chromatogram. Since detailed knowledge of the MWD may be required to correlate with physical properties of polymers, any artificial oscillations must be eliminated to make the methods of chromatogram interpretation completely effective.

7. THE METHOD OF BALKE AND HAMIELEC

Balke and Hamielec (8) have recently proposed a method that corrects separately for symmetrical and skewed band broadening and avoids oscillations in corrected MWD's. The method requires three GPC calibrations:

1. Molecular weight versus elution volume using narrow standards.
2. Resolution factor h versus elution volume to correct for symmetrical band broadening (narrow or broad standards can be used).
3. Skewing factor sk versus elution volume for skewed band broadening (narrow or broad standards can be used).

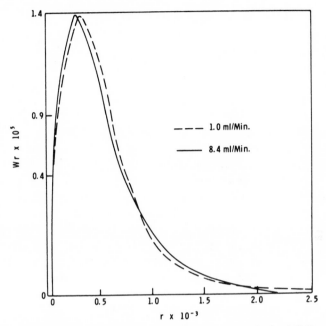

FIG. 13. Comparison of MWD's corrected by the method of Balke and Hamielec and Tung's polynomial expansion method.

Techniques for calibrating for h and sk without using reverse flow are described in the literature $(8, 17)$.

When the above calibrations have been made, corrected molecular weight averages for an unknown sample can be calculated from the following equations

$$M_n(h,sk) = M_n(\infty) \left[1 + \frac{sk}{2} \right] \exp\ (+A/h) \tag{7}$$

$$M_w(h,sk) = M_w(\infty) \left[1 + \frac{sk}{2} \right] \exp\ (-A/h) \tag{8}$$

where $M_n(h,sk)$ and $M_w(h,sk)$ are the number- and weight-average molecular weights corrected for symmetrical band broadening with h and for skewing from the Gaussian shape with sk. $M_n(\infty)$ and $M_w(\infty)$ are the averages calculated from the chromatogram assuming perfect resolution (no band broadening). $A = (2.303/2C_2)^2$, where C_2 is the slope of the molecular weight calibration curve.

The corrected molecular weight averages are then used to find an effective linear calibration curve, which is used with the raw chromatogram to calculate a corrected MWD. The corrected MWD will have the correct number- and weight-average molecular weights, but there is no guarantee that higher molecular weight averages will be accurate (17). Because the raw chromatogram, rather than a corrected chromatogram, is used to calculate the MWD, any chromatogram noise or reading inaccuracies are less likely to be magnified into artificial oscillations.

Figure 13 shows a comparison of two corrected MWD's for the same sample (8). The MWD obtained at 1.0 ml/min flow rate was calculated using Tung's polynomial expansion method (3). At this flow rate, the standard chromatograms were close to Gaussian. The MWD obtained at 8.4 ml/min was calculated using the method of Balke and Hamielec (8). Although skewing was significant at this high flow rate and the correction for band broadening was large, the MWD agrees very well with the MWD at 1.0 ml/min and does not have any inconsistent oscillations. A more severe test of the method of Balke and Hamielec would be obtained by treating a high flow rate chromatogram for a known mixture with a multimodal MWD to see if the method can resolve the peaks and give the correct MWD.

8. CONCLUSIONS

Tung's original polynomial expansion method works well when the chromatograms of single species are Gaussian. However, when they are skewed, and when the correction for band broadening is large, the corrected molecular weight averages are too low and the MWD's exhibit inconsistent oscillations. Two more recent methods by Tung are computationally faster and more accurate than his original polynomial expansion method. They also allow variable resolution factors. One of these also allows the use of a nonsymmetrical single species shape. Further evaluation of these more recent methods is required.

The method of Smith uses a log normal shape or two Gaussian halves for the single species to account for skewing. Although these work better than the Gaussian shape, they have proven to be inadequate to completely account for skewing. Inconsistent oscillations in the corrected MWD's were observed when skewing was significant and corrections were large.

The method of Hess and Kratz is based upon a dispersion model which predicts a nonsymmetrical single species shape. This method has not been properly evaluated due to computational difficulties with the matrix inversion technique for solving for species amounts.

The method of Pickett, Cantow, and Johnson uses observable shapes for narrow standards to represent the single species shapes. This method appears to account reasonably well for the effects of skewing on molecular weight averages. However, many MWD's exhibit inconsistent oscillations.

The method of Balke and Hamielec corrects separately for symmetrical band broadening and for skewing. The GPC is calibrated for molecular weight, resolution factor, and skewing factor using standards. Corrected molecular weight averages are first found and these are used to find an effective molecular weight calibration curve with which to calculate the corrected MWD from the raw chromatogram. This method appears to adequately account for skewed band broadening and does not generate inconsistent oscillations in the MWD. Further evaluation is necessary for more complex MWD's.

The inconsistent oscillations in the MWD's observed with most of the methods appear to be caused by GPC noise and limited accuracy in reading trace heights, rather than by the mathematical techniques involved. If this is the case, these oscillations might be eliminated by reducing instrument noise, improving the accuracy of height readings, and by smoothing the height readings before calculation. Further work is required in this area.

Acknowledgment

The author is indebted to Professor A. E. Hamielec for providing his GPC notes which were presented at a two-day short course at Washington University, St. Louis, April 25–26, 1969.

REFERENCES

1. G. I. Taylor, *Proc. Roy. Soc.*, **A225**, 473 (1954).
2. F. W. Billmeyer, Jr., and R. N. Kelley, *J. Chromatogr.*, **34**, 322 (1968).
3. L. H. Tung, *J. Appl. Polym. Sci.*, **10**, 375 (1966).
4. L. H. Tung, J. C. Moore, and G. W. Knight, *J. Appl. Polym. Sci.*, **10**, 1261 (1966).
5. L. H. Tung, *J. Appl. Polym. Sci.*, **10**, 1271 (1966).
6. J. H. Duerksen and A. E. Hamielec, *J. Polym. Sci., Part C*, **21**, 83 (1968).
7. J. H. Duerksen and A. E. Hamielec, *J. Appl. Polym. Sci.*, **12**, 2225 (1968).
8. S. T. Balke and A. E. Hamielec, *J. Appl. Polym. Sci.*, **13**, 1381 (1969).
9. L. H. Tung, *J. Appl. Polym. Sci.*, **13**, 775 (1969).
10. W. N. Smith, *J. Appl. Polym. Sci.*, **11**, 639 (1967).
11. M. Hess and R. F. Kratz, *J. Polym. Sci., Part A-2*, **4**, 731 (1966).
12. H. E. Pickett, J. R. Cantow, and J. F. Johnson, *J. Polym. Sci., Part C*, **21**, 67 (1968).

13. O. Levenspiel, *Chemical Reaction Engineering,* Wiley, New York, 1962.

14. J. H. Duerksen, Ph.D. Thesis, McMaster University, 1968.

15. P. E. Pierce and J. E. Armonas, *J. Polym. Sci., Part C,* **21,** 23 (1968).

16. S. T. E. Aldhouse and D. M. Stanford, Paper presented at the 5th International GPC Seminar, London, May, 1968.

17. A. E. Hamielec, GPC notes presented at a two-day short course at Washington University, St. Louis, April 25–26, 1969.

18. L. H. Tung, *ACS Div. of Petrol. Chem. Preprints* **15,** No. 1 (February, 1970).

II
REVIEWS OF THEORY AND EVALUATION METHODS IN GPC

Separation Mechanisms in Gel Permeation Chromatography

W. W. YAU, C. P. MALONE, and H. L. SUCHAN

ENGINEERING PHYSICS LABORATORY,
E. I. DU PONT DE NEMOURS & CO., INC.
WILMINGTON, DELAWARE 19898

Summary

This paper presents flow rate studies, vacancy chromatography, and a static mixing experiment. Data obtained on an unpacked column (a straight tube) and on a column packed with nonporous glass beads are also reported. The results reveal that peak dispersion in GPC arises mainly from lateral diffusion in the stationary phase (permeation in and out of the porous substrate) and from lateral diffusion in the mobile phase. GPC peak separation is mainly dominated by the process of steric exclusion. Pore size distribution data obtained on Bio-Rad porous glass are shown to illustrate the preference of random coil theories over theories of the equivalent sphere in the interpretation of steric exclusion of flexible polymers. The data are discussed in terms of Herman's diffusion theory and Cassasa's exclusion theory.

INTRODUCTION

Since the development of gel permeation chromatography (GPC) as a means to determine molecular weight distribution of flexible polymers, considerable interest has been shown in studying the separation mechanism. An understanding of the basic mechanism of GPC has great importance as a guide for such studies of practical interest as the improvement of separation efficiency, the correction for peak dispersion (1), and the development of a universal calibration curve (1).

Models have been proposed to explain GPC in terms of separation by flow (2), separation by restricted diffusion (1), and separation by steric exclusion (1). Since all of these postulated processes may occur in a GPC experiment, interpretation based on one model alone is not

105

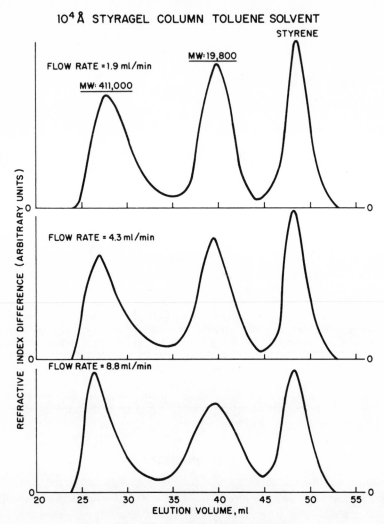

FIG. 1. Effect of flow rate on GPC curve shape.

sufficient to explain fully both the dispersion and the separation of GPC peaks. In the following, some experimental results are presented to show the relative importance of each of the postulated processes to GPC peak dispersion and GPC peak separation.

GPC data reported in this work were obtained either on a Waters Associates apparatus with a differential refractometer or on a GPC unit with the Du Pont Model 400 photometric analyzer as an ultraviolet detector. The polystyrene standards of narrow molecular weight distribution ($\bar{M}_w/\bar{M}_n < 1.10$) used in the experiments were obtained from Pressure Chemical Company, except for MW 4800, which was obtained from Waters Associates.

PEAK DISPERSION

The GPC elution curves obtained at three flow rates for a composite solution of styrene and two polystyrene standards are shown in Fig. 1. Peak dispersion changes with molecular size and with flow rate. As illustrated in Fig. 1, peak dispersion in GPC is greater at higher flow rates and for species of higher molecular weight except for those eluted near the void volume. The dispersion of peaks near the void volume, such as the peak of MW 411,000, will be discussed later. These observations imply that it is not longitudinal diffusion (in the flow direction) but lateral diffusion that is responsible for the dispersion of GPC peaks. In the case of longitudinal diffusion, the dispersion would be smaller for species of higher molecular weight (smaller diffusion coefficient) and would decrease with increasing flow rate (decreasing retention time). On the other hand, the results can be very well understood in terms of lateral diffusion processes, such as extra-column dispersion, permeation, and lateral diffusion in the mobile phase, which are described in the following paragraphs.

The characteristics of the extra-column dispersion are illustrated in Fig. 2, which shows the elution peak of styrene and that of polystyrene of MW 1.8×10^6 after passing through a tubing of small diameter. The dispersion of these peaks can be explained as the result of the velocity profile in the flow stream. The difference between the two curves is caused by the difference in the lateral diffusion rate between the two systems. As a solute band travels through the tubing, it becomes increasingly distorted due to the velocity profile. The center portion of the band travels faster, but it is less distorted than the outer portion near the wall of the tubing. In case of negligible lateral diffusion such a distorted band is expected to give an elution peak with a sharp front and a long tail, such as that observed for the polystyrene peak shown in Fig. 2. The distortion of the solute band also creates concentration gradients in the radial direction of the tubing. This concentration gradient is negative at the leading edge of the band; therefore, the solute molecules tend to diffuse from the center of the tubing to the slow-moving region near the wall. The reverse is true at the tailing edge of the band. This implies that a fast rate of lateral dif-

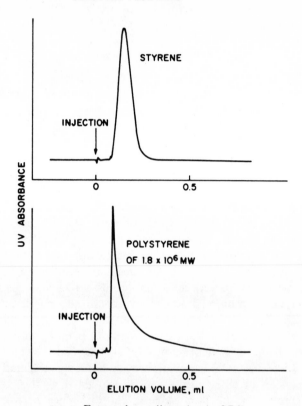

STRAIGHT TUBING, 0.018"I.D., I METER LONG
CHLOROFORM SOLVENT
0.5 ml/min FLOW RATE

FIG. 2. Extra-column dispersion in GPC.

fusion tends to give a symmetrical elution peak, and this seems to be
the case for the styrene peak shown in Fig. 2.

By comparing Figs. 1 and 2, it is obvious that GPC peak dispersion
is not explained by the capillary model (2) proposed to describe GPC
separation. The dispersion predicted by such a model, as one may
visualize by extrapolating the results in Fig. 2 to large retention
volume, would be orders of magnitude larger than what is observed
in Fig. 1. This suggests that the packing in a GPC column must have
sufficiently distorted the flow stream to prevent the development of a
persistent velocity profile in the column.

To prove the above hypothesis, GPC elution curves of the styrene

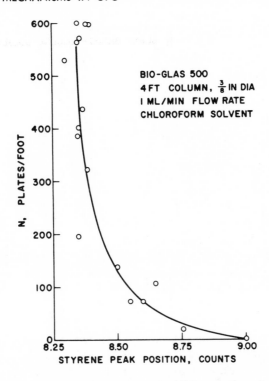

FIG. 3. Effect of packing density on GPC peak dispersion.

solution were obtained on a column which was repeatedly packed with different amounts of Bio-Glas 500 glass beads. At each packing density, the number of theoretical plates per unit column length (*3*), N, was calculated according to the approximate formula, $N \cong 4Ve^2/LW^2$, where Ve is the peak elution volume, L is the column length, and W is the peak half-width. Figure 3 shows how N decreases with increasing peak elution volume, i.e., with decreasing packing density. This is what one would expect from the disrupture of a persistent velocity profile mentioned in the previous paragraph.

Flow rate data obtained on columns packed with smooth, nonporous glass beads were used by Kelley and Billmeyer (*4*) to explain the mechanism of mobile phase dispersion in GPC. These authors showed that the experimental results were in good agreement with a coupling theory they developed [similar to that of Giddings' theory (*3*)], in which they interpreted the lateral diffusion as being caused by the velocity nonuniformity across the column cross section. The results

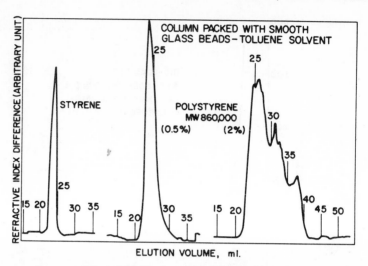

FIG. 4. Mobile phase dispersion in GPC.

of this study confirm that mobile phase dispersion plays an important role in GPC. It contributes a great deal to GPC dispersion because of the low diffusion coefficients of the polymer molecules.

The magnitude of mobile phase dispersion is illustrated in Fig. 4, which shows the elution curves of styrene and the polystyrene samples of MW 860,000 obtained on a column packed with smooth glass beads. The curve obtained for the polystyrene sample at 2% solution concentration is included in Fig. 4 to demonstrate that the so-called "overloading effect" can also happen in a column of nonporous packing. This would suggest that, whatever the causes of such an effect may be, it should not be considered in terms of oversaturation of the porous volume as the word "overloading" would imply. Comparing Figs. 1 and 4, one sees that the width of curves shown in Fig. 4 are of comparable magnitude to, yet are appreciably smaller than, those shown in Fig. 1. This indicates that permeation as well as mobile phase dispersion should be considered to fully account for GPC peak dispersion.

Both the extent and the rate of permeation are the important factors dictating the amount of dispersion caused by permeation. For this reason the stochastic model (5), which is derived on the basis of the extent of permeation only, is inadequate to describe GPC dispersion. The prediction of this model, viz., that dispersion increases with increasing retention volume, is not substantiated by the GPC results. A complete description of permeation dispersion was recently

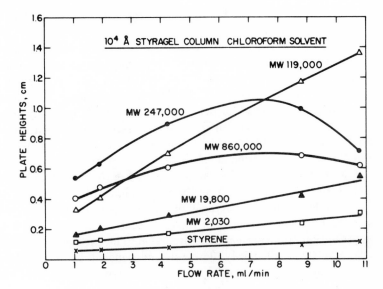

FIG. 5. Effect of flow rate on GPC peak dispersion.

developed by Hermans (*6*). In the following paragraphs the principal implications of his theory are briefly described and are compared with the results of the flow rate study on a 10^4 Å Styragel column which is shown in Fig. 5. This figure shows the dependence of the peak dispersion of several polystyrene samples of different molecular weight (MW) and the styrene solution, in which the plate height (*3*) of the elution peak, approximated by the quantity $LW^2/4Ve^2$, is plotted versus flow rate, where the symbols L, W, and Ve are the same as defined previously.

In the case of fast permeation, Herman's theory predicts [Eq. (38) in Ref. *6*] that the mean square fluctuation in retention volume should be proportional to u/kD_s. (Where u, k, and D_s are the symbols used by Hermans to represent the average flow velocity, the ratio of the concentration in the mobile phase versus that in the stationary phase, and the solute diffusion coefficient in the stationary phase, respectively. One may notice that the ratio k is a parameter to express the extent of permeation. It is equal to $1/K$, where K is the often-used symbol of the distribution coefficient.) Physically, this means the following: (*a*) at a fast diffusion rate, the peak dispersion, or plate height, should increase linearly with increasing flow rate (which is indeed observed for the curves of styrene and polystyrene of MW, 2,030, 19,800, and

119,000 shown in Fig. 5); (b) the dependence of peak dispersion on MW should be governed by the linear relationship between W^2 and $1/kD_s$. Since D_s decreases, yet k increases, with increasing MW, one should expect peak dispersion caused by permeation to increase with increasing MW until the product kD_s reaches a minimum, then to decrease with further increasing MW as k becomes increasingly large. In the extreme case, when no permeation occurs, k is infinitely large, the contribution of permeation dispersion is zero and the peak width should be affected only by the mobile phase dispersion. [This may explain the lower dispersion level of the peak of 860,000, relative to that of MW 247,000 and 119,000, shown in Fig. 5. The molecules of MW 860,000, which are eluted at the void volume of the column, are totally excluded from the Styragel packing in the column. The peak is less dispersed since it does not have, in contrast to the other ones, the contribution from permeation dispersion. The fact that the curve of MW 860,000 is relatively flow rate-independent is in agreement with the above reasoning, since it has been reported (4) that, for highly dispersed peaks, the mobile phase dispersion becomes flow rate-independent as a consequence of the coupling effect.]

For insufficient permeation rates, Hermans' theory predicts a highly dispersed and skewed elution peak. Insufficient permeation rate is defined here as the experimental condition under which the solution molecules do not have sufficient time to establish equilibrium between the mobile and the stationary phases. Such a condition is more likely to be true at high flow rates and for large solute molecules. From the concentration profile of such an elution peak [given in Eq. (26) in Ref. 6], can show that, under nonequilibrium condition, W^2 should vary linearly with D_s^2/k^4u^2, i.e., W should decrease with increasing flow rate. (This seems to be the reason for the decline of the curve of MW 247,000 in the high flow rate region shown in Fig. 5.) These results indicate that nonequilibrium is not realized under the normal operating condition of GPC. It becomes noticeable only at very high flow rate and for samples of high MW.

PEAK SEPARATION

Diffusion models (1) have been proposed that assume peak separation in GPC is caused by the nonequilibrium mechanism mentioned in the previous paragraph. In view of the results discussed above, it is obvious that such a model would not be adequate to account for the overall peak separation in GPC. The rate of permeation would

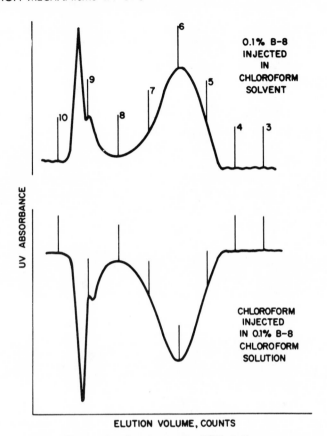

FIG. 6. Conventional and vacancy GPC elution curves.

affect peak separation only at high flow rate and for samples of high MW. This is reflected in the fact that GPC peak positions, except the ones of high MW, are virtually flow rate-independent.

A capillary model (*2*) has recently been proposed to explain GPC peak separation. The model assumes that GPC separation is the result of the capillary velocity profile in combination with a wall effect that causes the larger solute molecules being more populated near the center of the flow stream, therefore having a larger average flow velocity. Experimental evidence against such a model is provided in the results shown in Figs. 4 and 6. The fact that the styrene and the polystyrene peaks both elute near the void volume of a column packed with smooth glass beads (see Fig. 4) shows that the velocity profile in the interstitial spaces does not provide the separation capability.

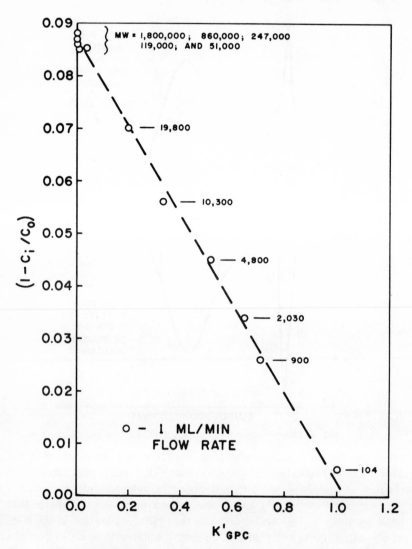

FIG. 7. Comparison of polymer-gel mixing data with GPC data.

The fact that the vacancy elution curve (7), obtained by an injection of solvent into the flow stream of the polymer solution, is characteristic of the polymer and not of the solvent (see Fig. 6), is also in direct contradiction to the capillary model. The conventional (top curve) and the vacancy (bottom curve) GPC elution curves shown in Fig.

6 were obtained on a 10^4 Å Styragel column for a polystyrene standard, designated B-8, obtained from Dow Chemical Company. These results indicate that the porous nature of the GPC packing, which is neglected in the capillary model, is the essential element of the separation capability in GPC.

The results discussed so far have indicated that kinetic processes contribute only in minor ways to the peak separation in GPC. This suggests that an equilibrium mechanism, viz., the extent of permeation, must be the origin of GPC peak separation. Direct experimental evidence for this contention is provided by a static experiment of polymer–gel mixing (8). The result of such an experiment is illustrated by Fig. 7 with data obtained on Bio-Glas 200 Å glass beads. Figure 7 shows that there is a change from an initial concentration C_i to a final concentration C_0 when a polymer solution is mixed with dry porous material. This concentration change is a function of MW of the polystyrene samples; therefore, it too depends on the distribution coefficient K'_{GPC}. K'_{GPC} is defined as $(V_e - V_0)/(V'_t - V_0)$, where V_e and V'_t are the GPC elution volumes of the polystyrene and the styrene peaks, respectively, and V_0 is the void volume of the GPC column. The linear relationship between $(1 - C_i/C_0)$ and K'_{GPC} shown in Fig. 7 indicates that the separation achieved in this GPC column is due to a distribution of the solute molecules between the mobile and the stationary phases which closely approximates the equilibrium condition.

These results demonstrate that GPC separates primarily by the extent to which the solute molecules can permeate the porous packing. Several theories based on steric exclusion have been proposed to explain the effect of the size of a flexible polymer molecule on the extent of permeation. The earlier models (1) assume that the exclusion effect of a flexible polymer molecule can be approximated by that of a rigid sphere with a radius equivalent to the radius of gyration of the molecule. As de Vries et al. (4) have pointed out, such a model is not adequate to explain GPC peak separation of flexible polymer molecules since the shape of the GPC calibration curve for flexible polymer molecules is different from the pore size distribution curve of the packing. Such a comparison is given in Fig. 8. The dashed curve is the pore size distribution curve of Bio-Glas 500 Å porous glass, and the data points identified by $\triangle$ are the GPC results obtained on a column packed with the same glass.

Models of steric exclusion based on thermodynamic reasonings have recently been proposed. The theory for rigid molecules was developed

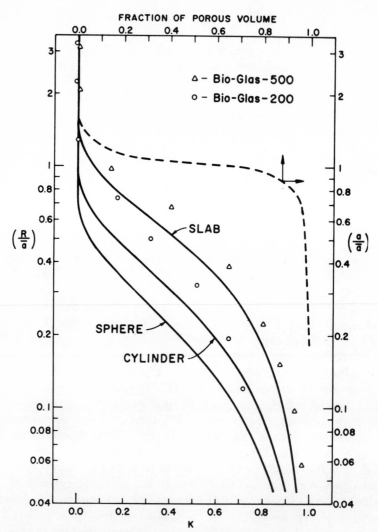

FIG. 8. Comparison of GPC data with pore size distribution curve and with Casassa's theoretical curves.

by Giddings et al. *(10)*. The theory for flexible polymer molecules was developed by Casassa *(11)*. An approximate treatment of the problem was given in the stochastic model *(5)*. Cassassa explained the decrease in the extent of permeation with increasing MW of the flexible polymer molecules as a consequence of the decrease in the conformational freedom of such a molecule in the pores of the GPC packing. For a pore of either sphere, cylinder, or slab shape, he derived the theoretical expression for the distribution coefficient (K) of the solute molecule as a function of the radius of gyration of the molecule (R) and the size of the pore $(\bar{a})$ [Eqs. (2), (3) and (4) in Ref. *11*]. The solid lines in Fig. 8 show the theoretical curves predicted by the theory (from top to bottom: slab, cylinder, sphere). They compare well with the GPC results obtained on a column packed with Bio-Glas 500 Å (plotted by $\triangle$) and with Bio-Glas 200 Å (plotted by $\bigcirc$). Obviously, Casassa's theory describes the shape of the GPC calibration curve much better than the earlier equivalent sphere models. This implies that the curvature of the GPC calibration curve is very much determined by the fluctuating nature of the polymer molecule. Therefore, the extent to which the calibration curve can be flattened to give better peak separation by improving the sharpness of the pore size distribution of the packing is limited.

Pore size distribution measurements were provided by the American Instrument Company. The average pore radius $(\bar{a})$ is 63.2 Å for Bio-Glas 200 Å as determined by nitrogen desorption, and 210 Å for Bio-Glas 500 Å as determined by mercury intrusion.

REFERENCES

1. K. H. Altgelt, in *Advances in Chromatography*, Vol. 7 (J. C. Giddings and R. A. Keller, eds.), Dekker, New York, 1968, pp. 3–46.

2. E. A. Di Marzio and C. M. Guttman, *J. Polym. Sci., Part B*, **7**, 267 (1969).

3. J. C. Giddings, *Dynamics of Chromatography. Part 1. Principles and Theory*, Dekker, New York, 1965.

4. R. N. Kelley and F. W. Billmeyer, Jr., *Anal. Chem.*, **41**, 874 (1969).

5. J. B. Carmichael, *J. Polym. Sci., Part A-2*, **6**, 517 (1968).

6. J. J. Hermans, *J. Polym. Sci., Part A-2*, **6**, 1217 (1968).

7. C. P. Malone, H. L. Suchan, and W. W. Yau, *J. Polym. Sci., Part B*, **7**, 781 (1969).

8. W. W. Yau, C. P. Malone, and S. W. Fleming, *J. Polym. Sci., Part B*, **6**, 803 (1968).

9. A. J. de Vries, M. Le Page, R. Beau, and C. L. Guillemin, *Anal. Chem.*, **39**, 935 (1968).

10. J. C. Giddings, E. Kucera, C. P. Russell, and M. N. Myers, *J. Phys. Chem.*, **72**, 4397 (1968).

11. E. F. Casassa, *J. Polym. Sci., Part B*, **5**, 773 (1967).

Gel Permeation Chromatography and Thermodynamic Equilibrium*

EDWARD F. CASASSA

MELLON INSTITUTE AND DEPARTMENT OF CHEMISTRY
CARNEGIE-MELLON UNIVERSITY
PITTSBURGH, PENNSYLVANIA 15213

Summary

Although gel permeation chromatography is firmly established as a technique for investigating heterogeneity in synthetic polymers, the character of the chromatographic process—whether it is dominated by diffusion. by flow effects, or by an equilibrium partitioning of polymer between the mobile phase and the micropores in the column packing—is still disputed. General chromatographic theory supports the idea that under ordinary experimental conditions the equilibrium distribution of a solute determines the position of its elution peak in the chromatogram. Statistical mechanical calculations of distribution coefficients for linear and branched polymer chains and idealized. pores of simple geometry lead to predictions in good accord with some experimental findings.

GEL PERMEATION AND CHROMATOGRAPHIC THEORY

In the decade or less since gel permeation chromatography (GPC) was first proposed as a general means for separation of synthetic polymers according to molecular weight, suitable instrumentation has become widely available and the method has attained status as the most popular one for analytical polymer fractionation. Although the practical success, first with compact biological macromolecules and more recently with typical flexible chain polymers, that has amply demonstrated that separation is effected according to molecular size has also spurred inquiry into the physical basis of the separation, the

* Presented at the ACS Symposium on Gel Permeation Chromatography sponsored by the Division of Petroleum Chemistry at the 159th National Meeting of the American Chemical Society, Houston, Texas, February, 1970.

mechanism responsible remains a matter of some dispute. Theoretical treatments have been variously concerned with diffusion processes (*1*, *2*), hydrodynamic effects (*2–5*), and thermodynamic equilibrium (*6–11*). Perhaps the only point of general agreement is that the mechanisms proposed are not mutually exclusive: the question is not what effect occurs but rather which one is dominant (*1, 2, 4, 12*).

It is attractive to suppose that a process at or (more precisely) near thermodynamic equilibrium is operative simply because a rigorous theoretical treatment will be inherently simpler than for a mechanism limited by a transport process. Furthermore, even if such optimism should be quite unjustified, it can be argued that the equilibrium situation retains significance as the asymptotic limit to which real behavior must tend when a chromatographic column is operated at extremely small flow rate.

To gain some idea of the applicability of an equilibrium model for GPC, we can turn to the general theory of chromatography. Of the several ways of treating chromatographic processes, the stochastic theory proposed by Giddings and Eyring (*13*), elaborated upon by Giddings (*14*) and by McQuarrie (*15*), and applied to GPC by Carmichael (*16*) is particularly appropriate. The original theoretical model is designed to describe adsorption chromatography. Hence, the description proceeds in terms of a column filled with a granular packing material bearing identical surface sites (constituting a stationary phase) capable of adsorbing a solute reversibly from solution. The interstices between granules, initially filled with solvent, constitute the mobile phase. At a certain moment, a narrow band of solution is introduced at the top of the column. This is followed by more solvent as liquid is withdrawn from the bottom of the column at a steady rate and the solution zone passes down through the column, exchanging solute with the stationary phase. Three assumptions are made: (a) that a molecule of a particular solute species in the liquid phase has a certain fixed probability k_1 per unit time of being adsorbed on a surface site, (b) that a molecule on the surface has a fixed probability k_2 per unit time of escaping into the mobile phase, and (c) that there is no net diffusion of solute molecules in the direction of the column axis while they are in the mobile phase. The first assumption implies that the adsorption isotherm is linear—the solution is so dilute that solute molecules do not interact and there is no multiple adsorption on sites. It further implies that the absorption is not diffusion controlled. This requirement is obviously not physically realizable since the probability for adsorption of a given molecule can hardly be independent of its distance from a reactive site, but a time-averaged k_1 will be the same for each solute molecule if the residence time in the column is long enough to permit the molecule to undergo a large number of adsorptions and desorption steps. Assumptions (a) and (b) require that absorption sites do not interact—that occupation of a site does not affect k_1 and k_2 at any other site. Assumption (c) means that every solute molecule spends the same time t_0 in the mobile phase as it passes through the chromatographic column. The theory thus neglects various flow effects, as well as longitudinal diffusion, that in real columns contribute to the spreading of chromatographic peaks. The peak shape predicted by the theory is accounted for solely by the probability distribution of the total time t_s that a solute molecule spends absorbed as it pursues its course down the column, alternately in the mobile phase and in the stationary (adsorbed) phase. The object of the stochastic theory is to calculate this distribution as a function of parameters k_1, k_2, and t_0. If the rate of withdrawal of fluid from the column is held constant, $t_0 + t_s$ is proportional to the volume eluted when molecules experiencing a given t_s emerge, and the probability of t_s plotted against t_s for a solute species has the shape of the elution curve—i.e., concentration versus elution volume.

It will be recognized that the foregoing description of a model for adsorption chromatography requires only minimal verbal changes to apply equally to GPC. Now we have a column packing that contains, in place of adsorption sites, microscopic voids, for our purposes of the same order of size as macromolecules and (for simplicity) assumed to be of identical size and shape. The stationary phase is the volume V_i of the part of the column inside micropores and accessible to solvent. All we then have to do is replace the word "adsorption" by the phrase "entrapment in micropores" and recognize that in GPC, unlike adsorption chromatography, the solvent—consisting of small molecules and, therefore, most easily trapped in voids—is retarded in the column relative to macromolecular solutes.*

The mathematical problem is a standard one in probability theory. Interestingly, the result of the calculation is an elution curve that departs somewhat from a Gaussian—it has a positive skew. However,

---

*Since the name "gel permeation" has been canonized by general acceptance, we use it freely here to designate a kind of chromatography, without meaning to imply that the column packing must be a gel in any conventional sense of the word.

our concern here is not with the details of the theoretical curve shape but primarily with the position of the maximum in the peak along the time (or volume) axis. The result we require is (13)

$$V_e \approx V_0 + KV_i(1 - 3/2\bar{N}) \tag{1}$$

where V_e is the volume eluted when the peak maximum appears (henceforth called simply the elution volume) and V_0 is the volume of the mobile phase. Then, $V_0 + V_i$ is the total volume of the column except for the volume of the solid matter in the packing and the volume of "blind" pores that cannot be permeated by solvent. Certain constants are gathered in K,

$$K = k_1 V_0 / k_2 V_i \tag{2}$$

and

$$\bar{N} = k_1 t_0 \tag{3}$$

is the mean number of entrapments suffered by a solute molecule in its passage through the column. Equation (1) is an asymptotic relation for $\bar{N}$. Recalling that the mean residence time of a molecule in a GPC column is of the order of a half hour or longer in a conventional experiment, we surmise that this condition is adequately fulfilled. Equation (1) also implies that at the start of the experiment—the moment when withdrawal of liquid from the column begins—the solute is all in the mobile phase. However, the only difference for the alternative extreme of all solute adsorbed at the beginning is the unimportant replacement of $3/2$ in the second term by $1/2$ (13).

The obvious, but crucial, deduction from Eq. (1) is that if $\bar{N}$ is sufficiently large, we can write

$$V_e = V_0 + KV_i \tag{4}$$

without appreciable error. Consequently, under realistic experimental conditions, the elution volume V_e from a given column is expected to be insensitive to the flow rate, which of course determines the time t_0, hence $\bar{N}$.

As we expect intuitively, the dispersion of an elution peak does not share this insensitivity to flow rate. Another result of the statistical theory is that the standard deviation of t_s is proportional to $\bar{N}^{1/2}$ (i.e., to $t_0^{1/2}$) for large $\bar{N}$. However, the standard deviation of the ratio t_s/t_0 is asymptotically proportional to $1/\bar{N}^{1/2}$. These relations conform to the essential requirement in a chromatographic separation that increasing the column length at constant flow rate improves resolution of multiple peaks, even though each peak becomes broader with increasing t_0.

Equation (4) can be regarded as the basic relation for GPC. The constant K as defined in the theory is the ratio of the two rate constants for passage of solute to and out of the stationary phase; hence it is the equilibrium constant, the distribution coefficient, for the partitioning of solute between mobile and stationary phases. Just as in the distribution of a solute between dilute macroscopic phases, K is the ratio of solute concentrations in the two phases; and $-RT \ln K$ represents the free energy change for the process of transporting a mole of solute in its standard state in one phase to its standard state in the other. The possibility is thus afforded of applying the conventional methods of statistical mechanics to simple models for entrapment of solute in pores to deduce K and thus to predict elution volumes from Eq. (4) if V_0 and V_i are known. In general V_0 and V_i can be determined experimentally: the former represents the elution volume for solute molecules so large that they do not penetrate the pores appreciably $(K = 0)$, and $V_i + V_0$ is the elution volume for small molecules that penetrate the voids as easily as the solvent does (so that $K = 1$—as would be expected for radioactively labeled solvent).

As the preceding discussion indicates, equilibrium theory may yield important information about gel chromatography—information on the relation of molecular conformation and pore geometry to the elution volume. However, it cannot be expected to reveal anything about peak spreading since the dynamics of molecular entrapment that determine the individual rate constants k_1 and k_2 do not enter into the equilibrium calculations.

CALCULATION OF THE SOLUTE DISTRIBUTION

We have already stipulated that the solution passing through the chromatographic column must be so dilute that interactions between solute molecules are negligible. Then, the equilibrium constant K can be written formally as a ratio of configuration integrals for one molecule:

$$K = \frac{\int \cdots \int e^{-U_s\{q\}} d\{q\}}{\int \cdots \int e^{-U_0\{q\}} d\{q\}} \tag{5}$$

where $U_s\{q\}$ represents an energy associated with a set of spatial coordinates $\{q\}$ that defines a solute molecular conformation in the stationary phase and $U_0\{q\}$ represents the energy as a function of the requisite coordinates in the mobile phase.* For a spherical molecule appropriate coordinates would be the three needed to locate the center of mass; for a rigid asymmetric molecule, the coordinates of the center of mass plus the angles determining orientation; and for a flexible polymer chain, coordinates of the center of mass, or a chain end, plus coordinates of each segment relative to this reference point. The symbol $d\{q\}$ denotes a differential element of the configuration space of as many dimensions as are needed to describe a molecular conformation. The two multiple integrations are carried out over equal macroscopic volumes in physical space, e.g., a unit volume of solution and the same volume of space inside pores constituting the stationary phase (10).

Evaluation of the integrals in Eq. (5) is in general impossible, and further progress depends on placing drastic limitations on the energies $U_s\{q\}$. Here we let these energies have but two values, zero (or any fixed finite value) and infinity. When U_s is zero, the conformation is allowed; when U_s is infinite, the conformation is forbidden. Thus we represent the walls of a pore as rigid boundaries which a molecule cannot pass. Any conformation intersecting the boundary has infinite energy and is thereby excluded. Adsorption of solute, which would imply minimum energy for conformations contiguous to the boundary surface, is also excluded. For nonrigid molecules the assumption that all allowed conformations have the same energy means that intramolecular interactions are ignored. For flexible long-chain molecules, the case of particular interest, this means that the polymer–solvent system is at its characteristic "theta temperature," at which the second virial coefficient in the osmotic equation of state is zero (and at which phase separation occurs if the polymer is of infinite molecular weight) (18). At the theta point, the molecular conformation of a long-chain polymer can be described by random flight statistics, a fact that enormously facilitates mathematical analysis.

With the above limitations on permissible energies, the distribution coefficient K is expressed as a ratio of volumes in configuration space; and thus $R \ln K$ is a standard entropy change per mole of solute. For rigid spherical molecules, the ratio of configuration integrals reduces to a ratio of volumes in physical space. This straight-

---

* In more precise language, these energies are potentials of mean force (17) between solute molecules. They include implicitly the effects of solute–solvent and solvent–solvent interactions.

forward volume-exclusion picture of GPC has long been used to account for the elution behavior of compact biological macromolecules. For example, a spherical solute molecule of radius r does not interact with a wall when its center of mass is farther than a distance r from the wall, and it can approach no closer than r. Consequently, in a slab-shaped space between parallel (infinite) planes a distance $2a$ apart, the equilibrium constant is $(a - r)/a$, the volume "seen" by the sphere divided by the actual volume of a cavity. The same sphere in a long cylindrical cavity of radius a and inside a hollow sphere of radius a gives K equal, respectively, to the square and cube of $(a - r)/a$. Various geometrical situations for exclusion of rigid molecules have been investigated (6–8, 10, 19). The calculations are sometimes intricate but the principle remains the same; the calculated K represents the fraction of the void volume effectively available to solute.

The exclusion of a flexible polymer chain from part of the volume inside micropores is perhaps less obvious. Because such a molecule is allowed to assume any shape, however tortuous, no part of the actual volume of a pore is excluded from occupancy by any part of the polymer chain; but since configurations that intersect the walls of a void are disallowed, the number of permitted configurations per unit volume is still decreased by proximity to a wall. In other words, although the geometry of the situation is more complicated than for the rigid sphere model, it is still true that the region of *configuration* space available to the solute is reduced by the presence of an impenetrable boundary.

These ideas can be illustrated concretely by considering the least specific model for a micropore: whatever the shape of the pore, we let it be characterized simply by the ratio σ of its surface to its volume and assume that solute molecules intersect with the surface as if it were a plane of infinite extent. Therefore, a spherical solute of molecular radius r, rolling over the surface of any cavity, will have its center of mass excluded from a fraction σr of the volume of the cavity, and then according to Eq. (5), the distribution coefficient is given by

$$K = 1 - \sigma r \qquad (6)$$

Representing σ by $1/a$, $2/a$, and $3/a$, for slab-shaped, cylindrical, and spherical cavities, respectively, we obtain expressions for K in agreement with the exact results mentioned above for the limit $r/a \rightarrow 0$.

Turning to a linear polymer chain, we depict it by a trace of the path of a particle undergoing a series of random displacements; thus we seek an appropriate solution of the diffusion equation

$$\frac{\partial P_n(x,y,z)}{\partial n} = \frac{b^2}{6} \nabla^2 P_n(x,y,z) \tag{7}$$

where $P_n(x,y,z)$ $dxdydz$ is the probability of finding the nth step of the random walk within a differential volume element $dxdydz$ located at a point x,y,z. The step length is normally distributed and its rms value is b. For the problem under discussion we suppose that the random flight begins at a point $(x',0,0)$ within a semi-infinite region $x > 0$ bounded by a plane at $x = 0$. Imposing the boundary condition that P_n vanish at $x = 0$, we can solve the differential equation to obtain:

$$P_n(x,y,z|x',0,0) = (6/\pi nb^2) \exp\{-(3/2nb^2)(y^2 + z^2)\}$$
$$\times [\exp\{-(3/2nb^2)(x - x')^2\} - \exp\{-(3/2nb^2)(x + x')^2\}] \tag{8}$$

the probability that a random flight starting at $(x',0,0)$ arrives at the point (x,y,z) after n steps without encountering the boundary (20). Integration gives an error function,

$$P_{x'} \equiv \iiint P_n(x,y,z|x',0,0) \, dxdydz = \mathrm{erf}\,[(x'/2)(6/nb^2)^{1/2}] \tag{9}$$

where

$$\mathrm{erf}\, u = (2/\pi^{1/2}) \int_0^u e^{-t^2} \, dt \tag{10}$$

for the probability that a random flight beginning at a distance x' from the plane $x = 0$ and proceeding for n steps does not touch the boundary. Equivalently, $P_{x'}$ can be described as the fraction of all possible conformations of a random-flight polymer chain of n segments with one end at $x = x'$ in unbounded space that still remains available when an impenetrable boundary is placed at $x = 0$. The integral

$$x_c = \int_0^\infty (1 - P_{x'}) \, dx' = 2(nb^2/6\pi)^{1/2} \tag{11}$$

represents an effective distance characterizing the depletion of polymer chains at equilibrium near the boundary. That is, in terms of the mean solute concentration (precisely, the concentration of chain ends), the cavity is equivalent to a volume in an unbounded space that is smaller than the real cavity by a layer of thickness x_c adjacent to the wall. The distribution coefficient $K = K_1$ for the polymer species is then simply obtained by putting x_c in place of r in Eq. (6).

Just as in the case of rigid spheres, expressions for K_1 for flexible chain molecules derived in this way are correct limiting forms for molecules small compared to the dimensions of the cavity. Calculations not so restricted, of K_1 for random-flight chains in the slab, cylinder, and sphere cavities have been described elsewhere (9). The results are shown in Fig. 1 for comparison with the limiting behavior. It can be seen that the straight line representing the approximate K_1 for the slab crosses the curve for the exact relation at $K \approx 0.25$. There are similar intersections at much larger K for the cylinder and sphere relations, although this is not discernible on the scale of Fig. 1.

Since the mean-square radius of a random-flight linear chain of n segments is

$$R^2 = nb^2/6 \tag{12}$$

the quantity, $(nb^2/6a^2)^{1/2}$, where a is the radius of the cavity (half the separation of the planes in the slab model), is a convenient meas-

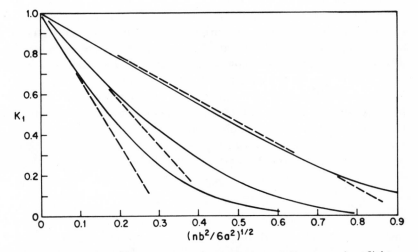

FIG. 1. Equilibrium constants for partitioning of linear random-flight polymer chains between a macroscopic solution phase and cavities of molecular dimensions. The solid curves (from top to bottom) are for slab-shaped, cylindrical, and spherical cavities. The abscissa is the ratio of the rms molecular radius to the cavity radius (half its thickness for the slab). The dashed lines are the corresponding limiting relations for the upper permeation limit, $K = 1$.

ure of the relative sizes of a polymer molecule and the cavity. For each pore geometry characterized by a single dimensional parameter a, it happens that K_1 will be a unique function of R/a. It can be seen from the plots, however, that these functions are quite different for the three cavity shapes. Although there is considerable interest in establishing a single combination of dimensional parameters to correlate (to some reasonable approximation) the equilibrium partitioning of any solute between a macroscopic solution phase and cavities of any geometry whatever (10), we shall pursue here only the more limited question of a comparison of the behavior of linear and branched polymer chains with respect to the same pore model. Thus the large absolute differences in K for a species in different types of pores are not our present concern.

Exact random-flight calculations would in general be very difficult for an arbitrary topology of chain branching. However, except for requiring lengthier computations, one model proves to offer no more real difficulty than do linear chains. This is what has been called the "regular star" model—a number of identical linear chain elements all joined by one end to a common branch point (21, 22). The procedure described above for a linear chain near a plane is carried out for one of the branches: the origin of the chain is taken as the branch point, which is placed at x', and P_x is calculated as before for a chain of n steps. Now, however, f chains of the same length are generated in the same fashion from the same origin; and thus $(P_{x'})^f$ is the probability that a conformation of an f-fold star with the branch point at x' will not touch the boundary at $x = 0$. Finally, the distribution coefficient is obtained:

$$K_f = 1 - \sigma \int_0 [1 - P_x')^f] \, dx'$$
$$= 1 - 2\lambda\psi(nfb^2/6a^2)^{1/2} \qquad (13)*$$

in which λ is a numerical factor dependent only on the cavity $(1, 2, 3,$ for the slab, cylinder, and sphere, respectively) and ψ is a function of f that has been evaluated by numerical integration ($\psi = 1/\pi^{1/2}$ for linear chains). A few values of ψ are listed in Table 1. As the discussion has already implied, Eq. (13) is a correct limiting form for a polymer chain with average dimensions small compared to the cavity; it gives the first two terms of an exact series in powers of $(nfb^2/6a^2)^{1/2}$

---

* This relation was written incorrectly in Eq. (13) of Ref. *11*, with an extra factor $f^{1/2}$ in the last term.

TABLE 1

Effect of Chain Branching on Permeation Equilibrium for K_f near Unity

f	g	ψ	$\psi/g^{1/2}$	ν
1, 2	1.0000	0.5462	0.5642	
3	0.7778	0.5415	0.6140	0.163
4	0.6250	0.5178	0.6634	0.182
6	0.4444	0.4775	0.7163	0.206
8	0.3838	0.4458	0.7603	0.220
12	0.2361	0.3997	0.8226	0.239

and hence, the correct initial slope of K as a function of this variable for a given f. As in the case of linear chains, rather more complicated calculations give K for chains that are not small compared to the radius of the cavity (*11*).

CORRELATION OF RESULTS AND COMPARISON WITH EXPERIMENT

Equation (13) permits us to investigate the effect of chain branching on K_f. Qualitatively, it is apparent (and expected) that increasing f while keeping molecular weight fixed (keeping nf constant) increases permeation—because with increasing branching the molecule becomes more compact. Increasing n with f fixed decreases permeation since the molecular domain becomes larger; increasing f at fixed n also decreases permeation because the greater the number of branches, the greater is the chance that a conformation randomly generated from a given point inside a cavity will be interrupted by the walls of the cavity. The more important quantitative question is whether the dependence of K_f on branching can be correlated with any accessible measure of molecular size. The two most obvious choices are inadequate. If the total mass (or number of statistical segments) were the sole determinant, K_f would depend on nf/a^2 and ψ in Eq. (13) would be a constant; in fact, ψ decreases with increasing f. The rms molecular radius R is no more satisfactory as a correlating parameter. The effect of branching on mean molecular size is conventionally expressed by

$$g = (R_{br}/R_{lin})^2 \qquad (14)$$

the ratio of mean-square radii of a branched chain and the analogous linear chain with the same mass. For random flight chains, R_{lin} is given by Eq. (12) and the proportionality factor g has been obtained for a number of branched models, including the regular star (21). If K depended uniquely on R_{br}/a, the ratio $\psi/g^{1/2}$, according to Eq. (13) would have to be constant; actually it increases, as Table 1 shows.

These unsuccessful analyses represent attempts to express K_f first as a function of $(nfb^2/a^2)g^0$ and then of $(nfb^2/a^2)g$. That the deviations with f are in opposite directions suggests applicability of an intermediate power of g. In the last column of Table 1, we give values of ν defined by

$$\pi\psi^2 = g^{2\nu} \tag{15}$$

It is evident that when f is small, ν is not far from 1/5 or 1/6. Accepting the latter value, we can combine Eqs. (13) and (15), and propose

$$K_f = 1 - 2\lambda\pi^{-1/2}(nfb^2/6a^2)^{1/2}g^{1/6} \tag{16}$$

as a general relation for K_f for linear and branched chains provided K_f is not far from unity. Calculations for star molecules in slab, cylinder, and sphere cavities for $K < 1$ confirm in these more general cases a dependence on $R_{lin}g^{1/6}/a$, as Eq. (16) indicates. Plots of K_f versus $(nfb^2/6a^2)^{1/2}g^{1/6}$ at constant f superpose quite well, at least for $f \leq 8$. A slightly larger power of g improves the superposition for larger f (11). It is likely that these conclusions need not be restricted to star molecules since the ratio g correlates physical properties for a variety of branched chains.

Taking ν as precisely 1/6 was arbitrary; but this value, in conjunction with theoretical relations for the intrinsic viscosity, brings our results into conformity with a recent empirical determinaton. If the intrinsic viscosity for a linear chain species is given (23) by the product of a universal hydrodynamic constant Φ and the ratio R^3/M (M being the molecular weight) and if, according to an approximate theory of Zimm and Kilb (24), the intrinsic viscosity for star molecules is related to that for linear chains of the same mass by

$$[\eta]_{br} = [\eta]_{lin}g^{1/2} \tag{17}$$

it follows that

$$([\eta]_{br}M/\Phi)^{1/3} = (nfb^2/6)^{1/2}g^{1/6} \tag{18}$$

If the location of an elution peak in a chromatogram from a GPC column is governed by the quantity on the right-hand side of Eq.

(16), any polymer having the same value of the product $[\eta]M$ should elute at the same point *from the same column*. On empirical grounds, Benoit et al. (*25*) have proposed just this relation.

Our equilibrium calculations presuppose a theta-solvent system. The extension to polymer in a good solvent—in which the chain is expanded beyond random-flight size owing to potentials of mean force $U_0\{q\}$ that depend on conformation—can be done to an adequate approximation, at the price of loss of elegance; however, there is some reason to believe that results for a random flight modified simply by use of the mean-square size of the real chain will not be seriously in error (*26*). Probably a more important shortcoming of the theory is the total neglect of adsorption of polymer on the column packing. It is very likely that this is a perturbing influence in many situations of practical interest.

It is important to note that the ideas proposed here are susceptible to experimental tests apart from verification of the predicted universality of $[\eta]M$ as a column calibration parameter. In the unlikely event that the real pore structure were sufficiently simply and precisely known, K for a polymer species of known molecular dimensions could be calculated *a priori* and compared with the apparent value deduced from elution measurements with the aid of Eq. (4). The column material coming closest to such idealized requirements is a special porous glass developed by Haller (*27*, *28*), which has been found by electron microscopy and mercury intrusion measurements to have pores of remarkably uniform cross section. Values of K determined from elution studies (*29*) of a series of nearly monodisperse linear polystyrenes from two columns of such glass with different pore sizes agree well with the theoretical relation for K versus R/a for the slab model (*9*). However, since the experimental estimates of a (by mercury intrusion) (*27*) may be subject to systematic error, the results can be taken as indicating no more than that the theory gives results of the right order.

Finally, it is important to remember that the validity of interpreting K in Eq. (4) as an equilibrium constant can be studied without regard to a particular model. The value of K can be obtained directly in a static experiment, by equilibrating an aliquot of polymer solution with the porous gel and determining the change in concentration in the supernatant liquid, provided that a calibration experiment with a solute species too large to penetrate the pores is also done. The equilibrium data can then be compared with results from column elution. In a recent study of this kind, Yau, Malone, and Fleming (*30, 31*) found agreement between static and chromatographic experiments for elution of polystyrene solutions from porous glass. With the more familiar cross-linked polystyrene gel as a column packing, deviations from equilibrium behavior were found— elution volumes exhibiting a dependence on flow rate that became more pronounced with increasing molecular weight.

Acknowledgment

This study was aided by the Air Force Materials Laboratory.

REFERENCES

1. W. W. Yau and C. P. Malone, *J. Polym. Sci., Part B,* **5,** 663 (1967).
2. G. K. Ackers, *Biochemistry,* **3,** 723 (1964).
3. K. O. Pedersen, *Arch. Biochem. Biophys., Suppl.* **1,** 157 (1962).
4. C. Gutmann and E. A. DiMarzio, *J. Polym. Sci., Part B,* **7,** 267 (1969).
5. F. H. Verhoff and N. D. Sylvester, *J. Macromol. Sci.—Chem.,* **A4,** 979 (1970).
6. J. Porath, *Pure Appl. Chem.,* **6,** 233 (1963).
7. T. C. Laurent and J. Killander, *J. Chromatogr.,* **121,** 317 (1964).
8. P. G. Squire, *Arch. Biochem. Biophys.,* **107,** 471 (1964).
9. E. F. Casassa, *J. Polym. Sci., Part B,* **5,** 773 (1967).
10. J. C. Giddings, E. Kucera, C. P. Russell, and M. N. Myers, *J. Phys. Chem.,* **72,** 4397 (1968).
11. E. F. Casassa and Y. Tagami, *Macromolecules,* **2,** 14 (1969).
12. R. L. Pecsok and D. Saunders, *Separ. Sci.,* **1,** 613 (1966).
13. J. C. Giddings and H. Eyring, *J. Phys. Chem.,* **59,** 416 (1955).
14. J. C. Giddings, *J. Chem. Phys.,* **26,** 169 (1957).
15. D. A. McQuarrie, *J. Chem. Phys.,* **38,** 437 (1963).
16. J. B. Carmichael, *J. Polym. Sci., Part A-2,* **6,** 517 (1968).
17. T. L. Hill, *Statistical Mechanics,* McGraw-Hill, New York, 1956, Chapter 6.
18. P. J. Flory, *Principles of Polymer Chemistry,* Cornell Univ. Press, Ithaca, N. Y., 1953, Chapter 12.
19. A. G. Ogston, *Trans. Faraday Soc.,* **54,** 1754 (1958).
20. H. S. Carslaw and J. C. Jaeger, *Conduction of Heat in Solids,* 2nd ed., Oxford Univ. Press, London, 1959. Chapter 10.
21. B. H. Zimm and W. H. Stockmayer, *J. Chem. Phys.,* **17,** 1301 (1949).
22. T. A. Orofino, *Polymer,* **2,** 295, 305 (1961).
23. P. J. Flory, Ref. 18, Chapter 14.
24. B. H. Zimm and R. W. Kilb, *J. Polym. Sci.,* **37,** 19 (1959).
25. H. Benoit, Z. Grubisic, P. Rempp, D. Decker, and J. G. Zilliox, *J. Chem. Phys.,* **63,** 1507 (1966).
26. D. J. Meier, *J. Phys. Chem.,* **71,** 1861 (1967).
27. W. Haller, *Nature,* **206,** 693 (1965).
28. W. Haller, *J. Chem. Phys.,* **42,** 686 (1965).

29. J. C. Moore and M. C. Arrington, International Symposium on Macromolecular Chemistry, Tokyo and Kyoto, 1966, Preprints VI-107.
30. W. W. Yau, C. P. Malone, and S. W. Fleming, *J. Polym. Sci., Part B,* **6,** 803 (1968).
31. W. W. Yau, *J. Polym. Sci., Part A-2,* **7,** 483 (1969).

Calibration of GPC Columns

HANS COLL

SHELL DEVELOPMENT COMPANY
EMERYVILLE, CALIFORNIA 94608

Summary

Calibration in gel permeation chromatography is reviewed with special reference to $M[\eta]$ as the parameter for universal calibration in the case of polymers

INTRODUCTION

The familiar event in gel permeation chromatography (GPC) is that the largest molecules emerge first from the column and the smallest ones last. Early in the development of GPC it was apparent that the peak elution volume (V) is not a function of molecular weight (M) of the solute species alone, but that molecular structure also plays a role. The task was then to find a formulation for the structural factors with the help of which M can be related to V.

A direct model calculation that starts with a determination of geometry and pore-size distribution of the gel is impractical, if not futile, even if a description could be given entirely in terms of configurational quantities (specific interactions are precluded). The GPC process and the geometry of the gel are in general too complicated for this approach.

The first step in a practical procedure is therefore to calibrate the column with a series of reference solutes usually belonging to a homologous series. In this manner one obtains the familiar plot of log M vs elution volume (cf. Fig. 1), the latter being identified with the volume of column effluent that corresponds to the peak maximum. For this purpose the reference solutes should be monodisperse, or at least exhibit a very narrow molecular-weight distribution (MWD),

135

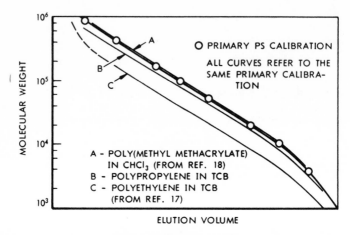

FIG. 1. Calibration curves.

because the position of the peak maximum cannot be directly related to any particular molecular-weight average [in the case of conventional polymers the position of the peak maximum will frequently range between $(M_n M_w)^{1/2}$ and M_w, depending on MWD (1)].

The calibration curve cannot be expected to hold if the material under investigation is structurally different from the reference compounds. This creates problems, particularly in the case of polymers, where it is difficult to obtain sharply fractionated samples. In fact, at the present time polystyrene (PS) standards (Pressure Chemical Company, Pittsburgh, Pa.; Waters Associates, Framingham, Mass; ArRo Laboratories, Joliet, Ill.) are the only readily-available polymeric reference standards for work in nonaqueous solutions.*

It is, therefore, desirable to find a way for transforming a primary calibration curve (as obtained with PS standard, for instance) in such a manner that it can be used with structurally different polymers.

EXTENDED CHAIN LENGTH AND MOLECULAR VOLUME

One of the first attempts to arrive at a useful calibration parameter was to correlate the extended chain length (L) of the solute molecules with the peak elution volume $(4,5)$. L can be determined by calcula-

---

* Balke et al. (2) have recently discussed calibration by means of polydisperse standards of known M_w and M_n using a computer search program. A procedure that uses a polymer of very broad but well-defined MWD spanning the molecular-weight range of interest has been described by Cantow et al. (3).

tion or from molecular models. This approach held some promise with oligomers, but was not satisfactory with polymers in general, although it represented an improvement over M as a calibration parameter.

The shortcomings of a calibration in terms of L are apparent if one accepts the GPC process to discriminate between molecular species on the basis of effective dimensions in solution. Thus, the concept of molecular volume as a universal parameter is a useful one. This has been shown for small molecules (6) provided specific solute–gel interactions (7, 8) and solvation effects (7, 9) on the elution volume can be discounted (specific effects greatly complicate the problem of universal calibration and will not be considered here).

Complications arise with all nonglobular macromolecules. A configurational effect (10) has to be taken into account if the molecules are rodlike. For flexible-coil molecules the concept of molecular volume, in the context of GPC, requires redefinition. The peak elution volume apparently, depends on an "effective" molecular volume operationally defined in terms of hydrodynamic parameters.

HYDRODYNAMIC VOLUME AS A UNIVERSAL CALIBRATION PARAMETER

Benoit and co-workers (11–13) found that the peak-elution volumes of fractions of a variety of chemically and structurally different polymers* conformed to a single curve if plotted against the product $M[\eta]$, where M is the molecular weight of the respective fraction and $[\eta]$ the intrinsic viscosity. Hence, $M[\eta]$ can be considered as the universal parameter, proportional to $R_H{}^3$, R_H being the viscometric hydrodynamic radius of the polymer coil,

$$M[\eta] = 10\pi N_A R_H^3/3 \qquad \text{(cgs)} \qquad (1)$$

(N_A designates Avogadro's number). Measurements in other laboratories on other polymer systems have generally confirmed this universality.† An exception are Meyerhoff's data on cellulose nitrate (a rather stiff coil) in tetrahydrofuran (19).

---

* Linear and branched PS in tetrahydrofuran (THF) at 25°C (11); PS, poly-(vinyl chloride), polybutadiene, poly(phenyl siloxane), PS-poly(methylmeth-acrylate copolymers in THF, 25°C (12, 13).

† PS, polyethylene in trichlorobenzene (TCB), 130°C, PS, polybutadiene in THF, 23°C (14). PS, polyisobutylene in TCB (15). PS, polyisobutylene in benzene and butanone/isopropanol at 25°C (16); PS, polypropylene in TCB at 135°C (17); see also Ref. 18.

Presumably there is nothing unique about $M[\eta]$ and R_H in the GPC process. It is not unlikely that, for instance, the Stokes radius of molecules obtained from diffusion measurements may serve equally well as a universal parameter, as this has indeed been proposed for biopolymers (20). (Preference is to be given to R_H because the intrinsic viscosity is much more readily determined than the diffusion coefficient.)

For linear flexible-coil molecules at least, a theoretical justification can be given for the universality of $M[\eta]$ as a calibration parameter:

1. The retardation of a solute molecule in its travel down the column is governed by the probability of the molecule entering into the pores of the gel. The probability of entry into a particular pore is given by the decrease of free energy associated with the volume restriction imposed on the molecule by the dimensions of the pore.

2. The change of free energy is to the largest extent configurational, the heat of mixing (with the solvent) in the deformation is only of minor importance (21).

3. It follows from Gaussian statistics that the dimensions of the pore and a single statistical parameter of the polymer coil [rms radius of gyration, or rms end-to-end distance, $(\bar{r}^2)^{1/2}$, for instance] suffice to describe the change of configurational free energy upon deformation (21). Here the theoretical work of Casassa (22, 22a) is of special importance. As a result, molecules of the same statistical dimensions should have the same emergence volume under a given set of experimental conditions.

4. Although the argument in Point 3 was restricted to equilibrium conditions, it should also hold, at least approximately, for dynamic processes in which the diffusion coefficient plays a role, since the latter should be equal for linear molecules having the same statistical dimension.

5. Finally, according to the Flory-Fox equation

$$[\eta] = \phi(\bar{r}^2)^{3/2}/M \qquad (\phi = \text{constant}) \qquad (2)$$

$M[\eta]$ is proportional to the cube of the statistical parameter.

This line of reasoning cannot be extended to branched molecules. Yet, calculations by Casassa (22) have shown that with reference to the theory of Zimm and Kilb (23) universal calibration in terms of $M[\eta]$ is still approximately correct for star-shaped molecules (the calculations were based on an equilibrium model). This has also been indicated by experimental results (11, 24). According to Casassa's treatment, the agreement seems to be fortuitous. Interestingly, GPC studies on the polypeptide benzyl-L-glutamate in dimethylformamide (25) show that even this rodlike (helical) molecule conforms to the present scheme of calibration. To which extent this result can be generalized remains to be seen.

Instead of the Flory-Fox equation—Eq. (2)—one may use the expression of Ptitsyn and Eizner (26). The universal calibration parameter then becomes $M[\eta]/f(\epsilon)$, where $f(\epsilon) = 1 - 2.63\epsilon + 2.89\epsilon^2$, and $\epsilon = (2a - 1)/3$, a being the exponent in the Mark-Houwink equation. Some arguments may be advanced in favor of including $f(\epsilon)$, but no clear distinction can be made on the basis of presently available GPC data (17).*

Substitution for $[\eta]$ by means of the Mark-Houwink equation, $[\eta] = KM^a$, immediately leads to an equation (17, 27) which transforms a primary calibration curve (obtained with polymer 1) for use with some other polymer (subscript 2)

$$\log M_2 = \frac{1}{1 + a_2} \log \frac{K_1 f(\epsilon_2)}{K_2 f(\epsilon_2)} + \frac{1 + a_1}{1 + a_2} \log M_1 \qquad (3)$$

K and a are the parameters of the respective Mark-Houwink equations. In certain cases these may be found in the literature, but they can usually be determined even if fractionated polymer samples are not available (17). (The transformation of the primary calibration curve may have to be carried out by segments if one set of Mark-Houwink parameters is insufficient for the whole range of molecular weights under consideration.)

An application of Eq. (3) may be illustrated by the example of polypropylene in trichlorobenzene at 135°C, PS serving as the primary calibration standard (17). The Mark-Houwink equations were determined as $[\eta] = 1.37 \times 10^{-4} M^{0.75}$ (dl/g) for polypropylene, and $[\eta] = 1.21 \times 10^{-4} M^{0.707}$ for polystyrene. From this, one calculates log $M_{PP} = 0.0496 + 0.975$ log M_{PS}. The displacement of the calibration curve for this example, and a few others, is shown in Fig. 1.

If $a_1 = a_2$, the calibration curves, log M vs V, are parallel. Therefore, if equality of the Mark-Houwink exponents can be anticipated for a particular pair of polymers in a given solvent, M_w/M_n, M_z/M_w,

---

* In the case of Meyerhoff's data on PS and cellulose nitrate ($a_1 = 0.74$, $a_2 = 1$) omission of $f(\epsilon)$ reduces the disagreement between results (19).

etc., for polymer 2, can be calculated from the chromatogram with reference to the primary calibration curve without a need for transformation (28). Furthermore, it is evident that the factor $f(\epsilon)$ can only be important if a_1 differs significantly from a_2.

M_1 and M_2 in Eq. (3) do not represent any particular averages of molecular weight. Thus, M_2 refers to the molecular weight of monodisperse samples provided M_1 and the Mark-Houwink parameters are valid for monodisperse polymer, as they should be.

DISCUSSION

It seems that the accuracy of calibration in terms of $M[\eta]$—or similarly, by Eq. (3)—is most likely impaired by unreliable Mark-Houwink parameters. Here a critical selection from published data is imperative. (Mark-Houwink equations based on number-average molecular weight should not be used because of differences between M_n and viscosity-average molecular weight, unless the sharpness of the polymer fractions has been documented.) More and better data for many polymer–solvent systems are needed.

One may expect universal calibration in terms of $M[\eta]$ to become unreliable in a molecular-weight range sufficiently low for substantial deviations from Gaussian coil statistics. But it should be kept in mind that the absolute magnitude of these deviations do not matter here; only the deviations of one polymer with respect to another are reflected in the calibration, which is definitely a mitigating factor.

Recently, Dawkins (18) has expressed some dissent, and he suggested that universal calibration should be based on the unperturbed dimensions of the polymer coil rather than on the dimensions of the expanded coil as implied in the $M[\eta]$ calibration. He correctly points out that present experimental data do not refute this hypothesis since practically all measurements which compare elution volumes of different polymers have been made in solvents where coil expansion was approximately the same for the polymers under investigation. For evidence Dawkins replotted the data of other investigators and added his own results on PS, poly(methyl methacrylate) and poly(dimethyl siloxane) in chloroform (18). Significantly, Meyerhoff's nonconformist cellulose nitrate (19) also falls on the same plot, in support of Dawkin's hypothesis. An explanation for the significance of the unperturbed dimensions is offered in terms of an interaction between the solute and polymer chains of the gel in the interior of the pores.

It should be possible to settle this argument by measurements under

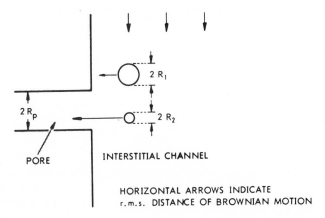

PORE

INTERSTITIAL CHANNEL

HORIZONTAL ARROWS INDICATE
r.m.s. DISTANCE OF BROWNIAN MOTION

FIG. 2. Schematic representation of capture of molecules by pore.

conditions such that the coil expansions of the respective polymers differ significantly from each other. Experiments of this kind should also reveal whether calibration in terms of $M[\eta]$ or $M[\eta]/f(\epsilon)$ gives more consistent results.

The theoretical discussions of the behavior of molecules in the GPC process have emphasized the equilibrium aspects between the moving phase (interstitial liquid) and the pores of the gel which represent the stationary phase. A few studies dealing with the dynamic behavior have been reported (*29–31*). Yet the following simple considerations suggest that arguments purely on the grounds of equilibrium effects are insufficient to describe exclusion from pores, the central phenomenon in our model of GPC.

Figure 2 represents schematically a wide interstitial channel and a slotlike pore of width $2R_P$. Consider then the progress of two solid spherical molecules with radii (smaller than R_P) in the channel (for the sake of simplicity we may assume uniform flow velocity, v, throughout the channel). Then the probability of entry of a molecule into the pore will first of all be determined by the flow-by time at the entrance to the pore,

$$t = 2(R_p - R)/v \qquad (4)$$

R being the molecular radius. Obviously, t is greater for the smaller molecule. Moreover, in this one-dimensional model, entry into the pore can only be brought about by lateral Brownian motion. Therefore, the probability of entry will further depend on the diffusion coefficient which again favors the smaller molecule, since $D \sim 1/R$. It follows that compact molecules, small enough to enter all pores of the gel, will still exhibit elution volumes depending on their sizes (*32*). In view of conventional flow velocities, gel dimensions, and diffusion coefficients of solute molecules, it appears that this dynamic exclusion effect should be quite significant. In the case of flexible coils, instead of compact molecules, one must further take into account configurational effects— as they apply to the equilibrium model (*22*)—in order to assess the overall probability of entry into a pore.

According to the present model, large molecules are only likely to be captured by a pore if they travel close to the surface of the gel, that is, at distances of the order of 100 Å or less).* This condition becomes less stringent for smaller molecules which diffuse more rapidly, but it is doubtful whether equilibrium between the moving and the stationary phase is ever approached under the conditions of a conventional GPC experiment.

At first it is somewhat surprising that flow rate seems to have only a rather insignificant effect on the peak elution volume (*32, 33*), unless the molecular weight of the solute is very high (*34*). But as Casassa and Tagami (*22a*) have pointed out, the equilibrium model is still applicable if nonequilibrium exists in the column. The only requirement is that a given molecule undergoes a large number of transfers between the moving and the stationary phase in the course of its passage down the column. Under this condition then, the elution volume that corresponds to the peak maximum should be virtually independent of flow rate.

It has also been suggested that dynamic effects which have to do with the flow pattern of solvent in the interstitial channels may play some role in the chromatographic separation (the assumption of uniform velocities in the channels is very unlikely to apply). One aspect of separation of molecules by laminar flow in the channels has been discussed by DiMarzio and Guttman (*35*).

A theory that takes equilibrium and dynamic effects into account has yet to be formulated. Nevertheless, it appears that the concept of universal calibration, as discussed before, remains valid at least in

---

* If we assume a pore width of $2R_p = 200$ Å, the molecular diameter $2R = 100$ Å, and a flow velocity of 0.1 cm/sec, one calculates by means of Eq. (4) $t = 10^{-5}$ sec; for a diffusion coefficient of 10^{-7} cm²/sec one then finds the rms distance of diffusion, corresponding to this time interval, as 141 Å (assuming diffusion in one dimension).

the case of linear flexible molecules, since molecules of the same statistical dimensions should exhibit the same dynamic and equilibrium behavior in the chromatographic process. Similarly, secondary effects, such as peak broadening, skewing, and concentration dependence of elution volume, should be approximately the same for all molecules of the same statistical dimensions.

CONCLUSION

In the absence of specific interactions, calibration in terms of $M[\eta]$ can be considered as universal with reasonable confidence if the polymer molecules are linear and randomly coiled. In the case of long-chain branching somewhat greater reservation is in order. The evidence that this calibration scheme applies to rodlike macromolecules in general is, at the present time, insufficient.

REFERENCES

1. H. L. Berger and A. R. Shultz, *J. Polym. Sci., Part A*, **2**, 3643 (1965).
2. S. T. Balke, A. E. Hamielec, and B. P. Leclair, *Ind. Eng. Chem., Prod. Res. Develop.*, **8**, 54 (1969).
3. M. J. R. Cantow, R. S. Porter, and J. F. Johnson, *J. Polym. Sci., Part A-1*, **5**, 1391 (1967).
4. J. C. Moore and J. G. Hendrickson, *J. Polym. Sci., Part C*, **8**, 233 (1965).
5. L. E. Maley, *ibid.*, **8**, 253 (1965).
6. W. B. Smith and A. Kollmannsberger, *J. Phys. Chem.*, **69**, 4157 (1965).
7. J. G. Hendrickson and J. C. Moore, *J. Polym. Sci., Part A-1*, **4**, 167 (1966).
8. J. Cazes and D. R. Gaskill, *Separ. Sci.*, **2**, 421 (1967).
9. *Idem., ibid.*, **4**, 15 (1969).
10. J. C. Giddings, E. Kucera, C. P. Russel, and M. N. Myers, *J. Phys. Chem.*, **72**, 4397 (1968).
11. H. Benoit, Z. Grubisic, P. Rempp, D. Decker, and J.-G. Zilliox, *J. Chim. Phys.*, **63**, 1507 (1966).
12. Z. Grubisic, P. Rempp, and H. Benoit, *J. Polym. Sci., Part B*, **5**, 753 (1967).
13. Z. Grubisic and H. Benoit, *C. R. Acad. Sci., Paris, Ser. C*, **266**, 1275 (1968).
14. K. A. Boni, F. A. Sliemers, and P. B. Stickney, *J. Polym. Sci., Part A-2*, **6**, 1567, 1579 (1968).
15. M. J. R. Cantow, R. S. Porter, and J. F. Johnson, *J. Polym. Sci., Part A-1*, **5**, 987 (1967).
16. J. C. Moore and M. C. Arrington, 3rd International GPC Seminar, Geneva, May, 1966.
17. H. Coll and D. K. Gilding, *J. Polym. Sci., Part A-2*, **8**, 89 (1970).
18. J. V. Dawkins, *J. Macromol. Sci.-Phys.*, **B2**, 623 (1968).
19. G. Meyerhoff, *Ber. Bunsenges. Phys. Chem.*, **69**, 866 (1965).
20. G. K. Ackers, *Biochemistry*, **3**, 723 (1964).
21. D. J. Meier, *J. Phys. Chem.*, **71**, 1861 (1967).

22. E. F. Casassa, *J. Polym. Sci., Part B,* **5,** 773 (1967); E. F. Casassa and Y. Tagami, *Polym. Preprints,* **9,** No. 1, 565 (1968).

22a. E. F. Casassa and Y. Tagami, *Macromolecules,* **2,** 14 (1969).

23. B. H. Zimm and R. W. Kilb, *J. Polym. Sci.,* **37,** 19 (1959).

24. L. Wild and R. Guliana, *J. Polym. Sci., Part A-2,* **5,** 1087 (1967).

25. Z. Grubisic, L. Reibel, and G. Spach, *C. R. Acad. Sci., Paris, Ser. C,* **264,** 1690 (1967).

26. O. B. Ptitsyn and Yu. E. Eizner, *Soviet J. Tech. Phys. (English Trans.),* **4,** 1020 (1960).

27. H. Coll and L. R. Prusinowski, *J. Polym. Sci., Part B,* **5,** 1153 (1967).

28. D. B. Bly, *Anal. Chem.,* **41,** 477 (1969).

29. W. W. Yau and C. P. Malone, *J. Polym. Sci., Part B,* **5,** 663 (1967).

30. J. J. Hermans, *J. Polym. Sci., Part A-2,* **6,** 1217 (1968).

31. R. N. Kelley and F. W. Billmeyer, Jr., *Anal. Chem.,* **41,** 874 (1969).

32. W. Haller, *J. Chromatogr.,* **32,** 676 (1968).

33. J. N. Little, J. L. Waters, K. J. Bombaugh, and W. J. Pauplis, *Proceedings, 7th International Seminar GPC* (Monte Carlo), October 1969.

34. W. W. Yau, H. L. Suchan, and C. P. Malone, *J. Polym. Sci., Part A-2,* **6,** 1349 (1968).

35. E. A. DiMarzio and C. M. Guttman, *J. Polym. Sci., Part B,* **7,** 267 (1969).

Data Treatment in GPC

L. H. TUNG

PHYSICAL RESEARCH LABORATORY
THE DOW CHEMICAL COMPANY
MIDLAND, MICHIGAN 48640

Summary

For high polymer samples, the peaks of individual components are not separable in GPC chromatograms. The interpretation of such chromatograms is different from that for other types of chromatograms. The outline of a computation procedure for treating such GPC chromatograms is given.

INTRODUCTION

In GPC chromatograms for monomeric compounds and oligomers, the individual components appear as separate peaks. These GPC chromatograms are interpreted in the same manner as that used in other types of chromatography. After the peaks are identified, the area under each of the peaks is measured to give the relative concentration of that component. For high polymer samples, the peaks of the individual components are no longer separable. The conversion of such chromatograms to the molecular weight distributions of the samples requires special considerations.

In the GPC manual by Waters Associates, an account for such a conversion has been given. In an instrument manual, however, the scope of discussion has to be limited. For many simple applications Waters' procedure is adequate, but to achieve the maximum accuracy for GPC, a more refined data treatment is required. For this reason many laboratories have adopted their own procedures for treating GPC chromatograms. The outline of a procedure used in our laboratory is given below. Hopefully, it will serve to illustrate the steps involved in the interpretation of the chromatograms for high polymer samples.

145

GENERAL CONSIDERATIONS

As mentioned above, complex computation is not always required in GPC. For instance, visual inspection of the GPC recorder traces is adequate to compare the relative breadth of distribution for some samples determined by the same GPC equipment using the same set of columns. Sometimes such information is all that one needs and no interpretation of any kind is necessary.

More often, however, it is desired to represent the distribution on a molecular weight scale. A knowledge for the relation between the elution volume and the molecular weight is needed for such a conversion. The distribution curves are usually normalized and the average molecular weights for the sample are usually calculated. Simple numerical integration steps are therefore involved in the computation.

For the most precise treatment of high polymer chromatograms, correction for instrumental spreading becomes necessary. This spreading is caused by many band spreading mechanisms in the instrument, and because of it, a high polymer chromatogram is a composite of the overlapping curves of all its components. The height of such a chromatogram no longer reflects the relative abundance of the component at the corresponding elution volume; it also depends on the abundance of the neighboring components. At the ends of the chromatograms there are curve portions representing components which do not even exist in the sample.

In order to correct for such spreading, the spreading characteristics for the instrument must be determined. This makes computations more complex and requires the use of a high-speed digital computer. Such a correction improves the accuracy of all chromatograms but is especially important when the distribution of the sample is narrow.

CONVERSION TO MOLECULAR WEIGHT DISTRIBUTION

In Waters' manual the integral molecular weight distribution is calculated from the chromatograms. To obtain the more demonstrative differential distribution curve from the integral distribution, numerical differentiation must be carried out. The chromatograms are thus integrated first and then differentiated. Such a process automatically takes into consideration the weighing factor involved in changing the scale from elution volume to molecular weight but at the same time sacrifices the accuracy which is potentially attainable by GPC.

The direct conversion of a chromatogram to differential distribution is more accurate. Let v represent the elution volume or count; M the molecular weight; $f_M(M)$ the differential molecular weight distribution function; and $f(v)$ the chromatogram

$$f_M(M) = f(v)(-dv/dM) \qquad (1)$$

The weighing factor, $-dv/dM$, is obtainable from the molecular weight calibration. In GPC such a calibration is usually plotted on a semilogarithmic graph with molecular weight on the logarithmic scale. The slope of the calibration on such a plot is $d \log M/dv$ and

$$dv/dM = \left(\frac{1}{M \times (d \log M/dv) \times 2.303} \right) \qquad (2)$$

In the special case where the relation between $\log M$ and v is linear, $d \log M/dv$ is a constant. When this relation is not linear, numerical differentiation of the calibration curve is needed.

It is often more convenient to represent the molecular weight distribution on a semilogarithmic plot. Let this distribution of log molecular weight be represented by $f_L(\log M)$, then

$$f_L(\log M) = -f(v)(1/[d \log M/dv]) \qquad (3)$$

or

$$f_L(\log M) = f_M(M)(1/2.303M) \qquad (4)$$

If $f(v)$ has already been normalized, the $f_M(M)$ and $f_L(\log M)$ computed from the above equations are also normalized. The average molecular weights can be calculated by using any one of the three f functions. For example, the weight-average molecular weight, M_w, is

$$M_w = \frac{\int M f_M(M)\, dM}{\int f_M(M)\, dM} = \frac{\int M f(v)\, dv}{\int f(v)\, dv} = \frac{\int M f_L(\log M)\, d \log M}{\int f_L(\log M)\, d \log M} \qquad (5)$$

MOLECULAR WEIGHT CALIBRATION

Coll (1) has discussed the calibration of molecular weight for GPC. Moore (2) and other early workers have assumed that the relation between the logarithm of molecular weight and elution volume is linear. In practice, however, only when a very small range of elution volume is used can the relation be assumed linear. The calibration curve is usually of the shape shown in Fig. 1. The deviation from linearity is the largest at the high and the low molecular weight regions. Many in-

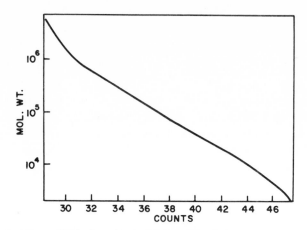

FIG. 1. Typical molecular weight calibration of GPC.

consistencies in GPC results can be traced to the improper use of the molecular weight calibration relation.

In our computation procedure the molecular weight calibration is represented by a polynomial. The calibration points obtained experimentally using standard samples are plotted first on a semilogarithmic graph paper. A smooth curve is drawn through the experimental points and extended to near interstitial volume on one end and the elution volume for monomers on the other end. The extrapolation of the calibration curve to the interstitial volume often is done in an arbitrary manner. Until very high molecular weight standard samples are available, such an uncertainty seems unavoidable. The polynomial is made to fit the smooth curve by the method of moments using the orthogonal Legendre polynomials. The best fit is selected from a set of polynomials with degrees ranging from 3 to 32. As the calibration curve does not contain abrupt slope changes, the selected polynomial usually is found to be indistinguishable from the plotted curve.

CORRECTION FOR INSTRUMENTAL SPREADING

The relationship between the experimental chromatogram, $f(v)$, and the chromatogram after the correction of instrumental spreading, $w(v)$, can be expressed by the convolution integral equation

$$f(v) = \int w(y)g(v - y)\, dy \tag{6}$$

where y is the elution volume under the integral sign and $g(v - y)$ is a function that describes the shape of the band spreading curve. Experimentally, low molecular weight compounds have been observed to give the Gaussian spreading in GPC, that is

$$g(v - y) = (h/\sqrt{\pi}) \exp\{-h^2(v - y)^2\} \qquad (7)$$

where h is a parameter describing the width of the spreading and h is related to the standard deviation, σ, of the Gaussian distribution by

$$h = 1/\sigma\sqrt{2} \qquad (8)$$

For high molecular weight species the shape of the spreading function cannot be determined directly because thus far there are no truly monodisperse high molecular weight polymer samples. It is indisputable that skewing does occur in the spreading for very high molecular weight polymer species (3), particularly at fast flow rates. The shapes of the chromatograms of the currently available high molecular weight narrow distribution polystyrene samples, however, cannot be used to estimate the extent of skewing nor can they even be used to judge whether skewing does occur at all. These samples are themselves skewed in the distribution. In our computation procedure the Gaussian spreading is assumed because using an incorrect degree of skewing may very likely introduce larger errors than using the simpler Gaussian spreading. Moreover, it has been demonstrated (4) that Gaussian spreading is a good approximation for polyethylene to a molecular weight at least as high as 460,000.

A number of methods (5) for solving the integral Eq. (5) have been published. The earlier methods have been evaluated by Duerksen and Hamielec (6). Our current computation procedure uses a method (7) that employs a fourth degree polynomial to fit the experimental chromatogram a section at a time. This method is fast and only in rare occasions does it give solutions with artificial oscillations.

CALIBRATION OF INSTRUMENTAL SPREADING

The proper use of the method for correcting instrumental spreading requires a precise calibration of the spreading characteristics of the instrument. Having made the Gaussian spreading assumption, we reduced this calibration to the determination of the parameter h in Eq. (7) as a function of the elution volume. This can be accomplished by a reverse-flow technique (4) using standard samples which are not truly monodispersed.

When a polydisperse sample is sent through a GPC column, its chromatogram is broadened by two processes, a desirable spreading

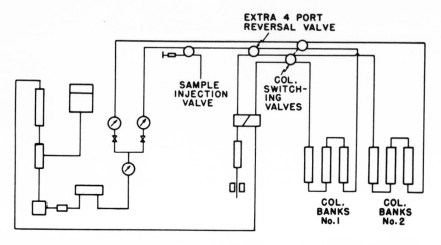

FIG. 2. Schematic GPC flow diagram with reverse-flow valve.

which separates the molecules according to their sizes and the undesirable instrumental spreading. If the elution of the sample is allowed to proceed to some part of the column and then the direction of flow is reversed, the size separation process is also reversed but the instrumental spreading continues to broaden the peak. The resulting chromatogram therefore reflects only the instrumental spreading.

To use the reverse-flow technique, a special four-port valve shown in Fig. 2 needs to be installed. In the experiments the peak position for the standard samples are predetermined. Then each sample is in-

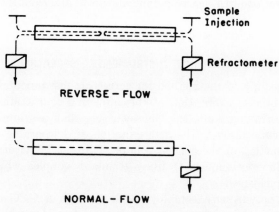

FIG. 3. Reverse-flow scheme.

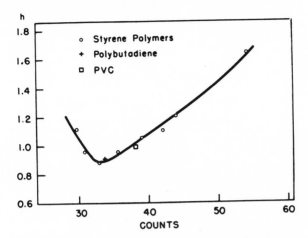

FIG. 4. Variation of the parameter h with elution counts.

jected twice. The first injection is made when the flow of the eluting solvent is in the normal direction. As the elution volume reaches one-half of that of the peak position of the sample, the flow is instantly reversed. The resulting chromatogram is used to compute the h parameter for the front half of the column. The process is repeated with a second injection of the sample when the flow is in the reverse direction. The chromatogram produced is then used to determine the h parameter for the second half of the columns. Figure 3 shows the reverse flow process schematically. The overall h for the sample is calculated from the equation

$$h = \sqrt{2}/\sqrt{(1/h_{\text{front}}^2) + (1/h_{\text{back}}^2)} \qquad (9)$$

The h values determined for our present GPC instrument are shown in Fig. 4. Two of the points in the figure represent values determined for a polybutadiene sample and a PVC sample. The rest of the points are values for standard polystyrene samples. All of the points are shown to follow a single curve. The spreading characteristics appears then to depend only on the elution volume regardless of the chemical composition of the polymer.

The reverse-flow method is tedious experimentally. Each sample requires two injections and the automatic sample injection device cannot be used for the injection. Hendrickson (8), Hamielec and Ray (9), and recently we (10) have proposed methods of computing h from the regular chromatograms normally required for the calibration of molecular weight to calibrate the instrumental spreading as well.

In the computation procedure we again represent the curve in Fig. 4 by a polynomial. The coefficients of the polynomial are determined in the same manner as that used for the molecular weight calibration.

DESCRIPTION OF THE COMPUTER PROGRAM

In our computer program the coefficients defining the polynomials for the two types of calibrations are fed to the computer just before the input for the actual chromatogram data sets. The computation for the correction of instrumental spreading is carried out first. The molecular weight distribution is then converted from the corrected chromatogram and also from the uncorrected chromatogram. Both the distribution of log molecular weight, $f_L(\log M)$, and the conventional differential distribution, $f_M(M)$, are calculated. The number-average, weight-average, and z-average molecular weights are calculated for both the corrected and uncorrected distributions.

If the instrumental spreading calibration is not fed to the computer before the data sets, only the results for the uncorrected chromatogram are calculated. If the molecular weight calibration is not fed in then, the correction for instrumental spreading is still carried out but the conversion to molecular weight is left out. If both calibrations are not

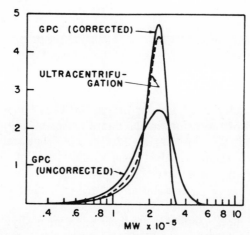

FIG. 5. Molecular weight distribution of a narrow distribution polystyrene sample.

fed to the computer, an error message will be printed. Various plotting options are also incorporated in the computation program.

Figure 5 shows the molecular weight distribution of a polystyrene sample calculated from the experimental chromatogram using our computer program and calibration procedures. Also shown in the figure is the distribution determined by sedimentation velocity measurement on a Spinco Model E ultracentrifuge. The agreement between the distribution calculated from the corrected chromatogram and the one determined by ultracentrifugation is within the uncertainty of the ultracentrifugation experiments. The uncorrected distribution from GPC is shown to be quite unsatisfactory for this sample. The sample has a M_w/M_n ratio of about 1.1. For broader samples the correction for instrumental spreading is less important. Our GPC unit consists of six columns with porosity ranging from 10^6 to 8×10^2 A. These columns are the high plate count type purchased from Waters Associates.

REFERENCES

1. H. Coll, *Separ. Sci.*, **5**, 273 (1970).
2. J. C. Moore, *J. Polym. Sci.*, **A2**, 835 (1964).
3. S. T. Balke and A. E. Hamielec, *J. Appl. Polym. Sci.*, **13**, 1381 (1969).
4. L. H. Tung, J. C. Moore, and G. W. Knight, *J. Appl. Polym. Sci.*, **10**, 1261 (1966).
5. L. H. Tung, Paper given in this symposium. To be published in *Separ. Sci.*
6. J. H. Duerksen and A. E. Hamielec, *J. Polym. Sci., Part C,* **21**, 83 (1968).
7. L. H. Tung, *J. Appl. Polym. Sci.*, **13**, 775 (1969).
8. J. G. Hendrickson, *J. Polym. Sci., Part A-2,* **6**, 1903 (1968).
9. A. E. Hamielec and W. H. Ray, *J. Appl. Polym. Sci.*, **13**, 1319 (1969).
10. L. H. Tung and J. R. Runyon, *J. Appl. Polym. Sci.*, **13**, 2397 (1969).

III

NEW DEVELOPMENTS IN GPC

The Overload Effect in Gel Permeation Chromatography*

J. C. MOORE

BASIC RESEARCH DEPARTMENT
TEXAS DIVISION
THE DOW CHEMICAL COMPANY
FREEPORT, TEXAS 77541

Summary

Overload phenomena in GPC are investigated with samples of narrow polystyrenes and short but efficient columns packed with Styragel. Viscous fingering is shown to be a leading cause of peak skewing and broadening. A correlation is proposed to define a safe operating range in terms of sample concentration, volume, and the average intrinsic viscosity of the solute polymer.

GPC presents an unusual problem in chromatography. The polymeric materials in the sample confer appreciable viscosity on their solutions, while a high resolution is demanded in a limited elution volume. Experience has shown that too much polymer in the sample causes not earlier but later elution of the peak. The curve starts up at the proper point but rises more slowly, and its return to baseline is delayed. There is a loss of resolution, particularly under the main part of the curve. This has been shown by collecting and rerunning the fractions (1). In such a case important details of the sample distribution may have been obscured. Attempts to extrapolate the mean values of the distribution (2, 3) to a zero load condition may be useful with very regular distributions, but it would be desirable to have a measure of a safe sample load to avoid undue loss of resolution.

This study suggests that a relative measure of this sample load

* Presented at the ACS Symposium on Gel Permeation Chromatography, sponsored by the Division of Cellulose, Wood, and Fiber Chemistry at the 159th National Meeting of the American Chemical Society, Houston, Texas, February, 1970.

effect may be based on the sample volume V, its total concentration C, and its average intrinsic viscosity $[\eta]$. The latter value may usably be obtained from a test chromatogram's center of area and a calibration curve for elution volume versus $[\eta]$, or the integration $[\eta]_{av} = \Sigma W_i [\eta]_i / \Sigma W_i$ may be performed.

Sample load effects should most easily be seen when column efficiency is high and a narrow-distribution sample is used, especially if the column is short. In this study two columns were used, each 4 ft $\times$ 3/8 in. (7.8 mm i.d.) in size. Column 17I was packed with 10^6 Å permeability Styragel (Waters Associates, Inc.), a fraction was used in which 80% of the particles were within the range 13–47 μ in diameter. Column 18W was packed with 3×10^4 Å permeability Styragel of size range 29–53 μ diameter. The eluting solvent was tetrahydrofuran at 25°C; column efficiencies measured with small benzene samples were 2360 and 1900 theoretical plates per foot, respectively, for the two columns including the spreading effect of the connecting tubing (about 70 in., 0.020 in. i.d.) and the Waters R-4 refractometer whose cell volume was 0.010 ml. Samples used were the narrow anionic

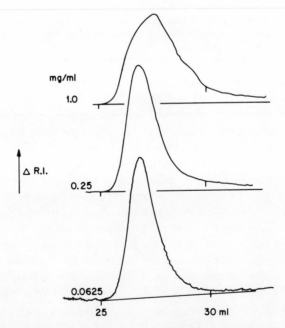

FIG. 1. Elution curves, Column 18w, 1 ml/min, 0.8 ml samples of polystyrene 14A in THF at three concentrations.

TABLE 1

Pressure Chemicals batch	$\bar{M}_w{}^a$	$\bar{M}_w/\bar{M}_n{}^a$	$[\eta]_{THF}$, 25°C
14A	1.8×10^6	1.2	3.8
6A	8.6×10^5	1.15	2.0
3A	4.1×10^5	1.06	1.2
1A	1.6×10^5	1.06	0.65

[a] Pressure Chemicals' property sheets.

polystyrenes obtained from Pressure Chemicals Inc., as shown in Table 1.

Elution curves from three concentrations of Sample 14A on Column 18W, in which the main part of the sample has little or no penetration of the gel, show in Fig. 1 that severe overloading has occurred in the highest concentration. The pattern of erratic delay is indicative of viscous fingering (4), an interstitial phenomenon. The drop in viscosity at the rear boundary of the sample has made the radial velocity profile

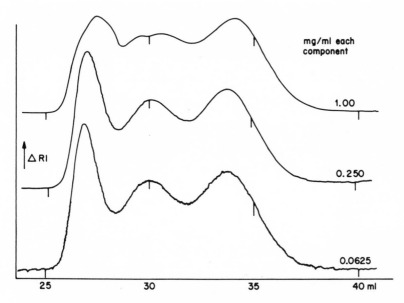

FIG. 2. Elution curves, Column 18w, 1 ml/min, 0.8 ml samples, polystyrenes 14A, 3A, and 1A in equal mixture, three concentrations.

unstable. In this zone it is easier for solvent to push through a slightly wider passage between grains than to maintain an even plug-type flow. Fingers of solvent may punch ahead while fingers of sample are temporarily delayed. As the sample becomes spread out, farther down the column the effect diminishes so that later eluting components may reflect the distortion effect in a lesser degree.

Elution curves from the same column showing the same samples with two lower molecular weights components added are shown in Fig. 2. Here it is evident that the distortion pattern is now largely super-imposed on the curve at the same elution volumes. If the distribution had been smoother, there might have been little evidence of the loss of resolution. It would be desirable to detect this in its early stages even with a broad distribution.

Since we observe a delay of material, it is interesting to test the intensity function that has been proposed as a detector of stagnancy in a residence time distribution (5, 6). The intensity function $L(V)$ measures the probability that a molecule still in the system may escape in the next volume increment.

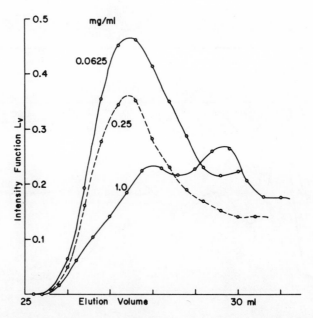

FIG. 3. Intensity function for elution curves in Fig. 1: Column 18w, 0.8 ml samples of polystyrene 14A at three concentrations.

$$L(V) = H(V)/[A - C(V)]$$

where H is the curve height and C is the cumulative area under the curve at the elution volume V, and A is the total area under the curve. This is the unnormalized form of Eq. (20) in Ref. 6. A decrease in this function would indicate stagnancy if the polymer sample were effectively mondisperse. In Fig. 3 values are shown as calculated from the three elution curves of Fig. 1. The sharp decrease shown is not related to viscous fingering since it is most evident at the lowest concentration. More likely, it may reflect the low-molecular weight tail of this sample which was capable of penetrating some of the gel pores. The sharply lower initial slope of the overloaded curve may be significant, but with these narrow samples the increased width of the curve is also sensitive. With the broader distributions in Fig. 2, the same relative pattern was followed by their calculated intensity functions. This may be more reliable than a direct measure of the initial slope of the chromatogram relative to total area.

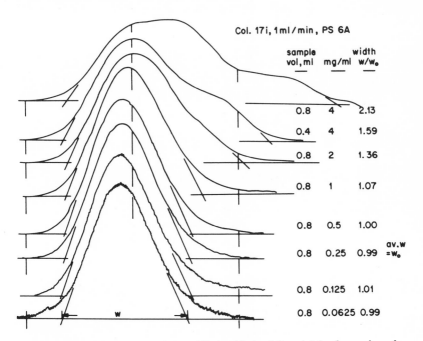

Col. 17i, 1 ml/min, PS 6A

sample vol, ml	mg/ml	width w/w₀
0.8	4	2.13
0.4	4	1.59
0.8	2	1.36
0.8	1	1.07
0.8	0.5	1.00
0.8	0.25	0.99 av.w =w₀
0.8	0.125	1.01
0.8	0.0625	0.99

FIG. 4. Elution curves, Column 17I, 1 ml/min, 0.8 and 0.4 ml samples of polystyrene 6A at seven concentrations, relative baseline width between tangent lines as shown.

TABLE 2

Peak Width Measurements under Varying Sample Loads

Col. No.	Sample Chemicals No.	Pressure Flow (ml/min)	Ml Baseline width between first and last tangent intercepts, at concentration (mg/ml)										
			0.0625	0.125	0.25	0.5	1.0	2.0	4.0	4.0	5.0	10.0	10.0
17i	14A	1.0	8.32	8.40	8.52	8.42	9.68	11.90	—	—	—	—	—
17i	14A	0.2	—	—	6.53	7.12	8.13	11.18	—	—	—	—	—
18w	14A	1.0	2.56	—	2.78	—	4.68	—	—	12.70	—	—	—
6A	17i	1.0	5.92	6.07	5.96	6.00	6.40	8.26	9.45[a]	12.70	—	—	—
6A	17i	0.2	—	—	—	4.46	5.11	6.96	—	10.61	—	—	—
18w	6A	1.0	3.88	—	3.80	—	4.30	4.84	5.35[a]	7.75	—	—	—
17i	1A	1.0	—	—	5.28	—	5.21	5.13	—	—	5.25	5.55[a]	8.20

[a] 0.4 ml sample, otherwise 0.8 ml.

It seemed desirable to study the onset of the delay phenomenon in more detail. A series of samples of three polystyrenes in the two columns were run at two flow rates, 1.0 and 0.2 ml/min, with two sample sizes, 0.8 and 0.4 ml, and a wide range of concentrations, as summarized in Table 2. Figure 4 illustrates the range of patterns encountered. At the lowest concentrations baseline drift and instrument noise limit the resolution. Then for a range the chromatogram is independent of sample concentration, with detector-recorder sensitivity adjusted to keep curve area constant. Above this range, gross distortion sets in and rapidly becomes severe. At the highest concentration overloading was sharply reduced by halving the sample volume, but it was reduced a little more by halving the concentration instead.

At lower molecular weights the safe and usable range is broader; at some higher molecular weight it would seem to disappear. For most samples the extremely high molecular weight components are minor in quantity but their accurate analysis is important.

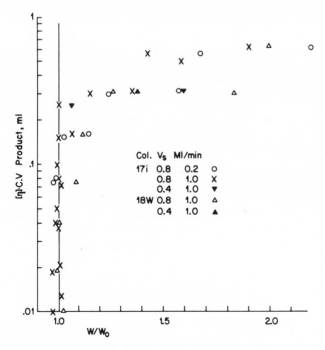

FIG. 5. Correlation of Table 2 data, relative peak broadening vs. solvated molecular volume of sample load.

Since intrinsic viscosity measures the ratio of molecular volume at infinite dilution to molecular weight, the product of sample concentration and intrinsic viscosity should give a ratio of solvated molecular volume to sample volume. Then multiplying this by sample volume gives a measure of load which implies that a given column can safely handle some total volume of solvated polymer molecules. Figure 5 shows Table 2 data treated in this way. With consistent units (dl/g $\times$ 0.1 $\times$ mg/ml $\times$ ml = ml) the safe limit for these columns appears to lie between product values of 0.05 and 0.10 ml.

It appears that much GPC work has been done with samples in the overload range, while long columns, broad distributions, and lower column efficiencies obscured the evidence. With improvements in column-packing techniques and in corrections for the system's spreading effects, it becomes possible to ask for faster and more accurate analyses, and these column load effects become more important. While comparison of a rerun cut with its expected distribution (7) is still considered the most revealing and fundamental technique, it seems desirable to direct attention to the causes of the gross distortions seen in this study, and if possible to find a simple expression to describe a safe operating range with a variety of samples as usually encountered.

REFERENCES

1. H. E. Adams, K. Farhat, and B. L. Johnson, *Ind. Eng. Chem., Prod. Res. Develop.,* **5**, 126 (1966).
2. M. J. R. Cantow, R. S. Porter, and J. F. Johnson, *J. Polym. Sci., Part B,* **4**, 707 (1966).
3. A. Lambert, *Polymer,* **10**, 213 (1969).
4. R. E. Collins, *Flow of Fluids Through Porous Materials,* Reinhold, New York, 1961, pp. 196–200.
5. P. V. Dankwerts, *Chem. Eng. Sci.,* **2**, 1 (1953).
6. P. Naor and R. Shinnar, *Ind. Eng. Chem., Fundam.,* **2**, 278 (1963).
7. L. H. Tung, J. C. Moore, and G. W. Knight, *J. Appl. Polym. Sci.,* **10**, 1261 (1966).

Gel Permeation Chromatography Using a
Bio-Glas* Substrate Having a Broad Pore Size Distribution†

A. R. COOPER, J. H. CAIN, E. M. BARRALL II,‡
and J. F. JOHNSON§

CHEVRON RESEARCH COMPANY
RICHMOND, CALIFORNIA 94802

Summary

Gel permeation chromatography has been performed using a porous glass of broad pore size distribution, which was subjected to hexamethyl-disilazane treatment. The elution volumes and peak widths of narrow molecular weight distribution polystyrenes using toluene as solvent have been determined. The physical characteristics of the porous glass have been studied by the methods of mercury porosimetry, nitrogen adsorption-desorption isotherms, and electron microscopy. The characteristics of this column packing material are compared with other packing materials in popular use.

INTRODUCTION

Recently a porous glass having a wide range of pore sizes, of suitable dimensions for the separation of high molecular weight polymers, became available to us (Bio-Glas BRX 85001, Bio-Rad Laboratories Richmond, Cal.). The elution volume–molecular weight relationship and the widths of the eluted peaks have not previously been obtained on a porous glass column with such a broad pore size distribution.

* Trademark name of a porous glass packing sold by Bio-Rad Laboratories, Richmond, California.

† Presented at the ACS Symposium on Gel Permeation Chromatography, sponsored by the Division of Petroleum Chemistry at the 159th National Meeting of the American Chemical Society, Houston, Texas, February, 1970.

‡ Present address: IBM Research Laboratory, San Jose, California 95114.

§ Present address: Department of Chemistry and Institute of Materials Science, University of Connecticut, Storrs, Connecticut 06268.

Thus, it became important to compare the gel permeation chromatography (GPC) characteristics of this packing with other available column packings to gain further insight into the mechanism of GPC. The porous glasses currently in use for GPC have either an intermediate pore size distribution (*1–3*) or an extremely uniform pore size (*4, 5*). Polystyrene gels have been shown by electron microscopy to have a very broad pore size distribution and, in general, have higher efficiencies than the porous glass columns (*6*). Porous silica beads which have recently been studied as a column packing have a distribution of pore sizes similar to the intermediate pore size distribution porous glasses.

EXPERIMENTAL

The glass as received contained a large distribution of particle sizes. For this study the 100–200 mesh fraction was separated using Tyler sieves and accounted for ~25% of the total weight. This fraction, after pretreatment under vacuum at 400°F for 2 hr, was treated with hexamethyldisilazane by refluxing in *n*-hexane for 6 hr. The water flotation test showed complete reaction with the surface hydroxyl groups. This was found to be a more convenient procedure than the vapor phase method (*7*). A column 64.5 in. long of ⅜-in. o.d. stainless steel tubing was packed dry and then pumped with solvent for 48 hr. Polystyrene fractions of narrow molecular weight distribution (MWD) were eluted from this column using toluene as solvent at room temperature using a chromatograph with a differential refractometer detector. Samples as 0.1% solutions were injected by displacement from a loop having a volume of 1.77 ml. Flow rates were 1.05 and 0.4 ml/min.

The siphon used for flow rate measurement was made according to the design of Gray (*8*) Type B nominal 5 ml which gives a volume dump of excellent constancy. This new siphon was calibrated for each flow rate used. The siphon chamber was saturated with solvent to eliminate evaporation errors (*9*).

Mercury porosimetry and nitrogen adsorption–desorption isotherm characterizations were carried out on samples that had been pretreated under vacuum for 2 hr at 400°F.

ELECTRON MICROSCOPY

A well-mixed 0.3 g portion of the crushed glass having particle sizes 100–200 mesh was mixed with an epoxy mounting resin prepolymer as described previously (3). The impregnated samples were cast into the shape required by the Ultratome II ultramicrotome. Numerous sections were cut and floated onto carbon-coated electron microscope grids. Those sections thinner than 1000 Å (by interference color) were chosen for observation. The samples and grids were coated with carbon in a vacuum sputtering apparatus.

No replication or shadowing was required as the electron contrast between the glass and the resin mounting was satisfactory for direct observation. The carbon coating was required to avoid the beam-induced sample charge. A Japan Electron Optics JEM 6-A electron microscope operated at 80 kV was used to examine the samples. The micrographs were made on Kodak medium projector slide glass plates developed in Kodak HRP high resolution developer. The magnification of the microscope at 80 kV was calibrated by the standard diffraction grating technique.

RESULTS

Physical Characterization

Table 1 lists the physical characteristics of the porous glass, which had been treated with hexamethyldisilazane, determined by mercury intrusion and nitrogen adsorption techniques. The calculations of pore radius are based on a cylindrical pore model. The cumulative pore volume and differential pore volume plots with respect to the logarithm of the pore radius are shown in Fig. 1.

The nitrogen desorption isotherm showed that there was very little pore volume of pore radius less than 21 Å. In Fig. 1 the differential pore volume distribution shows that there are some very large pores in this

TABLE 1

Mercury porosimetry	Pore volume, cc/g, 21.1 → 67,000 Å pore radius	0.47
Contact angle 140°		
Surface tension, 473 dynes/cm	Macro pore volume, cc/g, 500 → 67,000 Å pore radius	0.21
	Micro pore volume, cc/g, 21.1 → 500 Å pore radius	0.26
Nitrogen adsorption–desorption	BET surface area, m²/g	48.3
Isotherms	Liquid nitrogen micropore volume, cc/g, <500 Å pore radius	0.26

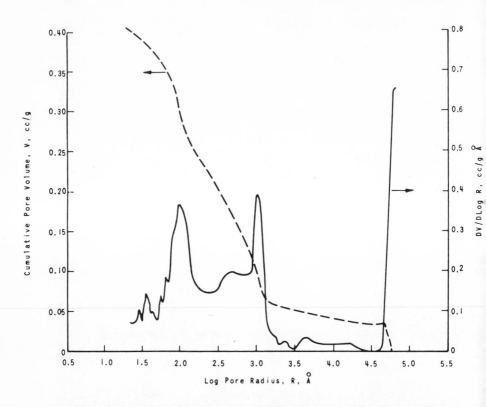

FIG. 1. Characterization of porous glass broad pore size distribution 100–200 mesh, disilazaned, by mercury porosimetry.

material. The peak at log $R = 4.8$ probably represents the filling of the interstitial voids in the bed of particles in the porosimeter cell. The mercury penetration data show that there are pores present in the material with radii of 17,800 Å. There is virtually a continuous distribution of pore radii down to the lower limit $\sim$25 Å of the porosimeter instrument. Two peaks on the differential pore volume distribution plot, corresponding to pore radii of 100 and 1000 Å, contribute the major portion of the total pore volume.

The gross optical appearance of the chips was the same as that seen previously for the narrower pore size distribution samples. The chips were obviously made by crushing a larger lump of glass. Figures 2–4 summarize the appearance of the thin sections in the electron microscope. The figures are positive prints taken from the glass negatives.

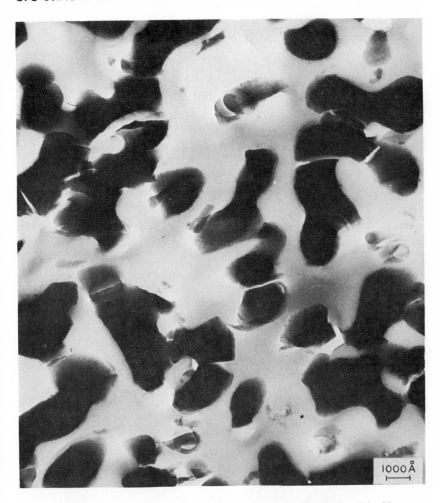

FIG. 2. Electron micrograph of broad pore diameter distribution Bio-Glas. Magnification of 100,000 diameters, sample mounted in epoxy resin with carbon coating. No metal shadowing. Positive print of glass negative. Dark areas represent glass walls.

The dark areas in the prints are the glass walls of the pores. The gray areas are the epoxy mounting resin. A few open areas exist where the resin did not penetrate or large glass walls dropped out. All channels appear to interconnect. The magnification of 100,000 diameters was verified with a diatom sample. Some optical gradation is noted where two layers of glass wall overlap due to sample thickness.

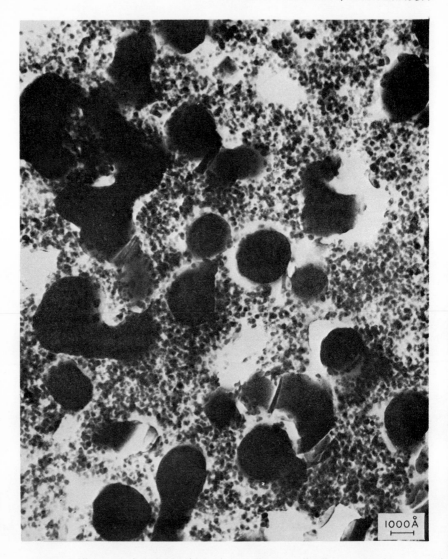

FIG. 3. Electron micrograph of broad pore diameter distribution Bio-Glas.

The field of Fig. 2 shows a group of open very large pores. In the same chip some of the smallest pores were noted. The pore diameters, measured on the perpendicular to a given dark surface (glass wall), range from 7500 to 250 Å. The average pore diameter was 3000 Å. All pores are very irregular. Materials with this large pore diameter were not seen in the previous study (*3*).

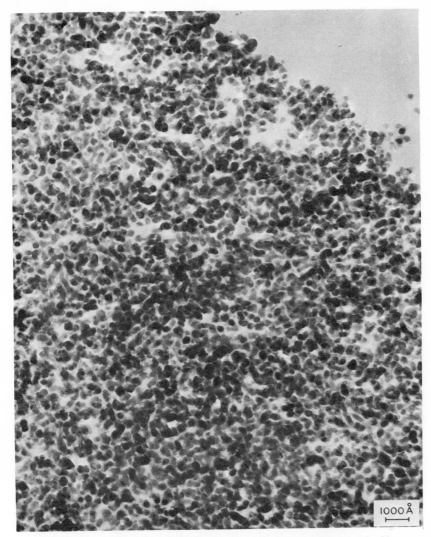

FIG. 4. Electron micrograph of broad pore diameter distribution Bio-Glas.

The field of Fig. 3 shows a material of about 100 Å pore diameter (granulations) which fills a structure similar to Fig. 2. This field was very near that shown in Fig. 2. The large, black voids represent the walls of the large matrix.

The relatively uniform field of Fig. 4 was taken at the edge of a chip. The average pore diameter is 200 Å with some large voids (filled with epoxy which precludes drop out) of 800 to 2000 Å diameter. Large

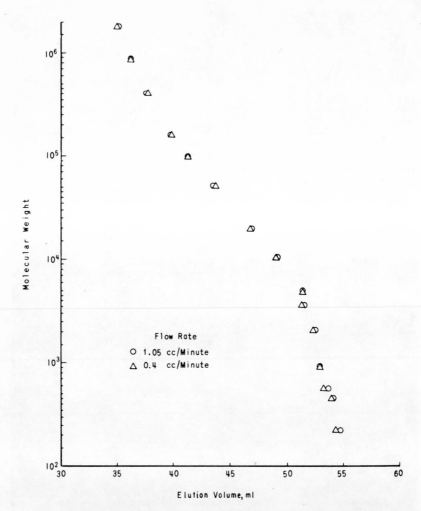

FIG. 5. Elution volume–molecular weight relationships for broad pore size
distribution porous glass.

glass walls are notably absent near the edges of about half of the chips
scanned.

The broad pore diameter distribution glass is a polydisperse material,
not a mixture of various pore diameter chips. This is very clear in
Fig. 3. Although it is not the purpose of this study to determine how
this porous glass was made, the impression gained from Figs. 2–4 is

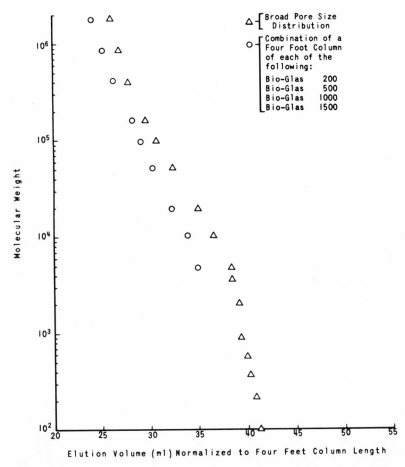

FIG. 6. Comparison of the elution volume–molecular weight relationships of the broad pore. Size distribution porous glass with a combination of narrower pore range porous glasses.

that a previously formed large pore diameter glass has been partially filled with a smaller pore diameter-forming glass. The effect is a product that appears to have undergone a series of treatments. Each chip contains a large number of microvolumes of differing pore sizes. If a broad pore size distribution substrate, mixed very intimately, has any real advantages in GPC, the Bio-Glas packing should exhibit the advantages fully.

GEL PERMEATION CHROMATOGRAPHY

The relationships between the elution volumes and the logarithm of the molecular weights of the polystyrene solutes at flow rates of 1.0 and 0.4 ml/min are shown in Fig. 5. The relationship exihibits the sigmoidal shape found for the other Bio-Glas substrates examined previously (*1*, *2*). The comparison between this column and four columns in series of varying but narrower pore size distribution Bio-Glas substrates, both normalized to 4-ft column length, is shown in Fig. 6. The curves are very similar in shape, the broad distribution column has a slightly flatter slope in the intermediate molecular weight region $0.5 \times 10^3 \rightarrow 1 \times 10^5$, and this exhibits greater selectivity with respect to the separation of the peak maxima of different molecular weights. This does not necessarily mean the column of broad pore size glass is more efficient, as this depends on the peak broadening also.

The elution volumes, peak widths, and the number of theoretical plates per foot for the polystyrene and hydrocarbon solutes at two different flow rates are recorded in Table 2. The column shows very little selectivity below molecular weights of 4000. Up to the highest molecular weights studied of 1.8×10^6 there is no flow rate dependence of elution volume for the flow rates studied. The number of theoretical plates, n, was calculated from the following equations:

$$n = \frac{1}{L}\left(\frac{V_e}{W_b/4}\right)^2 = \frac{1}{L}\left(\frac{V_e}{W_h/2.354}\right)^2 \tag{1}$$

where V_e is the elution volume to the peak maximum, W_b is the peak width at the base, and W_h is the peak width at half the peak height.

The variation in the peak widths with molecular weight at both flow rates studied shows an initial increase with molecular weight up to 20,000, then a decrease. The number of theoretical plates calculated from Eq. (1) is dominated by the elution volume, V_e; and the values decrease regularly with increasing molecular weight except for molecular weights greater than 160,000 at the faster flow rate where a fairly constant value is found. The column performs more efficiently at the slower flow rate, and the increase in the number of theoretical plates is due to a decrease in the peak width since the elution volumes are constant. At molecular weights above 860,000, it appears that no increase in efficiency is obtained by slowing the flow rate from 1.05 to 0.4 ml/min. This is quite different from the results obtained with the 860,000 molecular weight polystyrene eluting from a polystyrene gel

TABLE 2

Elution Volumes, Peak Widths, and the Number of Theoretical Plates of Polystyrene Solutes Eluted from a Broad Pore Size Distribution Porous Glass Column at Two Flow Rates

	Flow rate									
	1.05 ml/min					0.4 ml/min				
		Peak width (ml)		Number of theoretical plates per foot, n, measured at			Peak width (ml)		Number of theoretical plates per foot, n, measured at	
Solutes	Elution volume (ml)	At the base, W_b	At half the peak height, W_h	The base	The peak half height	Elution volume (ml)	At the base, W_b	At half the peak height, W_h	The base	The peak half height
$nC_{16}H_{34}$	54.69	8.40	4.99	126.1	123.6	54.37	7.85	4.62	145.0	145.2
$nC_{24}H_{50}$	53.97	8.60	5.08	117.6	116.5	53.91	8.18	4.74	131.1	135.1
$nC_{40}H_{82}$	53.66	9.77	5.71	91	92.2	53.24	8.77	5.16	111.3	111.3
Polystyrene, $\overline{M}_w$										
900	52.83	9.82	5.72	86.1	88.1	52.82	8.94	5.03	105.5	115.1
2,030	52.49	10.03	5.80	81.5	84.4	52.36	9.03	5.24	101.8	104.2
3,600	51.52	10.50	6.09	71.7	73.6	51.31	9.11	5.37	95.9	95.5
4,800	51.43	10.16	5.84	76.3	79.9	51.42	9.09	5.26	98.2	100.6
10,300	49.10	10.50	6.05	65.1	67.8	49.04	9.45	5.57	81.6	80.9
19,800	46.94	10.29	5.97	62.0	63.8	46.81	9.45	5.50	74.4	76.0
51,000	43.47	9.61	5.63	60.9	61.4	43.76	9.03	5.33	71.3	70.7
97,200	41.23	9.53	5.55	55.8	56.9	41.36	8.65	5.08	69.4	69.7
160,000	39.79	9.61	5.51	51.0	53.9	39.85	8.52	4.95	66.4	68.0
411,000	37.59	9.40	5.38	47.6	50.4	37.70	8.52	4.91	59.5	62.0
860,000	36.22	8.76	5.08	52.0	53.6	36.20	8.81	5.04	51.3	54.4
1.8×10^6	35.11	8.76	5.08	48.9	50.4	35.06	9.17	5.33	45.1	45.6

176 COOPER, CAIN, BARRALL, AND JOHNSON

column (10). There it was found that an increase in the number of theoretical plates per foot from 46 to 63 resulted from a similar change in flow rate. Moreover, it is interesting to note that the efficiency of the porous glass column is identical to that of the polystyrene gel column at this molecular weight.

The peak widths are largest for the samples that elute where the elution volume molecular weight relationship exhibits its smallest slope. This suggests that the discrete molecular weight range of the polystyrene samples may be influencing the total peak width. Assuming that the squares of the variances of the various processes occurring in the column are additive, the following equation may be written

$$W_{total}^2 = W_{inst}^2 + W_{MWD}^2 + W_{GPC}^2 \qquad (2)$$

where W_{total}, W_{inst}, W_{MWD}, and W_{GPC} represent the widths of the peak due to the total process, the instrumental broadening, the molecular weight distribution, and the gel permeation chromatographic process, respectively. The latter consists of several parts which can be divided into two groups, those processes occurring within the pores and those occurring in the interstitial volume. The contributions W_{total} and W_{inst} were measured directly, the contributions from W_{MWD} were calculated assuming a log normal molecular weight distribution and using the calibration curve to determine the volumes between which molecular weights corresponding to 5 and 95% weight composition of a particular sample would elute.

The values of the peak widths W_{GPC} corrected for instrumental broadening and sample molecular weight inhomogeneity at each flow rate are collected in Table 3. Also included in this table are values of the number of theoretical plates n_{CHR} resulting only from the chromatographic process and calculated from the equation:

$$n_{CHR} = \frac{1}{L}\left(\frac{V_e - 3}{W_{GPC}/4}\right)^2 \qquad (3)$$

The value of V_e is decreased by 3 ml, this being the holdup in the connecting tubings.

When the peak widths are corrected in this way, the values decrease fairly regularly with increasing molecular weight. This indicates that the efficiency of the separation process is not dependent upon the molecular diffusion coefficient as previously proposed (11–13).

TABLE 3

Correction of Experimental Peak Widths and Efficiencies for Instrumental Broadening and Sample Molecular Weight Inhomogeneity

| | | | Flow rate | | | |
| | | | 1.05 ml/min | | 0.4 ml/min | |
Solute	W_{inst}^2 (ml^2)	W_{MWD}^2 (ml^2)	W_{GPC}^2 (ml^2)	Number of theoretical plates/ft (n_{CHR})	W_{GPC}^2 (ml^2)	Number of theoretical plates/ft (n_{CHR})
$C_{40}H_{82}$	7.8	0	87.7	87.2	69.1	110.7

| Polystyrene | | | | | | | |
$\bar{M}_w$	$\bar{M}_w/\bar{M}_n$						
10,300	1.06	10.9	6.8	92.6	68.4	71.6	88.4
19,800	1.06	10.9	7.3	87.7	65.6	71.1	80.9
51,000	1.06	10.9	7.8	73.7	66.2	62.8	77.7
97,200	1.06	11.6	6.8	72.4	60.2	56.4	77.3
160,000	1.06	11.6	5.3	75.5	53.4	55.7	72.4
411,000	1.06	11.6	2.0	74.8	47.6	59.0	60.4
860,000	1.15	11.6	1.4	63.7	51.6	64.6	50.9
860,000	1.72[a]	11.6	2.3	62.8	52.4	63.7	51.6

[a] Value based on a membrane osmometer determination of the number-average molecular weight = 500,000 obtained in this laboratory.

CONCLUSION

The porous glass substrate was shown to be a truly polydisperse pore size material and not a mixture of glasses with smaller pore size distributions. The material has been shown to be an effective substrate for the separation of macromolecules. The efficiency of the porous glass column used was as good as that of a polystyrene gel column at molecular weights near to 1 million, but the efficiency of the porous glass column was inferior at lower molecular weights. It appears there is no particular preference between using a broad pore size distribution packing material in gel permeation chromatography or a mixture of different but narrower pore size distribution materials.

Acknowledgment

A portion of the work of one author was supported by National Science Foundation Grant GP14440.

REFERENCES

1. M. J. R. Cantow and J. F. Johnson, *J. Appl. Polym. Sci.,* **11**, 1851 (1967).
2. M. J. R. Cantow and J. F. Johnson, *J. Polym. Sci., Part A-1,* **5**, 2835 (1967).
3. E. M. Barrall and J. H. Cain, *J. Polym. Sci., Part C,* **21**, 253 (1968).
4. W. Haller, *Nature,* **206**, 693 (1965).
5. J. C. Moore and M. C. Arrington, *Intern. Symp. Macromol. Chem., Tokyo, 1966,* pp. vi–107.
6. K. H. Altgelt and J. C. Moore, *Polymer Fractionation* (M. J. R. Cantow, ed.), Academic, New York, 1967.
7. A. R. Cooper and J. F. Johnson, *J. Appl. Polym. Sci.,* **13**, 1487 (1969).
8. D. O. Gray, *J. Chromatogr.,* **37**, 320 (1968).
9. W. W. Yau, M. L. Suchan, and C. P. Malone, *J. Polym. Sci., Part A-2,* **6**, 1349 (1968).
10. A. R. Cooper, A. R. Bruzzone, and J. F. Johnson, *Polym. Preprints,* **10**, 1455 (1969).
11. M. LePage and A. J. DeVries, *Third International Seminar on Gel Permeation Chromatography,* Geneva, Switzerland, 1966.
12. W. B. Smith and A. Kollmansberger, *J. Phys. Chem.,* **69**, 1457 (1965).
13. J. G. Hendrickson, *J. Polym. Sci., Part A-2,* **6**, 1903 (1968).

High Resolution Gel
Permeation Chromatography—Using Recycle*

KARL J. BOMBAUGH and ROBERT F. LEVANGIE

WATERS ASSOCIATES, INC.
FRAMINGHAM, MASSACHUSETTS 01701

Summary

High resolution in gel permeation chromatography has been accomplished by use of long columns, since commercial rigid gels afford relatively fixed capacity ratios and since plate numbers of equivalent columns are additive. To overcome the high cost and high pressure requirement of long columns often required to resolve discrete species, high resolution gel permeation chromatography may be attained in commercial GPC equipment by recycle operation through the reciprocating pump, GPC column, and detector. However, peak width increases with the number of cycles (ν). Since the contained volume of the closed recycle system is constant, as ν increases the peak width (W) of the distribution will eventually exceed the volume of the system and peak overlap will occur. This presentation considers the increase in W and provides a method of "flush and draw off" to prevent sample overlap. Analytical and preparative scale separations were investigated, using both small and macromolecules. The effects of sample load and flow rate on resolution with recycle operation were investigated.

INTRODUCTION

Gel permeation chromatography, developed by Moore as a method of fractionating macromolecules, has also been extended to small molecules by a number of workers (2–8). Most of this work was done either at low resolution, or with species affording large alpha prime

*Presented at the ACS Symposium on Gel Permeation Chromatography sponsored by the Division of Petroleum Chemistry at the 159th National Meeting of the American Chemical Society, Houston, Texas, February, 1970.

values where alpha prime is equal to $V_2/V_1 \geq 1.1$, where V_2 and V_1 equal the elution volume of the two solutes.

To resolve species with alpha prime (relative retention) values less than 1.1, an increase is required, either in plate number of the system or in the capacity ratio (K') of the gel

$$K' = \frac{V_e - V_0}{V_0}$$

where V_e is the elution volume of the solute and V_0 is the interstitial (dead) volume of the column.

Commercial rigid gels offer a fixed K' of 0.8 to 1.2. Therefore, high resolution is accomplished in GPC by increasing the plate number (N). In an earlier work it was shown that high resolution was attainable in GPC by "brute force," using multiple gel columns in series at high pressure (3). This was possible since plate numbers of equivalent columns were additive (1). It was further shown that similar results could be attained by multiple use of columns in commercial GPC equipment by means of recycle operation through the reciprocating pumps, GPC columns, and detectors (9). The initial work was done with a narrow distribution pair of single species to demonstrate feasibility and to optimize the system for minimal band spreading. However, with broad distributions, a method of "draw off" was needed to prevent overlap, since band width increases with effective column length. Because the inlet from the solvent tank was left open during the recycle operation to prevent cavitation in the pumps, it was also necessary to devise a procedure to flush the inlet tee as well as the interconnecting tubing to prevent reinjection of minute quantities of sample which diffuse into the tees and remain in the interconnecting lines.

Recycle is an effective method of obtaining high resolution at heavy load for preparative separations of both small and macromolecules. Therefore, consideration of the effect of sample load on resolution was studied at both constant concentration and constant sample volume. The effect of flow rate on cycle requirement for resolution was also considered.

EXPERIMENTAL

Apparatus and Procedure

A Waters Associates ALC 100, adapted for recycle operation, was used to fractionate Triton X100, Triton X45, and C_{22-28} alpha olefins.

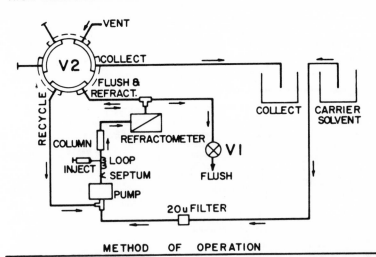

FIG. 1. Schematic diagram of recycle operation.

(A mixture of paraffins was used to evaluate the flush and draw off procedure described below.) Triton X100 was fractionated through six cycles at 3.1 and 0.5 ml/min. Five columns, 4 ft × 3/8 in., containing 60 and 100 Å Poragel were used to fractionate the smaller molecules. A schematic diagram of the recycle system and a mode of operation is shown in Fig. 1. The effect of sample load on resolution with recycle operation was investigated, using the new Waters Associates Chromatoprep. This is a new preparative scale liquid chromatograph, designed for recycle operation. The unit can accommodate four columns 2½ in. in diameter by 4 ft in length. Pump capacity is 10 to 120 ml/min. Sample injection is made through the solvent pump. The fraction collector employs an optically indexed 40-port valve controlled by an automatic time based programmer. A schematic diagram of the unit is shown in Fig. 2.

To determine the effect of sample load on resolution, equal parts polystyrene 51K and 10.3K were fractionated at constant volume (100

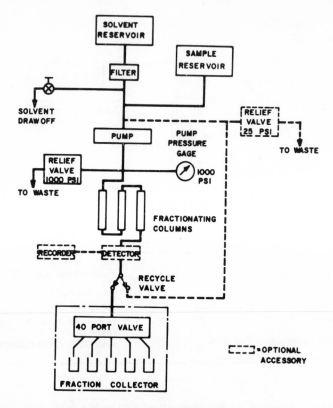

FIG. 2. Schematic diagram of Chromatoprep.

ml), varying concentration as shown in Table 1, and at constant concentration (10 mg/ml), as shown in Table 2. Resolution values were calculated for each cycle, using the equation

$$R = 2\left(\frac{V_2 - V_1}{W_1 + W_2}\right)$$

Resolution at various flow rates was determined at a 1-g load using a 100-ml sample at a concentration of 10 mg/ml.

RESULTS AND DISCUSSION

Peak Spread—System Capacity

Peak width (W) in GPC is proportional to the effective length (L) of the column by the relationship

$$L = 16H \left(\frac{V_r}{W}\right)^2$$

where V_r = retention volume and W is measured at the peak base. With equivalent columns in recycle operation, height equivalent to a theoretical plate, H may be considered constant since, with recycle operation, L and subsequently V_r, is a function of the number of cycles. W increases with the number of cycles (v) by the relationship

$$W^2 = a\frac{V_r^2}{v} = a\frac{v_n^2}{v} = av$$

or

$$W = W_0 \sqrt{v}$$

where W_0 is the peak width in milliliters of the first cycle.

Since the contained volume of the closed recycle system is constant as v is increased, the peak width of the distribution will eventually exceed the volume of the system and peak overlap will occur. To prevent overlap a "draw off" procedure is needed. The procedure is illustrated in Fig. 3, where Triton X45 is run through six cycles. The

TABLE 1

Resolution at Various Sample Loads Using Constant Volume (100 ml)

		Resolution		
g	Concentration (mg/ml)	Cycle 1	Cycle 2	Cycle 3
1.0	10	1.06	1.31	1.47
2.0	20	0.59	1.13	1.29
3.5	35	0.34	0.77	1.14
5.0	50	0.25	0.54	0.92

TABLE 2

Resolution at Various Sample Loads Using Constant Concentration (10 mg/ml)

		Resolution		
g	Volume (ml)	Cycle 1	Cycle 2	Cycle 3
1.0	100	1.06	1.31	1.47
2.2	220	0.70	1.20	1.34
3.5	350	0.39	0.91	1.14

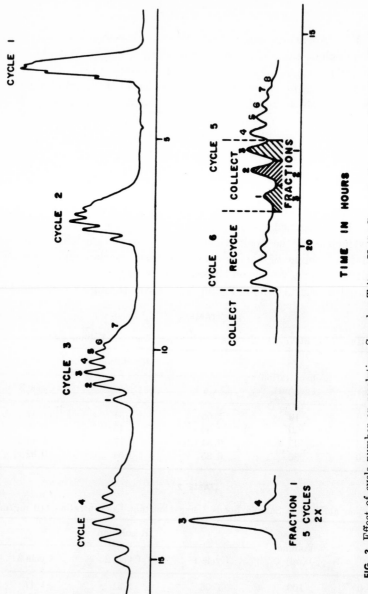

FIG. 3. Effect of cycle number on resolution. Sample: Triton X-45. Concentration: 50%. Injection: 30 μl. Solvent: THF. Flow rate: 0.48 ml/min. Columns: Stragel 60 Å (15 ft).

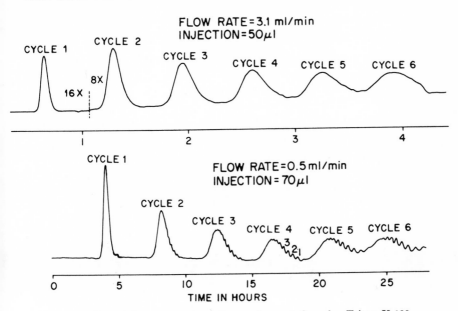

FIG. 4. Effect of flow rate on capacity requirement. Sample: Triton X-100. Concentration: 50%. Solvent: THF. Columns: Styragel 100 Å (8 ft); 60 Å (12 ft).

low molecular weight end is resolved first. Peak 1 could be removed after the third cycle, but in this work was removed along with Peaks 2 and 3 during the fifth cycle, just before overlap occurs from the high molecular weight end of cycle six. (After removal of the resolved low molecular weight species, the high molecular weight portion is recycled additional cycles until resolution is virtually complete.) Also shown is a chromatogram of Peak 3 (Fraction 1) as obtained by recycling five times on the same system. Relative areas indicate that Fraction 1 contained 92% of Peak 3 and 8% of Peak 4.

System Capacity—Flow Rate

Since W increases with v, it is essential to provide enough system capacity to get the desired resolution at the low molecular weight end of the distribution before peak overlap occurs. This is illustrated in Fig. 4. The top chromatogram of Triton X100 was obtained at 3.1 ml/min; the bottom chromatogram at 0.5 ml/min. At the high flow rate, six cycles were completed in slightly over 4 hr. However, peak overlap resulted before evidence of discrete peaks was observed. At the

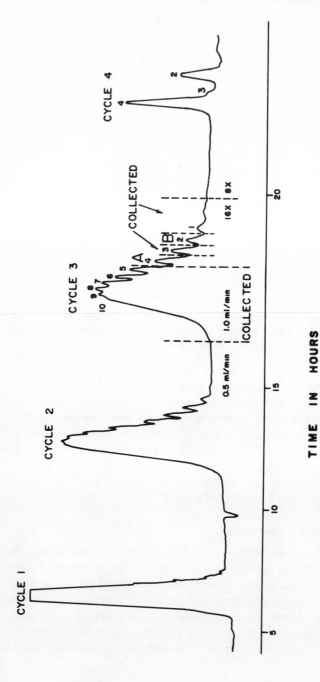

FIG. 5. Effect of system capacity on cycle requirement. Sample: Triton X-100. Concentration: 50%. Injection: 70 μl. Solvent: THF. Columns: Styragel 100 Å (8 ft) and 60 Å (24 ft).

low flow rate a small, low molecular weight peak is observed in the first cycle, which required 4½ hr. Clear evidence of individual species is seen in the second and fourth cycles (19 hr) are completed before overlap occurs. Peaks 1 through 3 could be removed during the fourth cycle, permitting additional cycles without overlap. In this manner additional components from the low molecular weight end of the distribution could be drawn off and recycling continued until resolution is complete and all species are isolated.

By increasing the columns from five (20 ft) to eight (32 ft), fewer cycles are necessary (and somewhat higher flow rates may be used) as is illustrated in Fig. 5. At this condition resolution is observed in the first cycle. Note that virtual baseline resolution of Peaks 2, 3, and 4 was attained on the third cycle. During Cycle 3, Peaks 1, 3, and 5 through 10 were drawn off. Peaks 2 and 4 remained in the system and were run through Cycle 4. High purity material remained, as only a trace of Peak 3 is seen between Peaks 2 and 4.

When the distribution is narrow, fewer columns are required, or more cycles can be run without draw off. The separation of C_{22-28} alpha olefins, using five columns as shown in Fig. 6, was run through five cycles without concern for overlap. Components are removed from both sides of the distribution until only Peak 2 remains.

Effect of Sample Load on Resolution, Using Large Diameter Columns (2½ in.)

The effect of sample load on resolution, as determined at a constant injection volume of 100 ml, is shown in Table 1, and the effect of sample load on resolution at constant concentration (10 mg/ml) is shown in Table 2. A plot of R vs. v for these data shows that poor resolution is obtained at >2 g with one cycle. However, at three cycles a 5-g load is resolved. At either constant volume or constant concentration, a 3.5-g load shows slightly higher resolution at three cycles than a 1-g load with a single cycle. In terms of through-put, therefore, a 3.5-g load at three cycles would yield more product than three separate 1 g injections run for one cycle. A comparison of constant concentration versus constant volume shows that better resolution is obtained at lower concentration (10 mg/ml) at two cycles. Lower resolution is obtained for the 3.5-g injection at a concentration of 35 mg/ml, probably due to viscosity effects. However, the points merge at three cycles.

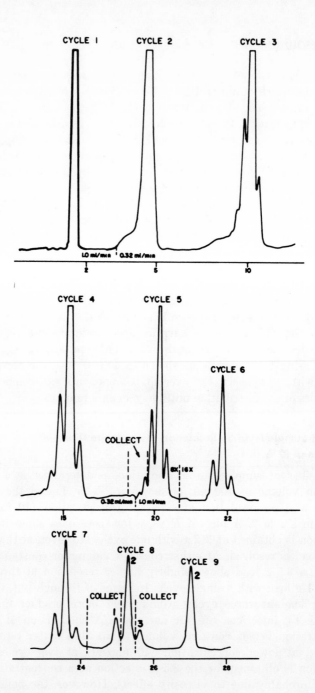

FIG. 6. Effect of sample distribution on capacity requirement. Sample: α-olefin mixture. Concentration: 60 mg/ml. Load: 90 mg. Solvent: toluene. Columns: Styragel 60 Å (15 ft).

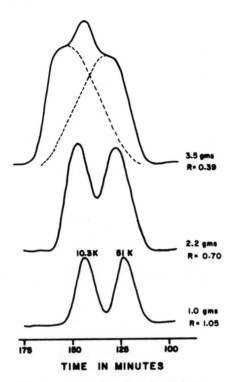

FIG. 7. Effect of load on resolution. Sample: Polystyrene mixture 51K + 10.3K (1 — 1). Concentration: 10 mg/ml. Injection: varied. Solvent: toluene. Flow: 14.4 ml/min. Column: Styragel 2.5 × 10⁴ Å (4 ft).

The effect of load on resolution is shown by Fig. 7, where chromatograms from three different sample loads are shown. The apparent trimodal distribution is the sum of the unresolved bimodal. The effect of improved resolution by recycle for the same sample load is shown in Fig. 8, where in the second cycle the apparent trimodal resolves into a bimodal distribution.

Effect of Flow Rate

The effect of flow rate on resolution for various cycle numbers is shown in Table 3. A plot of these data shows that higher resolution is obtained with three cycles at 121 ml/min than for one cycle at 14.4 ml/min. Therefore, for optimum resolution/time (R/t), the preferred modus operandi should employ maximum load and maximum flow rate, provided adequate system capacity is provided to prevent overlap.

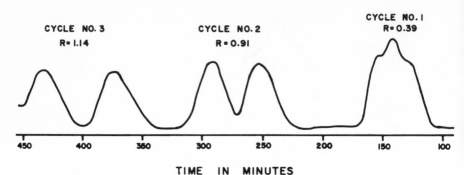

TIME IN MINUTES

FIG. 8. Effect of recycle on resolution at heavy load. Sample: polystyrene mixture $51K + 10.3K$ $(1-1)$. Concentration: 10 mg/ml. Injection 350 ml. Load: 3.5 g. Solvent: toluene. Flow: 14.4 ml/min.
Column: Styragel 2.5×10^4 Å (4 ft).

Adequate resolution of the species at the low end of the distribution
should be obtained before overlap occurs, if resolution is to be accomplished on a single injection. Where this is not practicable, it may
be necessary to use multiple fractionations, whereby the sample is
pre-cut sequentially into progressively more narrow fractions. By this
procedure, high resolution can be obtained, with broad distribution
over virtually any molecular weight range.

TABLE 3

Resolution at Various Flow Rates

	Resolution		
Flow rate	Cycle 1	Cycle 2	Cycle 3
14.4	1.06	1.31	1.47
29.6	0.95	1.30	1.40
59.1	0.81	1.03	1.11
92.0	0.69	1.04	1.13
121.0	0.57	0.95	1.09

REFERENCES

1. J. C. Moore, *J. Polym. Sci., Part A,* **2,** 835 (1964).
2. K. J. Bombaugh, W. A. Dark, and R. F. Levangie, *Z. Anal. Chem.,* **236,** 443
(1968).

3. K. J. Bombaugh, W. A. Dark, and R. F. Levangie, *Separ. Sci.,* 3, 375 (1968).

4. J. Cazes and D. R. Gaskill, *Separ. Sci.,* 2, 421 (1967).

5. J. G. Hendrickson and J. C. Moore, *J. Polym. Sci., Part A-1,* 4, 167 (1966).

6. J. G. Hendrickson, *Anal. Chem.,* 40, 49 (1968).

7. J. L. Spell, *Fourth International GPC Seminar,* Waters Associates, Framingham, Massachusetts, 1967.

8. W. Heitz, B. Bömer, and H. Ullner, *Makromol. Chem.,* 121, 102 (1969).

9. K. J. Bombaugh, W. A. Dark, and R. F. Levangie, *J. Chromatogr. Sci.,* 7, 42 (1969).

Gel Permeation Chromatography with High Loads*

KLAUS H. ALTGELT

CHEVRON RESEARCH COMPANY
RICHMOND, CALIFORNIA 94802

Summary

High loads on GPC columns usually lead to poor efficiency because of steep viscosity gradients. A great difference in density between solution and solvent can also cause excessive band broadening. However, under certain conditions good separations are achieved with loads of 150 mg/ 100 cc column volume and higher. Two mechanisms are proposed to explain this phenomenon. Secondary exclusion is caused by obstruction of pores to larger molecules by the more rapidly diffusing small molecules. It takes place predominantly with molecules of less than 2000 molecular weight in small pore gels. Incompatibility is caused by repulsive interaction between solute molecules and the polystyrene gel. It is observed with solutes which are chemically quite different from polystyrene, e.g., with polyvinyl acetate, and in a low to intermediate molecular weight range.

THE EFFECT OF LOAD ON COLUMN EFFICIENCY

Generally speaking, GPC columns should not be overloaded. High sample viscosity and ensuing pressure gradients lead to drastic drops of column efficiency. This was first pointed out by Flodin in 1961 (1) and was confirmed by many others. The sample load is restricted by two limitations: (a) the sample viscosity should not be different from the solvent viscosity by a factor greater than 2 (1–3); (b) the sample volume must be small since it increases linearly the width of a zone, i.e., the peak width (4–5). Taking these two effects into account, most researchers limit their sample size to about 15 mg/100 cc column volume.

*Presented at the ACS Symposium on Gel Permeation Chromatography, sponsored by the Division of Petroleum Chemistry at the 159th National Meeting of the American Chemical Society, Houston, Texas, February, 1970.

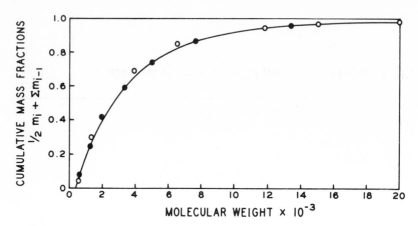

FIG. 1. Molecular weight distribution of a polybutene determined by GPC (○) and gradient elution chromatography (●).

On the other hand, we obtained surprisingly good separations with loads 10 times as great (*6*). Figure 1 compares the molecular weight distribution of a polybutene sample obtained by GPC with that obtained by gradient elution chromatography. The agreement is very good. Column load in the GPC separation was 1 g polybutene in 10 cc benzene + 10% methanol on a column with 600 cc total volume. Such good separation with high load is not a contradiction to previous findings but primarily a consequence of the low molecular weight of our samples (1200) as opposed to those of Flodin's (*1*) and others (100,000–2,000,000). Because of the relatively low intrinsic viscosities of our samples, we could use 10% solutions, in some cases even 20% solutions (*7*).

SECONDARY EXCLUSION

Some of our separations with highly loaded columns were so good that we searched for an explanation beyond that of undistorted flow due to low sample viscosity. Thus we came upon a separation mechanism in GPC which we named "secondary exclusion" (*5, 8*).

To understand the secondary exclusion mechanism, we must examine how the sample molecules diffuse into and out of the gel pores. The diffusion rate of a molecule in a gel depends not only on its size as it does in bulk liquid, but furthermore, on the ratio of the molecular size to the pore size. This is illustrated in Fig. 2. If we put a concentrated solution of small and large molecules onto a gel, then the small ones

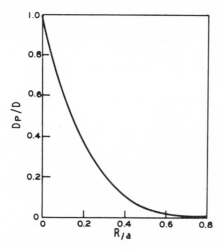

FIG. 2. Dependence of diffusion coefficient, D_p. of a spherical molecule in a pore on the ratio of particle radius, R, to pore radius, a. D is the diffusion coefficient in free solution. Reproduced from Ackers (9).

will rapidly diffuse into the available pores. The large ones, since they diffuse more slowly, will find more pores occupied. The probability of their diffusing into an occupied pore will be reduced depending on the reduction of available pore size and on the motion of the smaller molecule in the pore. If the small molecule is moving away from the oncoming larger molecule, it will have no effect. If it is moving toward it or lateral to it, it will obstruct the other molecule. In case of obstruction, the larger molecule will move on until it finds an unobstructed pore. Thus, it is effectively excluded from a pore which otherwise would have been accessible. This is secondary exclusion.

We tested our hypothesis with mixtures of a straight chain hydrocarbon, octacosane (C_{28}), and a narrow polystyrene standard of 900 molecular weight in chloroform on a polystyrene gel with an exclusion limit of about 3000 molecular weight. The materials were chosen such that (a) both molecular weights were low enough to permit high loads at moderate viscosities, (b) the low molecular weight sample consisted of molecules distinctly larger than the solvent molecules, and (c) the large molecules were small enough to fit into about 30% of the total gel pore volume. Figure 3 shows the elution pattern at low loading conditions. The limiting peaks, i.e., the first and the last ones, indicate the void volume, $V_0 = 44$ ml, and the total available column volume, $V_0 + V_i = 100$ ml, respectively. At this concentration, 15 mg/column

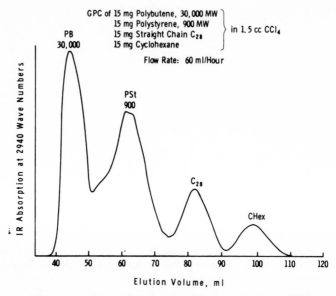

FIG. 3. Elution pattern of Polystyrene 900 and of octacosein under low load conditions.

volume (100 ml), the elution volume of the octacosane was 82 ml and that of the PSt 900 was 62 ml. Figure 4 demonstrates what happened at high loads, viz., 300 mg C_{28} and 50 mg PSt 900. The elution volume of PSt 900 shifted from 62 to 47 ml, i.e., almost all the way to V_0. Most of the C_{28} eluted with a peak at 65 ml leaving only a shoulder at its previous elution volume of 82 ml. At this high load not only the polystyrene had experienced secondary exclusion, but also part of the octacosane. An interesting detail is the fact that in this case the polystyrene had been injected not together with, but after, the octacosane. Thus it had literally to overtake the C_{28} molecules.

While I believe I have demonstrated the existence of the effect of secondary exclusion, I must point out that it seems to be restricted to a relatively narrow range of experimental conditions. We were not able to see it in gels of greater pore size. This restriction to small pores may perhaps be explained by the structure and "roughness" of the pore walls. Close examination of a polystyrene gel, Fig. 5, reveals that its walls consist of dense clusters which resemble grapes. The large pores are much less confined in that they have more connections with neighboring pores than do the smaller ones. In the extreme, the smallest

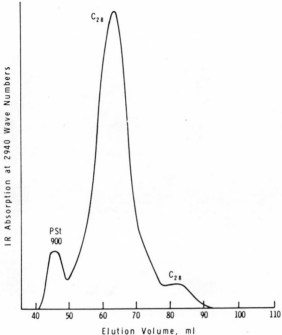

FIG. 4. Elution pattern of Polystyrene 900 and of octacosein under high load conditions.

pores are just crevices or holes in a wall and have just one opening. In pores with many openings a molecule has that many more ways of leaving the pore and thus not being in the way of an oncoming molecule.

In another paper given at this Symposium (*10*), Albaugh et al. also describe peak shifts toward V_0 in cases of high loads with no appreciable loss in separation power.

INCOMPATIBILITY

Another effect that may be encountered under conditions of high load is additional exclusion due to incompatibility between sample and gel. Incompatibility of two types of polymers has been described by various authors (*11–15*). It is caused by the low entropy of mixing

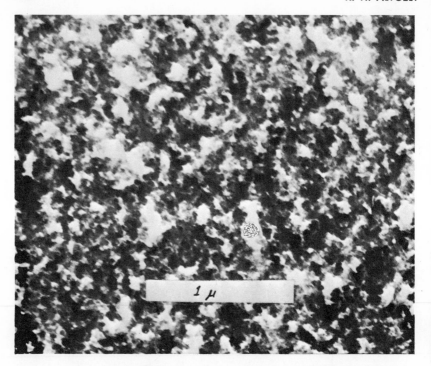

FIG. 5. Electron micrograph of large pore polystyrene gel.

between different macromolecules which cannot overcome even relatively weak positive heats of mixing and which thus leads to positive free energies of mixing. As a consequence, mixing does not take place, i.e., phase separation occurs at sufficiently high polymer concentration. The occurrence of phase separation is only dependent on the structure, size, and concentration of the two polymers; it is not affected by the solvent. Phase separation due to incompatibility takes place in good as well as in poor solvents.

In GPC one polymer is the gel, the other the sample. Here phase separation means complete exclusion. Complete exclusion would be an extreme case and is unlikely to be encountered in practice. However, partial exclusion due to incompatibility (*3, 6*) does take place and can be demonstrated. The calibration curve in Fig. 6 was established with polystyrene and with straight chain polymethylenes (C_6–C_{32}). The two sets of points representing high and low concentrations of polyvinyl acetate fractions are shifted toward smaller elution volumes.

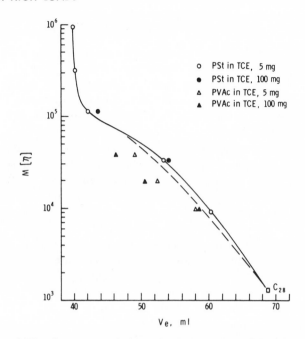

FIG. 6. Calibration curve of the polystyrene gel used for the present investigation. The shift of the PVAc points to the left is interpreted as ·incompatibility between the PVAc and the polystyrene gel.

This shift cannot be due to swelling or chain stiffness since the ordinate in our plot is not the molecular weight but the molecular size in terms of $M[\eta]$ (5, 16). Also, it can not be secondary exclusion in this case as the concentration dependence of the polystyrene samples on this gel was normal and opposite to that of the polyvinyl acetate fractions, see Fig. 6. An effect arising from repulsive interaction between solute and gel, i.e., incompatibility, seems to be the only explanation under these conditions.

The effect of incompatibility too is restricted to a limited set of conditions. While the molecular weight of the sample must be low enough to permit high loads, it must, on the other hand, be high enough to lower the entropy of mixing sufficiently to overcome the enthalpy term in the equation

$$\Delta F = \Delta H - T\Delta S$$

According to Fig. 6 the effect is quite small in the case of the 500 mol wt polyvinyl acetate and can be expected to be negligible at lower molecular weights for this polymer–gel combination.

VISCOSITY AND DENSITY EFFECTS

The detrimental effect of viscous samples in columns was described in the early days of GPC by Flodin (*1*). Less well known is the tailing that is caused by high viscosity and originates in tubings of the kind used for connections and sample loops in GPC systems. This is the subject of other presentations in this Symposium (*17, 18*) and need not be elaborated here.

Highly concentrated samples may be quite different in density from the solvent. Examples are hydrocarbons in chlorinated solvents. In such cases, broad peaks may result from two causes. First, fingering due to steep density gradients may occur in the column bed. We had occasion to observe this with colored samples such as azulene in carbon tetrachloride. Again, less well known is the second cause, viz., the layering which can take place in tubings. Initially, we used $\frac{1}{8}$-in. tubing for our sample loops. With colored samples, e.g., asphalt solutions, we could see how chloroform or carbon tetrachloride did not displace the solution as a plug but instead took off layer after layer starting at the bottom and slowly working its way to the top of the horizontally positioned sample loop. This problem was easily solved by replacing the $\frac{1}{8}$-in. by $\frac{1}{16}$-in. tubing.

CONCLUSIONS

Reasonably good separations can be achieved with over-loaded columns if care is taken to avoid relative viscosities above 2 and to minimize density effects. Secondary exclusion and incompatibility enhance separation but may lead to great errors if calibration curves are used indiscriminately. Secondary exclusion predominates in the low molecular weight range, incompatibility at higher molecular weights. It should be possible to exploit both effects for improving separations on a preparative scale.

Acknowledgment

I thank Dr. M. J. R. Cantow for providing his data on gradient elution chromatography for comparison with the GPC separation shown in Fig. 1.

REFERENCES

1. P. Flodin, *J. Chromatogr.*, **5**, 103 (1961).
2. J. L. Waters, *Polym. Preprints*, **6**, 1061 (1965).

3. K. H. Altgelt and J. C. Moore, *Polymer Fractionation* (M. J. R. Cantow, ed.), Academic, New York, 1967.

4. J. J. van Deemter, F. J. Zuiderweg, and A. Klinkenberg, *Chem. Eng. Sci.,* **5,** 271 (1956).

5. K. H. Altgelt, *Advances in Chromatography,* Vol. 7 (J. C. Giddings and R. A. Keller, eds.), Dekker, New York, 1968.

6. K. H. Altgelt, *Makromol. Chem.,* **88,** 75 (1965).

7. O. L. Harle and K. H. Altgelt, Unpublished Work.

8. K. H. Altgelt, Paper presented at the Pacific Conference on Chemistry and Spectroscopy, San Francisco, November 1968.

9. G. K. Ackers and R. L. Steeve, *Biochim. Biophys. Acta,* **59,** 137 (1962).

10. E. W. Albaugh, P. C. Talarico, B. E. Davis, and R. A. Wirkkala, *Separ. Sci.,* **5,** 801 (1970).

11. A. Dobry and F. Boier-Kawenoki, *J. Polym. Sci.,* **2,** 90 (1947).

12. R. L. Scott, *J. Chem. Phys.,* **17,** 297 (1949).

13. H. Tompa, *Trans. Faraday Soc.,* **45,** 1142 (1949).

14. H. J. Hildebrand and R. L. Scott, *The Solubility of Nonelectrolytes,* Reinhold, New York, 1950.

15. P. J. Flory, *Principles of Polymer Chemistry,* Cornell Univ. Press, Ithaca, New York, 1953.

16. Z. Grubisic, P. Rempp, and H. Benoit, *J. Polym. Sci., Part A-2,* **5,** 753 (1967).

17. A. Ouano and J. A. Biesenberger, *Amer. Chem. Soc. Div. Petrol. Chem. Preprints,* **15**(1), (February, 1970).

18. W. W. Yau, C. P. Malone, and H. L. Suchan, *Separ. Sci.,* **5,** 259 (1970).

Fast Gel Permeation Chromatography*

JAMES N. LITTLE, JAMES L. WATERS, KARL J. BOMBAUGH,
and WILLIAM J. PAUPLIS

WATERS ASSOCIATES, INC.
FRAMINGHAM, MASSACHUSETTS 01701

Summary

A systematic study of the major factors influencing fast analysis by gel permeation chromatography is presented. The study included the effects of (a) solvent flow rate [1–35 ml/min], (b) sample concentration [0.05–0.5%], (c) sample molecular weight [41–411,000 mol wt] and (d) particle size of column packing [10–42 μ]. The effect of the operating temperature at the high flow rates was also investigated.

INTRODUCTION

In a previous paper (1) we reported the effect of several operational parameters on separations by gel permeation chromatography (GPC). This study included the effects of (a) solvent flow rate, (b) sample concentration, (c) sample molecular weight, and (d) particle size of column packing. In the present study, we extended the upper pressure limit of the system to 2200 lb/in.² and evaluated these effects on column performance. The effect of the operating temperature at the high flow rates was also investigated.

EXPERIMENTAL

In these studies a modified Waters Associates Gel Permeation Chromatograph, Model 200—operated at room temperature and 80°C—was used with toluene as the solvent (see Fig. 1). The GPC-200 was

*Presented at the ACS Symposium on Gel Permeation Chromatography sponsored by the Division of Petroleum Chemistry at the 159th National Meeting of the American Chemical Society, Houston, Texas, February, 1970.

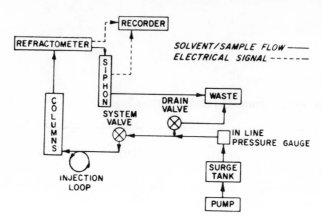

FIG. 1. Modified GPC-200 flow diagram.

modified to operate at an upper pressure limit of 2200 lb/in.^{2} by (a) replacing the standard sample and in-line gauges with 3000 lb/in.^{2} gauges; (b) filling the reference side of the detector with toluene and then closing the reference valve; (c) inserting a tee with two Hoke-needle valves after the surge tank, so that one line leads to the column system and the other to a drain; (d) substituting a variable relief valve (1600–2400 lb/in.^{2}) for the one in the instrument, and (e) disconnecting both injection port valve and the column switching valve. For the fast flow rates (up to 35 ml/min), a Milton-Roy Industrial Mini-pump with ¼ in. diameter plunger was used.

Samples were injected through a Swagelok fitted loop of 4 ft of 0.040-in. i.d. tubing (1 ml volume) inserted into the system ahead of the columns. A high-speed recorder was used for the high flow rates.

When operating at 80°C, an auxiliary preheater was used which consisted of three empty 4 ft × 0.305 in. i.d. stainless steel GPC columns installed in the oven and placed just before the sample injection loop. This ensured that the toluene solvent would be at the operating temperature (80°C) when passing through the analytical test columns. The sample injection loop was mounted outside the oven and connected to the system through two holes in the oven wall.

The procedure for injecting samples was the same as described previously (1). The siphon was not used at the high flow rates, since continuous siphoning occurred. Elution volumes and peak widths of test samples were determined by measuring the flow rate of solvent with a graduated cylinder and relating this to measured distances on the chart paper record.

Two sets of Styragel columns—as described in Table 1—were used in these tests.

TABLE 1

GPC Test Column Characteristics

Test column sets	Number of columns	Column dimensions		Styragel packing	
		Length (feet)	i.d. (inches)	Porosity ($\mathring{A}$)	Particle size (μ)
1	2	4	0.305	1.5×10^5	10–15
2	2	4	0.305	1.5×10^5	37–42

Samples in concentrations of 0.05 and 0.5% (w/v) were prepared by dissolving polystyrene standards (mol wt 411,000 and 19,850), acetonitrile (CH_3CN) and orthodichlorobenzene (ODCB) in toluene. The molecular weight of these test samples ranged from 41 to 411,000. A constant sample volume of 1 ml was used for all test samples.

RESULTS AND DISCUSSION

Effect of Flow Rate on Elution Volume

Elution volume was found to be independent of flow rate from 0.1 to 35 ml/min (see Fig. 2). The data for flow rates up to 12.5 ml/min are taken from our earlier study (1). Thus, the effect of permeation remained invariant to flow for both large and small molecules [polystyrene (mol wt = 411,000) and acetonitrile, respectively]. At the highest flow rate (35 ml/min), there appears to be a slight tendency for the more concentrated samples (0.5%) to be eluted earlier than the dilute samples (0.05%).

The data here and that in our earlier study are in disagreement with the work of Yau et al. (2), who found that elution volumes of the high molecular weight compounds decrease with increasing flow rate. They attributed this to a nonequilibrium distribution of the polymer molecule between the mobile and stationary phases at high flow rates (up to 10 ml/min).

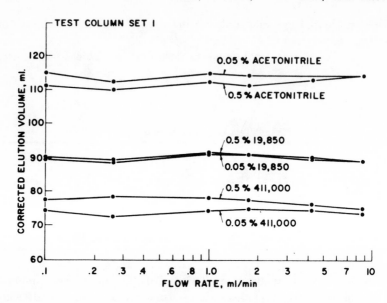

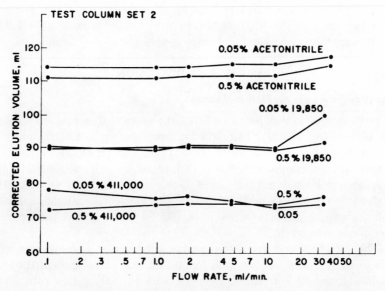

FIG. 2. Elution volume as a function of flow rate and column-packing particle size. (Acetonitrile and polystyrene standards used in all tests under same operating/analytical conditions.)

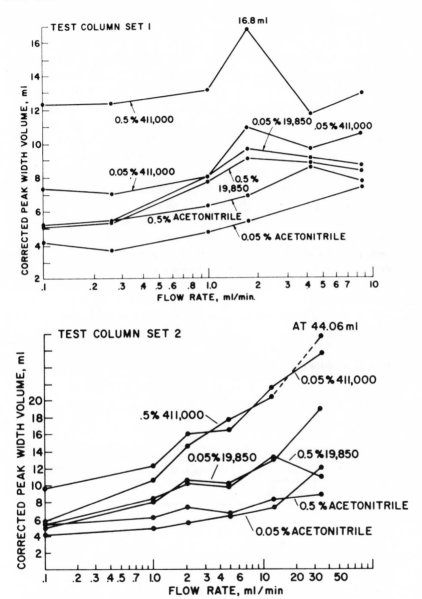

FIG. 3. Peak width as a function of flow rate and column packing particle size. (Acetonitrile and polystyrene standards used in all tests under same operating/analytical conditions.)

Effect of Flow Rate on Peak Width

The increase in peak width with flow rate was found to be less than would be anticipated from the Van Deemter equation (3, 4) (see Fig. 3). With Column Set 1, the peak widths of all samples only doubled, as the flow rate increased 87 times (0.1 to 8.7 ml/min). On Column Set 2, the sample peak widths increased approximately 3–4 times as the flow rate increased 350 times. Since the height equivalent to a theoretical plate (H) is a function of the square of the peak width, the value of H increased 10–16 times as the flow rate increased 350 times. Two of the data points (0.05% 19,850 and 0.05% 411,000) at the 35 ml/min flow rate were calculated from chromatograms with a sloping baseline and may be in error by as much as 20%.

As reported earlier (1), "viscous fingering" of high molecular weight compounds apparently does not occur at high flow rates. At 35 ml/min the high molecular weight compounds at concentrations of 0.5% gave symmetrical peak shapes and without delayed elution.

The results here extend the workable upper pressure limit of the Styragel column packing material to 2200 lb/in.^{2}. At 2200 lb/in.^{2} there was no change observed in column permeability with pressure and no indication that the gel network was collapsing or compacting.

Effect of Flow Rate on Resolution

Mixtures of polystyrene 411,000 and 19,850 were injected at 35 ml/min. The resolution obtained is shown in Table 2 and compared with

TABLE 2

GPC Resolution as a Function of Flow Rate and Sample Concentration (Column Set No. 2 and Same Operating/Analytical Conditions Used Previously)

Test sample[a]	Sample concentration (%)	Flow rate (ml/min)	Resolution
411,000 + 19,850	0.05 + 0.05	1.0	1.75
		5.0	1.33
		12.5	0.99
		35.0	0.66
411,000 + 19,850	0.50 + 0.50	1.0	1.43
		5.0	1.19
		12.5	0.89
		35.0	0.63

[a] Mixture of two polystyrene standards with weight-average molecular weights of 411,000 and 19,850, respectively. Test samples dissolved in toluene.

that found previously at slower flow rates. A 30% loss in resolution was observed when the flow rate was tripled (from 12.5 to 35 ml/min). Also, resolution did not decrease at a faster rate when the more concentrated (0.5%) mixture was used.

Effect of Temperature

Column Sets 1 and 2 were tested at 80°C using flow rates of 8.7 and 35 ml/min, respectively. These flow rates were the maximum attainable at room temperature and 2200 lb/in.^{2}. A comparison of the elution volumes and peak widths of the samples is shown in Table 3. Elution volumes of the test compounds were found to decrease at the higher temperature. The compounds may have a tendency to relax or uncoil at the higher temperature and occupy a larger volume. Therefore, a greater proportion of the pores may exclude the compounds and give earlier elution. Peak width volumes were also found to decrease

TABLE 3

Comparison of Elution Volumes and Peak Widths for Standards at 23 and 80°C

Sample	80°C elution volume	Peak width	23°C elution volume	Peak width
Test Set No. 1, 10–15 μ column set, flow rate 8.7 ml/min				
0.05% ODCB	97.95	5.04	100.98	5.95
0.5% ODCB	98.64	6.47	100.64	6.62
0.05% Acetonitrile	104.48	11.25	114.83	7.69
0.5% Acetonitrile	103.82	7.48	114.47	7.74
0.05% 19,850	87.03	8.97	89.42	8.70
0.5% 19,850	87.43	9.46	89.26	8.36
0.05% 411,000	71.30	10.30	73.88	10.60
0.5% 411,000	74.94	10.70	75.40	12.90
Test Set No. 2, 37–42 μ column set, flow rate 35 ml/min				
.05% ODCB	102.45	7.80	105.00	11.70[a]
0.5% ODCB	102.10	6.74	103.25	8.26
.05% Acetonitrile	109.19	7.09	117.73	12.39[a]
0.5% Acetonitrile	111.67	7.09	115.32	8.95
.05% 19,850	91.46	15.60	101.90	11.02[a]
0.5% 19,850	85.79	15.24	92.26	19.62
.05% 411,000	73.03	24.82	76.42	44.1[a]
0.5% 411,000	69.48	23.11	74.30	25.8

[a] Calculated from a sloping baseline.

TABLE 4

Resolution as a Function of Flow Rate and Temperature on Test
Column Set No. 2

	Elution volume	Peak width volume	Resolution
A. 12.6 ml/min, 23°C, 775 lb/in.²			
Mixture 0.5% 411,000	74.37	22.25	0.89
0.5% 19,850	90.52	14.21	1.88
0.5% Acetonitrile	112.12	8.81	
B. 35 ml/min, 23°C, 2200 lb/in.²			
Mixture 0.5% 411,000	76.77	26.16	.63
0.5% 19,850	92.95	25.47	1.30
0.5% Acetonitrile	115.32	8.95	
C. 35 ml/min, 80°C, 1350 lb/in.²			
Mixture 0.5% 411,000	75.1	26.8	.84
0.5% 19,850	92.0	13.47	1.83
0.5% Acetonitrile	110.2	6.38	

at the higher temperature. This would result from decreased viscosity
of the test compounds and solvent at 80°C, resulting in better mass
transfer.

The combined effects of temperature and flow rate on resolution for
Column Set 2 are shown in Table 4. As shown, resolution dropped 30%
on increasing the flow rate from 12.5 to 35 ml/min. By raising the
temperature from 23 to 80°C, the original resolution is almost achieved
while decreasing the analysis time by a factor of 3. Figure 4 shows
the chromatogram of the mixture run at 35 ml/min and 80°C.

The combined effects of particle size and flow rate on resolution for

TABLE 5

Resolution as a Function of Particle Size and Temperature for Test
Column Sets Nos. 1 and 2

System		Time (min)	R	Pressure (lb/in.²)
A. 37–42 μ, 8.7 ml/min	411,000	12	1.16	400
room temp	19,850			
B. 10–15 μ, 8.7 ml/min	411,000	12	1.56	2200
room temp	19,850			
C. 10–15 μ, 8.7 ml/min	411,000	12	1.66	2200
80°C	19,850			

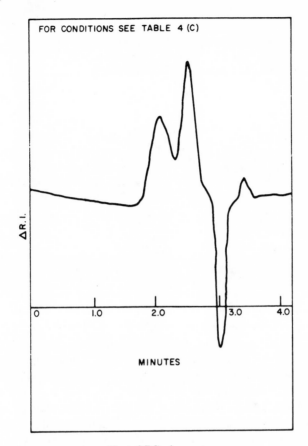

FOR CONDITIONS SEE TABLE 4 (C)

MINUTES

FIG. 4. Fast GPC chromatograms.

both column sets are shown in Table 5. At the same flow rate (i.e., same analysis time) and temperature, Column Set 1 (10–15 μ) gave better resolution (35%), but at the sacrifice of operating at over 5 times the pressure as Column Set 2 (37–42 μ). Only a 6% improvement in resolution was obtained when Column Set 1 was operated at 80°C. The much lower flow rate and smaller particle size of Column Set 1 tends to decrease the magnitude of the temperature effect.

Correlation of Data

The data obtained in this study and in the previous study represents a flow rate range from 0.1–35 ml/min, a particle size range from 10–

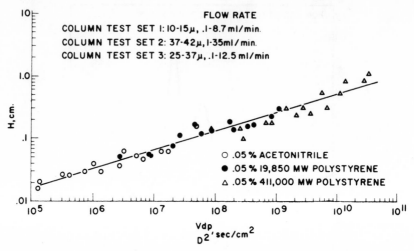

FIG. 5. H as a function of (Vdp/D^2).

$42\ \mu$ and test samples with diffusion coefficients from 10^{-5} to 10^{-7} cm²/ sec. Several theoretical models were considered in an attempt to correlate the data. Coupek and Heitz (5), using poly(vinyl acetate) and low cross-linked polystyrene gels, were able to correlate their data by plotting the reduced plate height (H/dp) versus the reduced velocity (Vdp/D). Our data did not fit this model, as the ordinate (reduced velocity) appeared to be particle size independent. Kelley and Billmeyer (6), using nonpermeating systems, found their data to fit a plot of plate height (H) versus reduced velocity (Vdp/D). A plot of our data as log H vs. log (Vdp/D) showed some correlation (the data formed a band), but a definite dependence on diffusion coefficient could be observed. We found that the square of the diffusion coefficient greatly improved the correlation and a plot of log (H) vs. log (Vdp/D^2) is shown in Fig. 5.

It is interesting to speculate as to why this model describes the data. In the Kelley and Billmeyer model $(H$ vs. $Vdp/D)$, for non permeating systems, the diffusion coefficient is to the first power. In their systems, only diffusion and spreading can occur in the interstitial spaces. In our work, this also occurs, but there is the added effect of permeation into and out of the gel pores. Thus, the data obtained here appears to show that when both processes are present, the square of the diffusion coefficient best describes the data.

CONCLUSIONS

In our studies of operational variables for fast gel permeation chromatography, we have found the following: (a) elution volumes are independent of flow rate for flow rates up to 35 ml/min; (b) for flow rates up to 35 ml/min, peak width volumes were found not to be excessive and were much less than predicted by Van Deemter's equation; (c) the effect of "viscous fingering" decreased at high flow rates resulting in improved peak symmetry; (d) loss of resolution at high flow rates is much less than expected from the Van Deemter equation; (e) increasing the operating temperature decreased peak widths and improved resolution; (f) a new model (H vs. Vdp/D^2) was found to best correlate the data.

Nomenclature

R = resolution
H = height equivalent of a theoretical plate (cm)
v = linear velocity (cm/sec)
dp = particular diameter (cm)
D = diffusion coefficient of sample in solvent (cm²/sec)

REFERENCES

1. J. N. Little, J. L. Waters, K. J. Bombaugh, and W. J. Pauplis, J. Polym. Sci., Part A-2, 7, 1775 (1969).
2. W. W. Yau, H. L. Suchau, and C. P. Malone, J. Polym. Sci., Part A-2, 6, 1349 (1968).
3. J. J. Van Deemter, D. J. Zuiderweg, and A. Klinkenberg, Chem. Eng. Sci., 5, 271 (1957).
4. J. L. Waters, J. N. Little, and D. F. Horgan, J. Chromatogr. Sci., 7, 293 (1969).
5. J. Coupek and W. Heitz, Makromol. Chem., 112, 296 (1968).
6. R. N. Kelley and F. W. Billmeyer, Paper presented at 6th International Seminar on Gel Permeation Chromatography, Miami Beach, 1968.

Extension of GPC Techniques*

G. MEYERHOFF

INSTITUT FÜR PHYSIKALISCHE CHEMIE
UNIVERSITÄT MAINZ
MAINZ, WEST GERMANY

Summary

Most of the common GPC gels exhibit pore sizes which are too small to separate very extended molecules, e.g., native cellulose in solution. Several gel types were tested with vinyl polymers of molecular weights up to 10^7 and with cellulose nitrate. Large pore size Styragel with acetone as solvent proved to be the most favorable gel system yielding effective separations for polymers with coil diameters in solution up to 4000 Å.

The evaluation of GPC runs usually requires a separate calibration procedure. We attempted to determine the molecular weight of the eluate directly as it leaves the column. This is done by a special automatic viscometer that allows measurement of the viscosity of small cuts (e.g., 0.5–1 ml) of the effluent volume. A set of six capillary viscometers are loaded and unloaded continuously. The details of the apparatus are described and examples of the performance reported.

INTRODUCTION

The lecture presented at the GPC Symposium in Houston, February 24, 1970, covered two areas, the search for a gel material of large pore sizes and the development of a viscometric technique allowing continuous determination of the eluate leaving a GPC column.

For the first purpose different gel types were tested with very extended molecules in solution, e.g., native cellulose nitrate and poly-

*Based on a lecture presented at the ACS Symposium on Gel Permeation Chromatography sponsored by the Division of Petroleum Chemistry at the 159th National Meeting of the American Chemical Society, Houston, Texas, February, 1970.

215

methyl methacrylate. The best resolution was obtained with large pore size Styragels of $>5 \times 10^6$ Å using acetone as solvent. Acetone is a nonsolvent for polystyrene. It slightly shrinks the volume of the gel particles compared with tetrahydrofuran. Measurements in acetone therefore require specially packed columns, which can only be used with nonsolvents for polystyrene, the base of Styragels. As a detailed description of the packing procedure and the resolution power of this and other column fillings appeared elsewhere (1), we restrict this paper to a description of the new viscometer system.

Gel permeation chromatography hitherto measured only the amount of polymer contained in the effluent volume and the chemical composition of the polymer. Both variables can be reported continuously by refraction and absorption measurements. For instance, a differential refractometer allows recording of the difference of the refraction indexes of sample and reference fluid as a function of time

$$n = f(t) \tag{1}$$

with additional markings of volumes counts. Since $n \propto c \propto w_{V_e}$ and, for constant flow rate, also $t \propto V_e$, the function

$$u_{V_e} = f(V_e) \tag{2}$$

is experimentally accessible. The term w_{V_e} is the normalized weight fraction of the polymer within the eluted volume V_e.

Since the desired molecular weight distribution is

$$w_M = f(M) \tag{3}$$

one has to establish a special calibration function with narrow samples of known molecular weight of the same polymer type

$$M = f(V_e) \tag{4}$$

which allows to transform w_{V_e} to w_M and V_e to M. Likewise, each single GPC elution curve contains a relation (4). Therefore, the potential of the GPC technique will not be fully realized until one can omit the separate calibration and determine the molecular weight of the polymer directly as a function of the eluted volume.

Recently a continuous method of molecular weight determinations on a flowing solution was given by Cantow (2), who used a light-scattering cell with two distinct angles of observation. This method implies the use of a θ-solvent to avoid extrapolation to zero concentration. The light-scattering apparatus together with a differential refractometer basically allows determination of w and M of solutes in column effluents.

There are now many techniques available which allow estimation of M_n by one of the colligative properties in a matter of minutes. In all of these cases equilibrium must be established, which is scarcely possible with a flowing liquid of changing composition.

A way to determine M by continuous viscometric measurements has been reported by the author (3). The situation seems to be somewhat more favorable than with the light-scattering technique. If η is measured continuously for small cuts $\triangle V_e$, the ratio of the viscosities of the polymer-containing effluent to that of the pure solvent, $\eta_r(V_e)$, can be calculated. The intrinsic viscosity

$$\lim_{c \to 0} \frac{\eta_r - 1}{c} = [\eta] \qquad (5)$$

may be obtained by any of the numerous extrapolation methods from η_r and c, the concentration measured with a differential refractometer. $[\eta]$ is converted to the molecular weight by the relation of Kuhn-Mark-Houwink-Sakurada:

$$[\eta] = KM^a \qquad (6)$$

This is a universal relation and has to be established only once for the elution solvent. It is not restricted to a special apparatus as are Eqs. (1) and (4).

DESCRIPTION OF THE GPC RECORDING UNIT

Usually the effluent from a GPC sample column passes a refraction cell and/or an absorption cell and then enters a siphon. When the siphon is emptied, a mark is produced on the chart recording the refraction difference and/or the absorbance. Since normally the flow rate is not completely constant, the distance from count to count varies slightly. The recorder trace often has to be corrected to obtain precise effluent volumes. Of course the correction can be avoided by applying a constant volume flow, which is possible by special regulating devices. It is more economical to use a drop counter such as that shown in Fig. 1. This schematic shows how the eluate passes the detectors for concentration and chemical composition before it

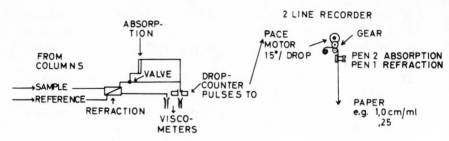

FIG. 1. Schematic diagram of detectors attached to GPC columns. Concentration is measured by a differential refractometer, chemical composition by an additional absorption spectrometer, and volume by a drop counter, which moves the chart drive of the recorder. The details of the viscometer are given in Fig. 2.

leaves the tubing as single drops which interrupt the light barrier in the drop counter and then enter a viscometer. Each drop renders a pulse, which is intensified to activate a pace motor. With our special setup, the motor axis is rotated 15 angular degrees per drop. A gear box slows down the rotary motion, so that the chart paper of the recorder moves, e.g., at a rate of 1.0 or 0.25 cm/ml, as a function of volume instead of time in any case.

The drop size depends on the surface tension, which varies slightly with polymer and polymer concentration. Since the concentration of the polymer in the effluent volume is low, the effect on the drop size and on the recorded volume is smaller than the error resulting from a very good siphon. Aside from this, it is practically impossible to use a siphon for volume counts appreciably smaller than the usual 5 ml, while a drop counter, if operated electronically, allows any reasonable volume counts to be recorded, e.g., 1 ml, 2 ml. A further advantage of a base line divided into volume units is that the evaluation and comparison of GPC runs with different flow rates are easily made.

Additional time counts are marked on the chart edge for control.

DESCRIPTION OF THE AUTOMATIC VISCOMETER

A special viscometer is needed to measure the viscosity of the effluent continuously. Many automatic viscometers which repeatedly measure the flow time of a liquid have been described and are commercially available, but they must be loaded and emptied by hand. Our viscometer has the advantage of automatic filling and unloading besides having automatic timing.

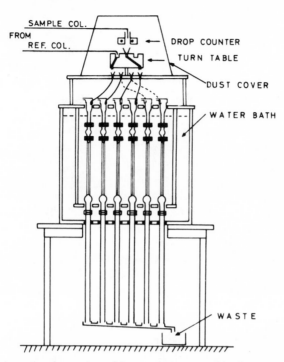

FIG. 2. Automatic viscometer with rotating turn table for consecutive filling of six viscometers. For details see text.

This is done with the arrangement demonstrated in Fig. 2. Six viscometers of free-flowing type are placed side by side in a thermostated water bath ($\Delta T < 0.003°C$). The volume of the viscometers is 0.5 ml; the capillary has a diameter of 0.03 cm and is bent backwards (below the plane of the drawing), thus accommodating a length of 17 cm and a medium driving height of 14 cm. The flow time for THF is >135 sec, while the filling time at a flow rate of 1 ml/min is less than 45 sec. This permitted omission of the valves at the outlet of each viscometer, which were previously described (1). The Teflon valves had been operated by two rotary magnets. They had been closed for filling only, but they had exhibited mechanical shock and leakage problems.

The timing of the six viscometers is performed by light barriers measuring the decrease of the intensity due to the falling meniscus. Two phototransistors start and stop an electronic counter for each viscometer. The counters' digits are printed out together with the number of the viscometer being timed and the hours and minutes of daytime. If the flow exceeds 200 sec, the capillary of the viscometer is likely to be partially blocked. Then an alarm is given and the viscometer must be replaced by a spare viscometer.

The liquid which has passed the capillary leaves the viscometer via a stainless steel tube connected to its glass outlet. The stainless steel tube is placed in a boring of the thermostat bottom and tightened by a rubber bearing, which allows the raising the viscometer until the glass steel connection is above the meniscus of the water bath. In this manner each viscometer can be separatively exchanged.

For the consecutive filling of the six viscometers, a rotating turntable is used. It is moved from 30° to 30° by a pace motor not shown in Fig. 2. The sample fluid, after passing a drop counter, enters a smaller funnel from which it runs down a stainless steel tube, while the reference fluid enters a circular groove of the turntable and flows down another tube. The exits of these tubings let the liquids enter two funnels from which they are led to the viscometer entrances through a small cap. Six of these funnels are arranged on a circle at 60° intervals. By this method two measuring rhythms are possible. If the viscometers are filled alternately with reference and sample fluid, η_r is measured with 2 ml intervals. If the reference line is not used, the viscometers are filled with the sample fluid only; and η_r can be determined in 1 ml intervals as demonstrated in Figs. 3 through 5.

The turntable is triggered by the meniscus of the liquid when it passes the upper light barrier of a viscometer. It moves with a time lag of 5 sec by 30°, placing the ends of the fluid lines between two funnels. About 5 sec after that, the counter is started. The rotary movement to the next full 60° position of the turntable, regulated by a clock, always brings the ending of the central funnel to the next viscometer at each full minute. Thus, the stepwise rotation is alternately controlled by the upper light barrier and an electronic clock. The 30° positions are necessary to avoid any overfilling of the viscometers. The 60° positions are enforced by a safeguard, so that an incomplete filling of one viscometer will not ruin the correct rhythm.

The drop counter, the turntable, and the viscometer entrances are placed below a dust cover, which also maintains a solvent saturated vapor around the drops.

TEST RUNS WITH POLYSTYRENE

For our test runs, 3 Waters Styragel columns $>5 \times 10^6$, 7×10^5, and 10^4 were used with tetrahydrofuran as a solvent. The filling loop contained 1.96 ml. Standard Polystyrene 1,800,000 from Pressure Chemical Company, Pittsburgh, was used with concentrations ranging from 1 to 4 mg/ml. In Fig. 3 both the concentration and η_r, the

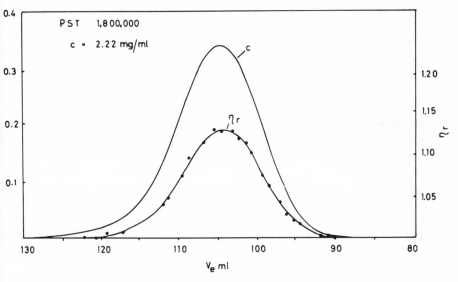

FIG. 3. Concentration and relative viscosity of a polystyrene with $M = 1,800,000$ after passing three Styragel columns with THF as produced by a differential refractometer and an automatic viscometer as a function of the effluent volume.

latter indicated by small open circles, are shown as a function of the effluent volume for a solution of $c = 2.22$ mg/ml. The η_r scale given at the right ordinate follows directly from the printed times of our tape. For the determination of the correct scale of the concentration, some precaution is necessary.

DETERMINATION OF CONCENTRATION

Usually it is sufficient to plot the difference of the refraction index and, hence, the concentration in arbitrary units. But for the calculation of $(\eta_r-1)/c$, it is essential to know the correct absolute value of the concentration. In this case it is not permissible to calibrate the differential refractometer in the usual way with static fillings of polymer solutions of known concentration. After passing a GPC column, the solvent contains small amounts of water and other low molecular weight impurities which exhibit a pronounced effect on dn/dc. This effect is as large as that due to negative and positive tails of a GPC run. Therefore, we measured the total area below the concentration line, which corresponds to (2.22 mg/ml) (1.96 ml) = 4.34 mg of polymer. This, together with the total volume over which the polymer is distributed, allows us to calculate the concentration scale given at the left ordinate. For this method it is essential to have a correct volume base line as produced by our drop counter-step motor combination.

The same column set and polymer were tested at three more concentrations. Figure 4 shows the recorder traces from the differential refractometer for $c = 4, 3, 2,$ and 1 mg/ml. The concentrations in the eluates ranged up to 0.47 mg/ml for the highest and up to 0.17 mg/ml for the lowest injected sample concentration. Concentrations >4 mg/ml exhibited an overload effect, characterized by a delayed second peak, viz., near $V_e = 117$ ml. The second peak starts to show up slightly at $c = 4$ mg/ml.

The areas under the peaks of Fig. 3 proved to be proportional to the concentration. This was not exactly the case when the usual timing method for the injection of different volumes was applied. Very

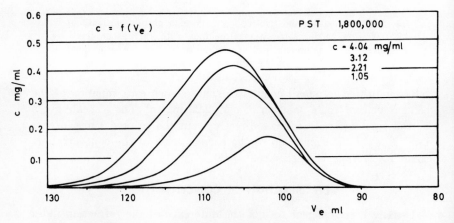

FIG. 4. Recorder traces of the differential refractometer for four different concentrations of polystyrene as a function of the effluent volume.

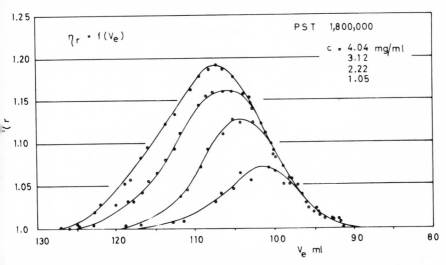

FIG. 5. Relative viscosities as a function of the effluent volume corresponding to the concentration curves given in Fig. 4.

likely some part of the polymer remains at the walls of the filling loop.
It is therefore preferable to use separate filling loops of the proper
volume to inject smaller volumes of solution (at Mainz 0.25, 0.5, 1.0,
and 2.0 ml). The injection valve stays open for at least the threefold
loop volume.

In Figure 5 the relative viscosities of the runs of Figure 4 are represented by open circles and connected by smoothed curves. Naturally
the deviations are more pronounced for the lowest concentrations.

CONCLUSIONS

Our measurements reveal the feasibility of direct determinations of
molecular weight in the eluates of a GPC apparatus by consecutive
viscosity measurements. Since the viscosity of a solution increases
with molecular weight and concentration, higher concentrations are
needed for polymers of lower molecular weight. As a rough measure,
the concentration may be increased inversely to the intrinsic viscosity
if two samples of similar distribution but different molecular weight
are tested.

The technique described here looks quite promising for further
automation of GPC and its evaluation procedures.

REFERENCES

1. G. Meyerhoff, *Makromol. Chem.*, **134**, 129 (1970).
2. H. J. Cantow, E. Seifert, and R. Kuhn, *Chem.-Ing., Tech.*, **38**, 1032 (1966).
3. G. Meyerhoff, *Makromol. Chem.*, **118**, 265 (1968).

Phase Distribution Chromatography (PDC) of Polystyrene*

R. H. CASPER and G. V. SCHULZ

INSTITUT FÜR PHYSIKALISCHE CHEMIE DER UNIVERSITÄT MAINZ
MAINZ, WEST GERMANY

Summary

Phase distribution chromatography is a new chromatographic method of fractionating polymers. The separation is achieved by partitioning the sample between the solvent and a polymer phase, in the case a noncrosslinked polystyrene of very high molecular weight, which is coated as a thin layer on small glass beads. The temperature must be held below the θ-temperature of the sample. The separation increases sharply with decreasing temperature. The theory of this method is presented and compared with experimental results. The agreement is very good.

INTRODUCTION

In ordinary gel permeation chromatography, the elution volume increases with decreasing sample molecular weight. It is easy to modify this method in such a way that this relation is inverted. For this purpose, only the cross-linking of the gel phase must be omitted. Then the high molecular species dissolve in the gel phase to a higher extent than those of lower molecular weight and therefore leave the column later. The theory of this effect seems to be simpler than that of GPC, it is similar to the Baker-Williams method (1, 2), but again considerably simpler.

EXPERIMENTAL

The main parts of the apparatus are sketched in Fig. 1. The sample column (a glass tube 100×3 cm) was filled with glass

* Presented at the ACS Symposium on Gel Permeation Chromatography sponsored by the Division of Petroleum Chemistry at the 159th National Meeting of the American Chemical Society, Houston, Texas, February, 1970.

225

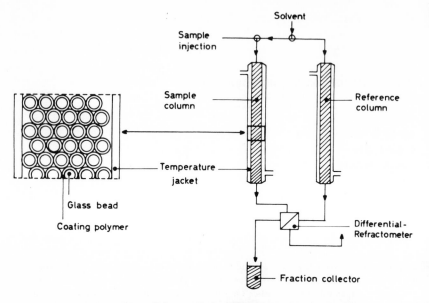

FIG. 1. Scheme of the PDC apparatus.

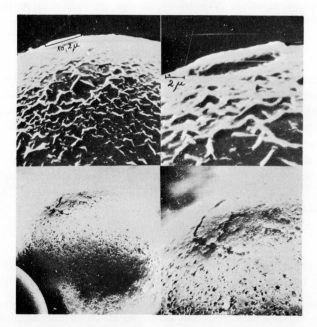

FIG. 2. Surface of the glass beads coated by PS (top) and uncoated (bottom).

beads (ballotines, diameter 0.1 mm) which were covered with a layer of a well-fractionated tritium-labeled polystyrene of very high molecular weight $(M_w = 8 \times 10^6)$. The coating polymer was practically insoluble in the solvent used in our experiments (cyclohexane below the θ-temperature of 35°C. The layer of the swollen polymer had a thickness of nearly 3000–4000 Å. It was extracted at 28°C until the solvent leaving the column contained less than 1 mg/liter (as measured by radioactivity). An electron micrograph of the coated and uncoated glass beads is shown in Fig. 2.

The separation efficiency of the column and its dependence on temperature is demonstrated by the experiments represented in Fig. 3.

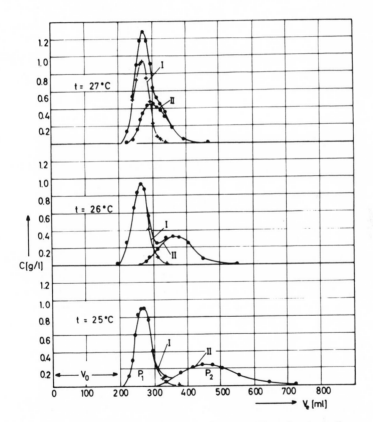

FIG. 3. Elution curve of a 1:1 mixture of two samples of anionically prepared polystyrenes; $\bar{M}_w = 135,000$ (I) and 415,000 (II) (300 mg in 20 ml cyclohexane).

A 1:1 mixture of two samples of anionically prepared PS was dissolved in 20 ml of cyclohexane and passed through the column with a flow rate of 10 ml/hr. The sample with the molecular weight 4.15×10^5 was labeled with tritium; thus, the amount of this polymer could be determined in each collected fraction. One can see that with decreasing temperature, the separation efficiency increases sharply as the elution volumes of the higher molecular weight samples become larger while those of low molecular weight remain nearly constant. Some preliminary experiments were published elsewhere (3).

THEORY OF THE METHOD

The elution volume V_e is controlled by the distribution coefficient $K(P)$ which defines the distribution of a species with the degree of polymerization P between the two phases

$$K(P) = \Phi_{gel}/\Phi_{sol} \qquad (1)$$

Φ_{gel} and Φ_{sol} are the volume fractions of the sample polymer P in the two phases. The ratio of the volumes V_{sol} and V_{gel} of the phases in the column is

$$r_v = V_{sol}/V_{gel} \qquad (2)$$

Applying the usual transport equation for a chromatographic process, we obtain

$$V_e = V_e^0[1 + K(P)/r_v] \qquad (3a)$$

and

$$K(P) = r_v(V_e/V_e^0 - 1) \qquad (3b)$$

where V_e^0 is the void volume of the column.

With these equations, one can predict the elution volume if the distribution coefficient is known from thermodynamic calculations or, in turn, one can calculate the distribution coefficient from the measured elution volume. The required condition of working close to equilibrium can be easily controlled by varying the flow rate. Equilibrium is established if the elution volume does not depend on the flow rate. This condition was maintained in our experiments.

Elution experiments were performed with some anionically prepared polystyrene samples of narrow molecular weight distributions in order to determine experimentally the relation of the distribution coefficient with the degree of polymerization and with temperature. The DP covered the range from 300 up to 4000, and the

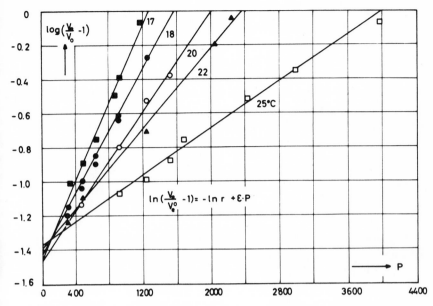

FIG. 4. Elution experiments Log $(V_e/V_e^0 - 1)$ as a function of the degree of polymerization P at different temperatures. Polystyrene in cyclohexane below the θ-temperature (35°C).

temperature was 10–18°C below the θ-temperature (35°C). Figure 4 shows that the relation between the elution volume V_e and the degree of polymerization P can be described by

$$\log [(V_e/V_e^0) - 1] = -a + bP \qquad (4)$$

Comparison of Eq. (4) with Eq. (3b) reveals that

$$a = \log r_v \qquad (5)$$

and

$$\log K(P) = bP \qquad (6a)$$

On the average, a has a value of 1.4; hence, $r_v \approx 30$. This agrees satisfactorily with independent swelling experiments performed with the high molecular polystyrene.

Equation (6a) is identical with the following relation which was first introduced by Brönsted (4)

$$K(P) = \exp (\epsilon P) \qquad (6b)$$

Equation (6b) had previously been used by one of us to evaluate experimental results of polymer fractionation in terms of its mechanism $(5, 6)$. A theoretical treatment of the Baker-Williams column was based $(1, 2)$ on the same equation. This treatment could be experimentally confirmed (2).

We call $v_{p,s}$ the volume of the polymer solution which is injected at the top of the column. The solution drains into the interstitial volume between the glass beads and comes in contact with a certain amount of the stationary gel layer of the volume $v_{p,g}$. Obviously the total volume which contains the sample is given by

$$v_p = v_{p,s} + v_{p,g} \qquad (7)$$

The composition of the two phases can approximately be described by a quaternary system consisting of the solvent and three polymer components with the molecular weights M_1, M_2, and M_3. The "polydisperse" low molecular weight polymer mixture is represented by M_1 and M_2, and the high molecular weight component by M_3. The "molecular weight distribution" of the low molecular weight mixture can be characterized by the ratio of mass m_1/m_2. Its total amount is $m_1 + m_2$.

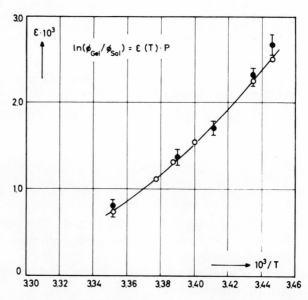

FIG. 5. Brönsted coefficient ϵ in Eq (6b) as a function of the temperature: $\bigcirc$, calculated according to Koningsveld; $\bullet$, experimental points (slope of elution curves in Fig. 4).

The dependence of the Huggins interaction coefficient χ on the temperature and on the concentration in the system cyclohexane polystyrene was measured by Koningsveld (7) and Rehage (9–11). Assuming that phase equilibrium has been established, the composition of the adjacent phases as well as the distribution coefficients for M_1, M_2, and M_3 can then be calculated on the basis of the Flory-Huggins relation for the free enthalpy of mixing of polymer solutions (8). The good agreement of measured and calculated ϵ values, as shown in Fig. 5, proves that our theory of the PDC method is complete and correct.

On this theoretical basis, model calculations were made, the results of which are listed in Tables 1–3.

Three conclusions can be drawn from Table 1:

(a) The high molecular sample of polystyrene which forms the gel phase 10°C below the θ-temperature is practically insoluble in the sol phase.

(b) The distribution coefficients $K(P)$ of the two lower samples (P_1 and P_2) of polystyrene are completely independent from their concentrations and also with good approximation from the ratio of their masses.

(c) At lower concentrations the elution volumes of the different

TABLE 1

Distribution Coefficient, Elution Volume, and the Ratio of Phase Volumes $r_v = v_{p,s}/v_{p,g}$ Calculated as a Function of the Mass of Three Samples of Polystyrene with Various Degrees of Polymerization, $P_1 = 1,300$; $P_2 = 4,000$; $P_3 = 80,000$. Total Interstitial Volume ca. 250 ml; Temperature 25°C. $v_{p,s} = 20.0$ ml; $m_3 = 124.5$ mg

m (mg)		$K(P) = \Phi_{sol}/\Phi_{gel}$			V_e/V_e^0		
1	2	1	2	3	1	2	r_v
0.5	0.5	0.380	0.0512	1.5×10^{-26}	1.093	1.690	28.3
5.0	5.0	0.380	0.0512	1.5×10^{-26}	1.096	1.711	27.5
50.0	50.0	0.380	0.0512	1.5×10^{-26}	1.119	1.889	22.0
500.0	500.0	0.381	0.0512	1.5×10^{-26}	1.178	2.319	14.8
1.0	9.0	0.379	0.0506	1.2×10^{-26}	1.097	1.727	27.2
3.0	7.0	0.380	0.0509	1.3×10^{-26}	1.096	1.720	27.3
5.0	5.0	0.381	0.0511	1.5×10^{-26}	1.096	1.711	27.5
9.0	1.0	0.382	0.517	1.8×10^{-26}	1.094	1.696	27.8

TABLE 2

Distribution Coefficients and Elution Volume as a Function of the Degree of
Polymerization at 25°C (Calculated for a Quaternary System).
$m_1 = m_2 = 5$ mg; $m_3 = 124.5$ mg; $v_{p,s} = 20$ ml

Degree of polymerization			Φ_{sol}/Φ_{gel}			V_e/V_e^0	
1	2	3	1	2	3	1	2
500	4000	80,000	0.69	0.0512	1.6×10^{-26}	1.05	1.71
1,300	4000	80,000	0.38	0.0511	1.5×10^{-26}	1.10	1.71
2,500	4000	80,000	0.16	0.0509	1.4×10^{-26}	1.24	1.72
3,500	4000	80,000	0.07	0.0506	1.2×10^{-26}	1.50	1.73
11,000	4000	80,000	3.0×10^{-4}	0.0502	1.0×10^{-26}	142.1	1.75

species are independent from each other. At higher concentrations
the elution volume is more or less influenced by changes of the relation r_v caused by the local enlargement of the gel phase.

In Table 2 the molecular weight of Component P_1 was varied up to
a factor of 22. One sees that the mutual influence of the components
at the conditions given here is very small and that neighboring degrees of polymerization (e.g., 3500 and 4000) are well separable. The
large difference of the distribution coefficients of the samples with DP
of 11,000 and 80,000, respectively, suggests that a polymer can be
used as the gel-forming component in the PDC column for a sample,
the molecular weight of which is smaller by less than one order of
magnitude.

Table 3 shows the very large influence of temperature.

TABLE 3

Calculated Distribution Coefficients of Three Degrees of Polymerization as a
Function of Temperature ($\theta = 35$°C)

T (°C)	Φ_{sol}/Φ_{gel}			V_e/V_e^0	
	DP = 500	4000	80,000	500	4000
17	0.28	4.3×10^{-5}	0	1.085	566
21	0.46	2.2×10^{-3}	0	1.061	17.97
23	0.57	1.2×10^{-2}	10^{-39}	1.055	3.684
25	0.69	5.1×10^{-2}	10^{-26}	1.053	1.709

CONCLUSIONS

The PDC method has two aspects: it allows determination of the thermodynamic parameters, e.g., distribution coefficients in phase equilibria. On the other hand, it is a new method for the determination of molecular weight distributions. It seems that the fractionating efficiency for polymers can be considerably improved if one makes use of a programmed temperature increase. Experiments in this direction are under way.

Acknowledgment

This work was supported by the Arbeitsgemeinschaft industrieller Forschungsvereinigungen (AIF).

REFERENCES

1. G. V. Schulz, P. Deussen, and A. G. R. Scholz, *Makromol. Chem.*, **69**, 47 (1963).
2. G. V. Schulz, K. C. Berger, and A. G. R. Scholz, *Ber. Bunsenges. Phys. Chem.*, **69**, 856 (1965).
3. R. H. Casper and G. V. Schulz, *J. Polym. Sci., Part A-2*, **8**, 833 (1970).
4. J. N. Brönsted, *Z. Phys. Chem. (A)*, **155**, 343 (1931); **168**, 381 (1934).
5. G. V. Schulz, *Z. Phys. Chem. (B)*, **46**, 137 (1940); **47**, 155 (1940).
6. G. V. Schulz and E. Nordt, *J. Prakt. Chem.*, **155**, 115 (1940).
7. R. Koningsveld and A. Staverman, *J. Polym. Sci., Part C*, **16**, 1775 (1967); *Part A-2*, **6**, 305, 325, 349, 367, 383 (1968).
8. R. H. Casper, Dissertation, University of Mainz, 1970.
9. G. Rehage and R. Koningsveld, *J. Polym. Sci., Part B*, **6**, 421 (1968).
10. G. Rehage and D. Möller, *J. Polym. Sci.*, **16**, 1787 (1967).
11. G. Rehage, J. Palmen, D. Möller, and W. Wefers, IUPAC Symposium on Macromolecules, Toronto, 1968.

Apparent and Real Distribution in GPC (Experiments with PMMA Samples)*

K. C. BERGER AND G. V. SCHULZ

INSTITUT FÜR PHYSIKALISCHE CHEMIE DER UNIVERSITÄT MAINZ
MAINZ, WEST GERMANY

Summary

Molecular weight distribution curves obtained by GPC are broadened if concentration and flow rate are fixed in the usual range. Therefore, the apparent nonuniformity U_{app} of the samples is larger than the real nonuniformity $U = (M_w/M_n) - 1$. For a number of fractionated and unfractionated samples of polymethyl methacrylate we determined M_n and M_w by osmotic, light-scattering, and viscosity measurements. Thus, the real value of U can be compared to U_{app} obtained by GPC at different concentrations and flow rates, v. The excess nonuniformity U_{exc} is evaluated as function of concentration c, polydispersity, molecular weight, and flow rate. For $c = 0$ and $v = 0$, U_{exc} is not far from zero. For standard conditions one certain value of the excess standard deviation of the elution volume allow calculation of U_{exc} for narrow and broader distributions and the obtaining of nearly correct values for the real nonuniformity U.

INTRODUCTION

The aim of this contribution is to determine experimentally the broadening effect of a GPC column. The best method for this purpose would be to pump samples with known molecular weight distribution through the column and to compare the uncorrected distribution curves with the known distributions. Unfortunately, there are no polymer samples for which the real distribution is precisely known. Therefore, we propose a simplified procedure which may be regarded as an approach.

* Presented at the ACS Symposium on Gel Permeation Chromatography sponsored by the Division of Petroleum Chemistry at the 159th National Meeting of the American Chemical Society, Houston, Texas, February, 1970.

235

The average values $\bar{M}_n = M_0\bar{P}_n$ and $\bar{M}_w = M_0\bar{P}_w$ of given samples are directly measured by osmotic pressure and light scattering or viscometry, respectively. On the other hand, these values of the samples are calculated from the elution curves by the so-called strip method: The elution curve is divided into small strips and, by using a calibration curve, the molecular weight averages are calculated according to the equations

$$\bar{P}_n = 1/\sum_i (m_i/P_i) \tag{1a}$$

and

$$\bar{P}_w = \sum_i (m_iP_i) \tag{1b}$$

m_i and P_i are the weight fraction and the degree of polymerization (DP) of the ith fraction, respectively. Comparison of the different values of $\bar{P}_n$ and $\bar{P}_w$ obtained by the two methods yields a direct measure of the broadening effect. For a quantitative comparison we introduce the nonuniformity (Uneinheitlichkeit) defined by (1)

$$U = (\bar{P}_w/\bar{P}_n) - 1 \tag{2}$$

Table 1 shows the directly measured values of $\bar{P}_n$ and $\bar{P}_w$ of 4 fractions and 1 unfractionated polymethyl methacrylate obtained by radical polymerization. The real values of the unfractionated sample, Polymer III, may be a little higher than listed in Table 1 because the osmotically determined value of $\bar{M}_n$ is too high, though by no more than 5–10%. Other samples of PMMA (see Table 2) were fractionated by the Baker-Williams technique in combination with the GPC technique. They have nonuniformity values between 0.95 and 1.00.

TABLE 1

Nonuniformity Data of PMMA Samples, as Determined by Osmotic Pressure ($\bar{P}_n$), Light-Scattering, and Viscosity Measurements ($\bar{P}_w$)

Sample	$\bar{P}_n \times 10^3$	$\bar{P}_w \times 10^3$	U, Eq. (2)
C.1 (fraction)	1.74	2.33	0.34
C.2 (fraction)	1.98	2.40	0.21
B.1 (fraction)	3.54	4.46	0.26
B.2 (fraction)	5.47	5.96	0.09
III (unfractionated)	1.73	3.25	0.9

FORMAL RELATIONS

If the experimental elution curve is Gaussian with the variance σ_v and the calibration curve is given by

$$\ln P = A - BV_e \tag{3}$$

(V_e = elution volume), the molecular weights follow a log-normal distribution with the variance $\sigma_P = B\sigma_v$. In this case U is given by (2)

$$U = \exp(\sigma_P^2) - 1 \tag{4}$$

The direct measurements of $\bar{P}_n$ and $\bar{P}_w$ allow calculation of the *real* values of U and σ_P according to Eqs. (1), (2), and (4). The strip method gives an *apparent* value U_{app}

$$U_{app} = \exp(B\sigma_v)^2 - 1 \tag{5}$$

which includes the broadening effect of the column.

Let $\sigma_{v,0}$ be the variance of the elution curve corresponding to the molecular weight distribution and $\Delta\sigma_v$ the contribution of the broadening effect of the column (axial dispersion), then

$$\sigma_{v,0}^2 = \sigma_v^2 - \Delta\sigma_v^2 \quad \text{and} \quad \sigma_P = B\sigma_{v,0} \tag{6a,b}$$

It follows from Eqs. (4), (5), and (6) that

$$\ln(1 + U_{app}) = \ln(1 + U) + (B\Delta\sigma_v)^2 \tag{7}$$

One can assume that the treatment of experimental errors as proposed here for Gaussian distributions is at least approximately correct for other distribution functions (3). To prove that assumption, the following experiments were performed.

EXPERIMENTS

We used a Waters GPC apparatus equipped with the following set of columns: 10^6, 10^5, 2×10^4, and 10^3. Figure 1 shows the calibration curves of 3 fractions with approximately the same values of U. Each sample was run at 5 concentrations. Figure 1 shows the considerable effect of the concentration on the position of the elution curves. Obviously, to obtain accurate values for the DP, the concentration must be standardized. It will be shown later that it is best to choose zero concentration.

As can be seen from Fig. 2, the elution curves of the fractions are much more sensitive to changes in concentration than those of the un-

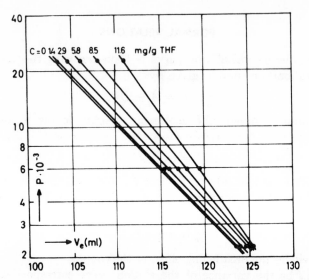

FIG. 1. Concentration dependence of calibration curves (3 fractions of PMMA).

fractionated samples. Obviously, due to their narrow distribution, the fractions migrate through the columns with comparatively high concentrations. They undergo greater dispersion which results in a wider spread of the elution curve. Therefore the concentration, c_m, at the maximum of the elution curve must be taken into account. Moreover, one can see from Fig. 2 that at higher concentrations the maximum of the elution curve is shifted to higher elution volumes and the curves are broadened. To obtain well defined values for both DP and U, an extrapolation is necessary.

A linear extrapolation of the DP (maxima of the curves) to $c = 0$ can be achieved by plotting log DP vs. concentration. The slope of the straight lines is a function of DP and U.

Figure 3 shows that the plot of $\log (1 + U_{app})$ vs. c^2 gives straight lines down to zero concentration. This plot corresponds to the equation

$$\log (1 + U_{app}) = \log (1 + U_{app}^0) + k_u c^2 \tag{8}$$

where k_u depends on the DP and the nonuniformity. It increases with increasing DP and decreasing U.

Comparing Eqs. (8) and (7), one could assume that the value U_{app}^0 is identical with U and that $\Delta\sigma_v$ is proportional to c. As Table 2 shows, this is not true. The extrapolated values U_{app}^0 are higher than the directly

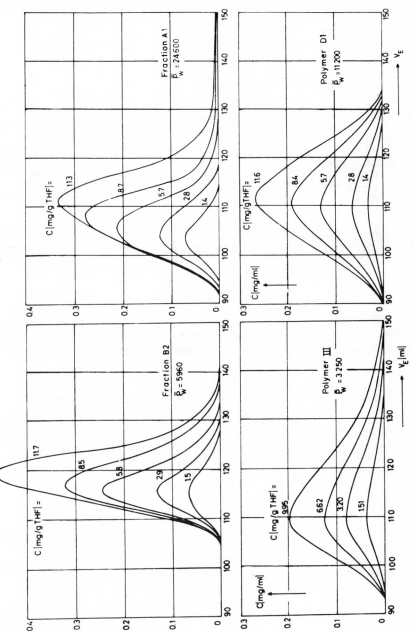

FIG. 2. Elution curves for 2 fractions and 2 unfractionated polymers of PMMA.

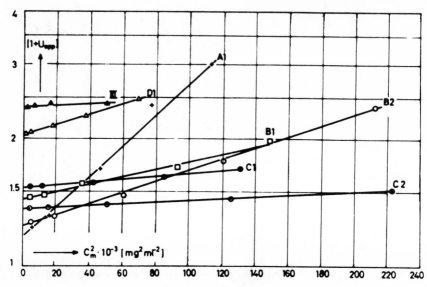

FIG. 3. Nonuniformity as a function of concentration.

measured values. However, the additional variance $\Delta\sigma_v$ is approximately constant regardless of the nonuniformity and of the molecular weight. Thus it is possible to determine the broadening effect of a combination of columns if some samples of known nonuniformity are available.

TABLE 2

Comparison of U^0_{app} (GPC) and U (determined by light scattering and osmotic pressure)

Sample	$\bar{P}_w \times 10^3$	U	U^0_{app}	$\Delta\sigma_v$ (ml)
C.1 (fraction)	2.33	0.34	0.53	3.08
C.2 (fraction)	2.40	0.21	0.37	3.19
B.1 (fraction)	4.46	0.26	0.44	3.19
B.2 (fraction)	5.96	0.09	0.25	3.29
				$\Delta\sigma_v = 3.2$
III.1 (polymer)	3.25	0.95	1.36	3.85
Mo.1 (polymer)	4.06	0.95	1.36	3.68
Mo.2 (polymer)	6.99	0.95	1.26	3.40
II.1 (polymer)	11.26	0.95	1.28	3.49
I.1 (polymer)	21.03	0.95	1.29	3.49
				$\Delta\sigma_q = 3.6$

The additional variance $\Delta\sigma_v$ comes close to zero by extrapolating the flow rate to zero as some preliminary experiments suggest (4). It seems that the extrapolated value $U_{app\ c} = 0/$flow rate $= 0$ is not far from the correct value of U.

A more detailed report has been published in *Die Makromolekulare Chemie* (5).

Acknowledgment

This work was supported by the Arbeitsgemeinschaft industrieller Forschungsvereinigungen (AIF).

REFERENCES

1. G. V. Schulz, *Z. Phys. Chem.* (B), **43**, 25 (1939).
2. H. Wesslau, *Monatsh. Chem.*, **20**, 111 (1956).
3. L. H. Tung, *J. Appl. Polym. Sci.*, **10**, 375 (1966).
4. G. Meyerhoff, *J. Polym. Sci.*, Part C, **21**, 67 (1968). W. W. Yau, H. L. Suchan, and C. P. Malone, *J. Polym. Sci.*, Part A-2, **6**, 1349 (1968). W. W. Yau, *J. Polym. Sci.*, Part A-2, **7**, 483 (1969).
5. K. C. Berger and G. V. Schulz, *Makromol. Chem.*, **136**, 221 (1970).

The Instrument Spreading Correction in GPC. I. The General Shape Function Using a Linear Calibration Curve *

THEODORE PROVDER† and EDWARD M. ROSEN

MONSANTO COMPANY
ST. LOUIS, MISSOURI 63166

Summary

A general shape function is proposed for describing the instrumental spreading behavior in gel permeation chromatography (GPC) columns due to axial dispersion and skewing effects. The general shape function contains statistical coefficients which describe the axial dispersion, skewing, and flattening of ideal monodisperse standards. A method denoted as the "method of molecular weight averages" is used to derive equations to correct GPC number- and weight-average molecular weights and intrinsic viscosities calculated from linear molecular weight calibration curves. The validity of these equations is experimentally verified with data for polystyrene, polybutadiene, and polyvinyl chloride polymers in tetrahydrofuran. The physical significance of the correction equations and their statistical coefficients is discussed in relation to the observed GPC chromatograms. Application of this shape function to the numerical Fourier analysis method for correcting differential molecular weight distribution (DMWD) curves is outlined. Also, a method is presented for obtaining corrected DMWD curves from a fitted molecular weight calibration curve corrected for instrument spreading by use of the hydrodynamic volume concept.

INTRODUCTION

Molecular weight distribution (MWD) curves calculated from gel permeation chromatograms are generally broader than the true or

* Presented at the ACS Symposium on Gel Permeation Chromatography, sponsored by the Division of Petroleum Chemistry at the 159th National Meeting of the American Chemical Society, Houston, Texas, February, 1970.

† Author to whom all correspondence should be addressed at: SCM Corporation, Glidden-Durkee Division, Dwight P. Joyce Research Center, 16651 Sprague Road, Strongsville, Ohio 44136.

243

absolute MWD curves due to instrumental spreading of the experimental chromatogram. Thus, the molecular weight averages calculated from the experimental chromatograms can be significantly different than the absolute molecular weight averages. The instrument spreading in gel permeation chromatography (GPC) has been attributed to axial dispersion (1) and skewing (2) effects. Several computational procedures (1, 3–10) have been reported in the literature to correct for these effects. In each method a specific shape for the chromatogram of an ideal monodisperse species or narrow MWD sample is assumed. The use of a Gaussian shape function as in Tung's (1) basic method may be adequate when skewing effects are absent. The methods of Hess and Kratz (4), Smith (3), and Pickett, Cantow, and Johnson (5) attempt to correct for skewing effects. Duerksen and Hamielec (2) have shown that these methods are inadequate where overloading and interaction between species occur, particularly under conditions of high concentrations, high flow rate, and at high molecular weights. In addition these methods were subject to oscillations in the corrected chromatograms and differential molecular weight distribution curves (DMWD) because of the problem of distinguishing between the noise and the data in the chromatogram, particularly at the tails of the chromatogram.

Recently, Balke and Hamielec (11) used the number- and weight-average molecular weights to obtain empirical skewing operators to successfully correct molecular weight averages obtained from skewed chromatograms. Also, recently Tung (12) has illustrated the usefulness of the Fourier transform method for correcting observed chromatograms with a Gaussian instrument spreading function. In this paper a general statistical shape function is proposed for describing the instrumental spreading behavior in GPC columns and is applied to linear calibration curves by a method denoted as the "method of molecular weight averages." The use of this function with the Fourier transform method is touched on briefly and will be discussed extensively in Part II of this series. Also, a method is presented for obtaining corrected DMWD curves from a fitted calibration curve which has been corrected for instrument spreading by use of the hydrodynamic volume concept.

EXPERIMENTAL

Gel Permeation Chromatography

A Waters Associates Model 200 Gel Permeation Chromatograph, fitted with five Styragel columns having nominal porosity designations of 10^6, 10^5, 10^4, 10^3, and 250 Å, was used for the analysis of molecular weight distributions. The average plate count of the columns at a solvent flow rate of 1 ml/min was 734 plates/foot with o-dichlorobenzene (ODCB). The columns were operated at room temperature, 24 ± 1°C, with Fisher Scientific Co. certified reagent grade THF (n_D^{25}, 0.888; bp, 64–66°C) used as the eluting solvent. The solvent contained 0.025 (w/v)% ditert-butyl-p-cresol which served as an antioxidant. The solvent flow rates were controlled at better than 1.00 ± 0.05 ml/min and 2.00 ± 0.05 ml/min. The degasser was operated at 55°C. The differential refractometer which had a 0.019 in. slit was operated on 8× and monitored the effluent streams at 42°C. To eliminate errors in elution volume measurement due to variations in the rate of solvent evaporation in the siphon tube (1 count = 5.003 ml at 1 ml/min, 1 count = 5.067 ml at 2 ml/min) a vapor feedback loop device, similar to that of Yau et al. (*13*), was installed. Samples were dissolved in degassed solvent taken from the GPC solvent reservoir and were filtered under N_2 pressure through 0.2 μ millipore filters. The samples were injected for 120 sec by means of the Waters Automatic Sample Injection System, ASIS. The variance in the reproducibility of chromatograms obtained from repetitive sample injections through the same loop was comparable to that obtained from repetitive sample injections between different loops. The GPC traces were digitally recorded at 20 sec intervals by means of the Waters Digital Curve Translator. Molecular weight averages, intrinsic viscosity, and normalized differential distribution curves were calculated on an IBM 360/65 computer according to the basic integral formulas given by Pickett et al. (*14*).

Calibration Standards

The calibration standards used were linear polystyrene standards (PS) from Pressure Chemical Co. (PC) and Waters Associates (W), linear polybutadiene standards (PBD) from Phillips Petroleum Co. (P), and linear polyvinyl chloride standards (PVC) from PC. The Waters Associates PS standards included some ultranarrow MWD recycle standards for which $\bar{M}_w/\bar{M}_n \leq 1.009$. The PS, PBD and PVC standards were injected via the ASIS at concentrations of 0.04, 0.05 and 0.075 (w/v) % respectively. The absolute number- and weight-average molecular weights and polydispersity ratios, designated respectively by $\bar{M}_n(t)$, $\bar{M}_w(t)$, and $P(t)$, and shown in Table 1, were those supplied by the vendor except where noted. The Mark-Houwink in-

TABLE 1

Absolute and Infinite Resolution Values of $\bar{M}_n$, $\bar{M}_w$, $[\eta]$, and P

Run	Standard	PEV	$10^{-3}\bar{M}_n(t)$	$10^{-3}\bar{M}_w(t)$	$[\eta](t)$	$P(t)$	$10^{-3}\bar{M}_n(\infty)$	$10^{-3}\bar{M}_w(\infty)$	$[\eta](\infty)$	$P(\infty)$
				Polystyrene, 1 ml/min						
180-175	PC-16A	40.15	0.578^a	0.636	—	<1.1	0.521	0.632	—	1.21
180-176	PC-15A	38.20	1.21^a	1.41	—	<1.16	1.20	1.57	—	1.31
180-177	PC-12A	37.50	1.73^a	1.85	—	1.07	1.68	2.04	—	1.21
180-178	PC-11A	36.60	3.18^a	3.53	0.0492	1.11	2.87	3.47	0.0496	1.21
180-179	W-4190042	34.55	9.70	10.3	0.106	1.06	7.86	9.59	0.103	1.22
180-180	W-4190039	33.25	19.65	19.85	0.172	1.01	16.01	18.67	0.163	1.17
180-181	W-4190041	31.09	49.0	51.0	0.331	1.04	—	—	—	—
180-182	W-41995	29.70	96.2	98.2	0.530	1.02	82.5	89.2	0.491	1.08
181-183	W-41984	28.50	164	173	0.781	1.06	139	161	0.747	1.15
181-184	NBS-705	28.30	170.9	179.3	0.802	1.049	144	178	0.797	1.232
181-185	NBS-706	27.85	136.5	257.8	1.006	1.889	119	284	1.052	2.377
181-186	W-4190037	26.25	392	411	1.441	1.05	271	407	1.417	1.50
181-187	PC-5A	25.60	404	507	1.568	1.26	348	531	1.713	1.53
181-188	W-490038	24.60	773	867	2.382	1.12	545	818	2.337	1.50
181-189	PC-14A	23.70	1610	1900	4.071	1.18	649	1160	2.964	1.79
181-190	W-61970	23.65	1780	2145	4.402	1.21	650	1147	2.944	1.77
184-207	W-27231	34.50	9.91	10.0^b	0.107	<1.009	8.84	10.13	0.108	1.15
184-208	W-27232	33.45	16.35	16.5^b	0.152	<1.009	14.31	15.63	0.144	1.09
184-209	W-27233	31.39	42.12	42.5^b	0.296	<1.009	37.81	41.57	0.287	1.10
184-210	W-27234	29.65	95.14	96.0^b	0.525	<1.009	82.94	90.18	0.495	1.09
184-211	W-27235	28.50	160.6	162^b	0.760	<1.009	136.8	157.8	0.738	1.15
184-212	W-27231	34.51	9.91	10.0^b	0.107	<1.009	8.92	10.25	0.109	1.15
184-213	W-27232	33.48	16.35	16.5^b	0.152	<1.009	13.88	15.09	0.141	1.09
184-214	W-27233	31.42	42.12	42.5^b	0.296	<1.009	38.03	41.45	0.287	1.09
				Polystyrene, 2 ml/min						
193-278	PC-16A	39.60	0.578^a	0.636	—	<1.1	—	—	—	—
193-279	PC-15A	37.60	1.21^a	1.41	—	<1.16	—	—	—	—

246

193-280	PC-12A	37.00	1.73[a]	1.85	—	1.07	1.75	2.12	—	1.23
193-281	PC-11A	35.90	3.18[a]	3.53	0.0492	1.11	2.93	3.62	0.0510	1.24
193-282	W-4190042	33.90	9.70	10.3	0.106	1.06	7.96	10.0	0.105	1.26
193-283	W-4190039	32.45	19.65	19.85	0.172	1.01	17.8	20.0	0.172	1.12
193-284	W-4190041	30.40	49.0	51.0	0.331	1.04	49.0	50.7	0.330	1.03
193-285	W-41995	28.95	96.2	98.2	0.530	1.02	90.4	100	0.541	1.11
194-286	W-41984	27.80	164	173	0.781	1.06	150	161	0.750	1.07
194-287	NBS-705	27.40	170.9	179.3	0.802	1.049	171	194	0.848	1.13
194-288	NBS-706	27.25	136.5	257.8	1.006	1.889	125	303	1.107	2.42
194-289	W-4190037	25.40	392	411	1.441	1.05	336	442	1.519	1.32
194-290	PC-5A	24.60	404	507	1.568	1.26	409	573	1.823	1.40
194-291	W-490038	23.80	773	867	2.382	1.12	853	880	2.477	1.03
194-292	PC-14A	23.10	1610	1900	4.071	1.18	834	1140	2.963	1.37
194-293	W-61970	23.00	1780	2145	4.402	1.21	924	1170	3.02	1.27
Polybutadiene,[c] 1 ml/min										
159-045	P-17M	32.01	16.1	17.0	—	1.06	12.4	18.3	—	1.64
159-046	P-170M	27.16	135	170	—	1.26	89.1	149	—	1.68
159-047	P-272M	26.10	206	272	—	1.32	145	303	—	2.09
159-048	P-332M	25.80	226	332	—	1.47	132	401	—	3.05
159-049	P-432M	25.22	286	432	—	1.48	127	493	—	3.88
Polyvinyl Chloride, 1 ml/min										
158-042	PC-V2	30.80	25.5	68.6	0.70	2.69	24.0	59.9	0.74	2.50
158-043	PC-V3	29.70	41.1	118	1.13	2.87	38.8	95.5	1.06	2.46
158-044	PC-V4	29.31	54.0	132	1.25	2.44	45.9	114	1.21	2.48

[a] The $\bar{M}_n(t)$ value used were those determined by Wachter and Simon (32) with an extended range vapor pressure osmometer. $\bar{M}_w(t)$. $\bar{M}_n(t)$ values were determined from the $\bar{M}_n(t)$ values and the vendor supplied $P(t)$ values.

[b] The $\bar{M}_w(t)$ values for the recycle PS-standards were read off the molecular weight calibration curve associated with the PEV. $\bar{M}_n(t)$ values were determined from the $\bar{M}_w(t)$ values and the vendor supplied $P(t)$ values.

[c] The ranges in per cent geometrical isomer content of the PBD-standards varied as follows: $43.5 \leq cis \leq 51.7$, $41.7 \leq trans \leq 49.1$ and $6.6 \leq vinyl \leq 8.4$.

trinsic viscosity-molecular weight relations used to obtain $[\eta](t)$ for PVC (15) and PS were, respectively,

$$[\eta]_{\text{THF},25°C}^{\text{PVC}} = 1.63 \times 10^{-4} \bar{M}_v^{0.766} \qquad 20,000 < \bar{M}_v < 170,000 \qquad (1)$$

and

$$[\eta]_{\text{THF},25°C}^{\text{PS}} = 1.60 \times 10^{-4} \bar{M}_v^{0.706} \qquad \bar{M}_v > 3000 \qquad (2)$$

Equation (2) was obtained via least-squares analysis of data supplied by ArRo Laboratories (16).

The calibration curves for PS and PBD standards were obtained by associating peak elution volume (PEV) with $\bar{M}_w$, while for the broad

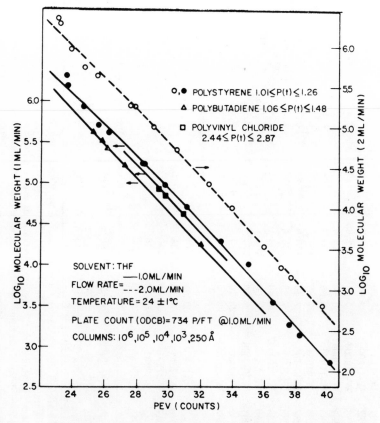

FIG. 1. Molecular weight calibration curves for polystyrene, polybutadiene, and polyvinyl chloride.

TABLE 2

Absolute and Infinite Resolution Values of $\bar{M}_n$, $\bar{M}_w$, $[\eta]$, and P

Run	Wt% composition[a]	$10^{-3}\bar{M}_n(t)$[b]	$10^{-3}\bar{M}_w(t)$	$[\eta](t)$	$P(t)$	$10^{-3}\bar{M}_n(\infty)$	$10^{-3}\bar{M}_w(\infty)$	$[\eta](\infty)$	$P(\infty)$
		Polystyrene Blends, Flow Rate = 2 ml/min							
196-302-2	0/50/50	64.9	74.6	0.431	1.11	61.9	73.7	0.428	1.19
196-303-2	50/50/0	28.1	35.4	0.252	1.24	26.9	36.8	0.258	1.37
196-304-2	50/0/50	32.6	59.0	0.351	1.79	30.6	61.8	0.363	2.02
196-305-2	75/25/0	23.1	27.6	0.212	1.18	21.8	29.2	0.219	1.34
196-306-2	25/75/0	35.6	43.2	0.291	1.18	34.3	42.6	0.289	1.24
196-307-2	33.3/33.3/33.3	36.7	56.3	0.344	1.51	34.6	56.9	0.347	1.65
196-308-2	16.7/33.3/50	48.8	69.4	0.404	1.42	47.1	70.5	0.409	1.50
196-309-2	50/33.3/16.7	29.4	43.3	0.285	1.45	27.8	44.4	0.290	1.60

[a] The wt% of each polystyrene standard in the mixture refers to the standards in the following order: W-4190039, W-4190041, and W-41995.

[b] The absolute values $\bar{M}_n(t)$, $\bar{M}_w(t)$, and $[\eta](t)$ of the blends were calculated from the absolute values of the individual components and the compositions of the blends.

PVC standards PEV was associated with $(\bar{M}_n \cdot \bar{M}_w)^{1/2}$. The number- and weight-average molecular weights, intrinsic viscosity, and polydispersity ratios calculated from the molecular weight calibration curves (shown in Fig. 1), assuming perfect or infinite resolution, are designated as $\bar{M}_n(\infty)$, $\bar{M}_w(\infty)$, $[\eta](\infty)$, and $P(\infty)$, and are listed in Table 1.

Polystyrene Blends

Some polystyrene blends composed of varying amounts of samples W-4190039, W-4190041, and W-41995 having a polydispersity range of $1.11 \le P(t) \le 1.79$ were injected via the ASIS at 0.04 (w/v)% for 120 sec. These samples were run through the columns at 2.00 ± 0.05 ml/min. The compositions of the blends and absolute and infinite resolution values of $\bar{M}_n$, $\bar{M}_w$, $[\eta]$, and P are shown in Table 2. The baseline adjusted raw chromatograms of these multimodal polystyrene blends are shown in Figs. 2 and 3.

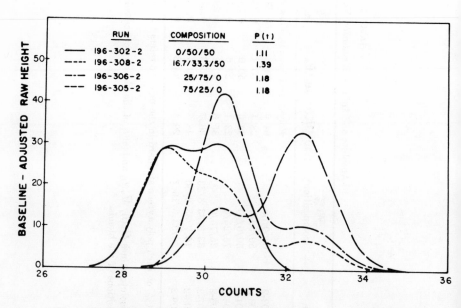

FIG. 2. Baseline-adjusted raw chromatograms of some polystyrene blends of samples W-4190039, W-4190041, and W-41995 at a flow rate of 2 ml/min.

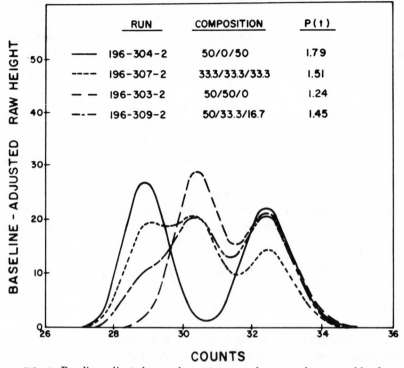

FIG. 3. Baseline-adjusted raw chromatogram of some polystyrene blends of samples W-4190039, W-4190041, and W-41995 at a flow rate of 2 ml/min.

THEORY

Method of Molecular Weight Averages

Tung (*1*) has shown that the normalized* observed GPC chromatogram, $F(v)$, at elution volume v is related to the normalized GPC chromatogram corrected for instrumental broadening, $W(v)$, by means of the shape function $G(v)$ through the relation

$$F(v) = \int_{-\infty}^{\infty} G(v - y)W(y)\, dy \qquad (3)$$

The kth molecular weight average associated with the corrected chro-

---

* The word "normalized" means that the area of the observed chromatogram is unity.

matogram is denoted by $M_k(t)$ (true or absolute value), and the kth molecular weight average associated with the observed chromatogram is denoted by $M_k(\infty)$ assuming perfect or infinite resolution. Then, the ratio $M_k(t)/M_k(\infty)$ is given by

$$\frac{M_k(t)}{M_k(\infty)} = \frac{\int_{-\infty}^{\infty} W(v)M(v)^{k-1}\,dv \Big/ \int_{-\infty}^{\infty} W(v)M(v)^{k-2}\,dv}{\int_{-\infty}^{\infty} F(v)M(v)^{k-1}\,dv \Big/ \int_{-\infty}^{\infty} F(v)M(v)^{k-2}\,dv} \tag{4}$$

where $k = 1, 2, 3, 4$ corresponds to the number-, weight-, Z- and $Z + 1$- average molecular weight, respectively.

A similar expression can be written for the ratio of the intrinsic viscosity associated with the corrected chromatogram, $[\eta](t)$, to the intrinsic viscosity associated with the observed chromatogram, $[\eta](\infty)$, assuming infinite resolution. This expression is given by

$$\frac{[\eta](t)}{[\eta](\infty)} = \int_{-\infty}^{\infty} W(v)M(v)^{\epsilon}\,dv \Big/ \int_{-\infty}^{\infty} F(v)M(v)^{\epsilon}\,dv \tag{5}$$

where ϵ is the exponent in the Mark-Houwink intrinsic viscosity-molecular weight expression.

For calibration curves which are linear in $\log_{10} M$ or curves which are linear in $\log_{10} M$ over the elution volume range of interest

$$M(v) = D_1 \exp(-D_2 v) \tag{6}$$

where $D_1, D_2 > 0$. Substituting Eq. (6) into (4) and (5) yields the following relations

$$\frac{M_k(t)}{M_k(\infty)} = \frac{\bar{W}[D_2(k-1)]/\bar{W}[D_2(k-2)]}{\bar{F}[D_2(k-1)]/\bar{F}[D_2(k-2)]} \tag{7}$$

$$\frac{[\eta](t)}{[\eta](\infty)} = \bar{W}[D_2\epsilon]/\bar{F}[D_2\epsilon] \tag{8}$$

where $\bar{W}$ and $\bar{F}$ are the bilaterial Laplace transforms of $W(v)$ and $F(v)$, respectively, and the square brackets indicate functionality. The general bilateral Laplace transform of $F(v)$ can be written as

$$\bar{F}(s) = \int_{-\infty}^{\infty} \exp(-sv)\left[\int_{-\infty}^{\infty} G(v-y)W(y)\,dy\right]dv \tag{9}$$

Application of the convolution theorem of the bilateral Laplace transform leads to

$$\bar{F}(s) = \bar{W}(s)\bar{G}(s) \tag{10}$$

The ratio of the absolute to infinite resolution molecular weight averages and intrinsic viscosity becomes

$$\frac{M_k(t)}{M_k(\infty)} = \bar{G}[D_2(k - 2)]/\bar{G}[D_2(k - 1)] \tag{11}$$

$$\frac{[\eta](t)}{[\eta](\infty)} = 1/\bar{G}[D_2\epsilon] \tag{12}$$

where

$$\bar{G}(s) = \int_{-\infty}^{\infty} \exp\left[-s(v - y)\right]G(v - y)\,d(v - y) \tag{13}$$

and $G(v - y)$ is the general instrument spreading function defined in Eqs. (14) and (15).

The Instrument Spreading Shape Function

Tung (1) has used a Gaussian function to describe the instrumental spreading. This shape function is inadequate when skewing effects are present. Smith's (3) use of a log-normal shape function is not of sufficient generality to account for skewing and has not been successful (2). The method of Pickett and co-workers (5), which uses the chromatogram shapes of narrow MWD standards, has been only partially successful since it overcorrects for skewing (2). The need for a non-Gaussian shape function has been empirically demonstrated rather extensively by Hamielec and co-workers (2, 11). Skewed chromatograms are produced under conditions of high flow rate (short residence times), increased viscosity due to high-molecular weight species, and/or column overloading as well as loss of resolution at the high- or low-molecular weight ends of the calibration curve.

A general normalized statistical shape function describing the instrumental broadening behavior in GPC columns is proposed and has the form

$$G(v - y) = G_T(v - y) + \sum_{n=3}^{\infty} \{(-1)^n A_n \mu^{n/2} G_T^n(v - y)/n!\} \tag{14}$$

where $G_T(v - y)$ and $G_T^n(v - y)$ denote the normal form of the Gaussian instrumental spreading function first used by Tung (1),

$$G_T(v - y) = (1/2\pi\mu_2)^{1/2} \exp\left[-(v - y)^2/2\,\mu_2\right] \tag{15}$$

and its nth order derivatives with respect to v, respectively.

The coefficients A_n are functions of μ_n, the nth order moments about the elution volume y of the instrument spreading function. The first two coefficients are of direct statistical significance and are

$$A_3 = \mu_3/\mu^{3/2} \tag{16}$$

$$A_4 = (\mu_4/\mu_2^2 - 3) \tag{17}$$

where the variance μ_2 is related to the resolution factor h by the relation

$$\mu_2 = 1/h \tag{18}$$

The resolution factor h defined by Eqs. (15) and (18) is equal to two times the resolution factor defined in the original Tung (1) formulation. In the limiting case of an ideal monodisperse standard or ultranarrow MWD standard for which $P(t) = 1$, the moments μ_n are also the nth order moments about the mean of the normalized observed GPC chromatogram.

The coefficient A_3 provides an absolute statistical measure of skewness. When $A_3 = 0$, the chromatogram of the ideal monodisperse standard is symmetrical about the elution volume y. When $A_3 > 0$, the chromatogram of the ideal standard has a long tail at high elution volumes and is skewed to high elution volumes or low molecular weights; and when $A_3 < 0$, the chromatogram has a long tail at low elution volumes and is skewed to low elution volumes or high molecular weights.

The coefficient A_4 provides a statistical measure of the flattening or kurtosis of the chromatogram of the ideal monodisperse standard and is related to instrumental broadening effects due to skewing and axial dispersion. The kurtosis coefficient measures the excess flatness or thinness of the chromatogram peak compared to that of a Gaussian curve. When $A_4 = 0$, the chromatogram is Gaussian in shape. When $A_4 > 0$, the chromatogram is leptokurtic, taller and slimmer than the Gaussian curve. When $A_4 < 0$, the normalized observed chromatogram is platykurtic, flatter or squat at the center of the curve than the corresponding Gaussian curve.

In principle, the coefficients A_n can be obtained from the moments about the mean elution volume of the normalized observed chromatograms of narrow MWD standards. However, the narrow MWD standards may have some natural skewness, flatness, and dispersion associated with their MWD's, and the computation of μ_3 and higher moments may be subject to large numerical errors. Furthermore, it is desirable to consider the coefficients A_n as parameters which can be mathematically manipulated. The coefficients A_n are functions of elution volume, but will be considered to be constant or slowly varying functions of elution volume in the integration in Eq. (13). In order to obtain the contributions to the coefficients A_n due solely to instrumental broadening, expressions will be derived for $\bar{M}_n(t)$, $\bar{M}_w(t)$, and $[\eta](t)$ in terms of h, μ_3, and μ_4. Subsequently, the values of h, μ_3, and μ_4 will be determined by fitting these expressions to the experimental values of $\bar{M}_n(t)$, $\bar{M}_w(t)$, and $[\eta](t)$.

The coefficients A_5 and higher are functions of higher order moments and/or products of lower order A_n coefficients. The higher order moments do not have a simple geometrical interpretation. The higher even order moments are further measures of dispersion or flatness and the higher odd order moments are further measures of skewness. A complete derivation of the general statistical shape function describing deviations from ideal Gaussian behavior by means of moment generating and cumulant generating functions has been described by Aitken (*17*) and others (*18, 19, 22*).

Evaluation of $M_k(t)/M_k(\infty)$ and $[\eta](t)/[\eta](\infty)$

The bilateral Laplace transform of $G(v - y)$ in reduced variable notation is

$$\bar{G}(s/\sqrt{h}) = \int_{-\infty}^{\infty} \exp{(-sx/\sqrt{h})}G(x)\,dx \tag{19}$$

where $x = (v - y)/\sqrt{\mu_2} = \sqrt{h}(v - y)$. The evaluation of Eq. (19) can be conveniently carried out by expressing the integrand in terms of Hermite polynomials. Using the relations (*20*)

$$\begin{aligned} H_n(x) &= (-1)^n \phi^n(x)/\phi(x) \\ \phi(x) &= (1/2\pi)^{1/2} \exp{(-x^2/2)} \end{aligned} \tag{20}$$

in conjunction with Eqs. (14) and (15), the general shape function becomes

$$G(x) = \phi(x)\left[1 + \sum_{n=3}^{\infty} (A_n/n!)H_n(x)\right] \tag{21}$$

The form of the general shape function in Eq. (21) is known as the Gram-Charlier series (*21–23*). Through the use of the generating function (*20*) for Hermite polynomials, the exponential term in the integrand of Eq. (19) becomes

$$\exp{(-sx/\sqrt{h})} = \exp{(s^2/2h)}\sum_{m=0}^{\infty} (H_m(x)/m!)(-s/\sqrt{h})^m \tag{22}$$

Substitution of Eqs. (21) and (22) into Eq. (19) leads to

$$\bar{G}(s/\sqrt{h}) = \exp(s^2/2h) \left\{ \sum_{m=0}^{\infty} (s/\sqrt{h})^m (1/m!) \left(\int_{-\infty}^{\infty} \phi(x) H_m(x) H_o(x)\, dx \right) \right.$$

$$\left. + \sum_{n=0}^{\infty} \sum_{m=3}^{\infty} (A_n/n!m!)(-s\sqrt{h})^m \left(\int_{-\infty}^{\infty} \phi(x) H_m(x) H_o(x)\, dx \right) \right\} \quad (23)$$

Using the orthonormality conditions (20) for Hermite functions,

$$\int_{-\infty}^{\infty} \phi(x) H_n(x) H_m(x)\, dx = \left. \begin{matrix} n! \\ 0 \end{matrix} \right\} \begin{matrix} n = m \\ n \neq m \end{matrix} \quad (24)$$

in Eq. (23) yields the following expression,

$$\bar{G}(s/\sqrt{h}) = \exp(s^2/2h) \left\{ 1 + \sum_{n=3}^{\infty} (A_n/n!)(s/\sqrt{h})^n \right\} \quad (25)$$

Substitution of the expression for $\bar{G}(s)$ into Eqs. (11) and (12) yields

$$\frac{M_k(t)}{M_k(\infty)} = \exp\left[\frac{-D_2^2(2k - 3)}{2h} \right] \frac{\left\{ 1 + \sum_{n=3}^{\infty} \left(\frac{A_n}{n!}\right)\left(\frac{-D_2(k - 2)}{\sqrt{h}}\right)^n \right\}}{\left\{ 1 + \sum_{n=3}^{\infty} \left(\frac{A_n}{n!}\right)\left(\frac{-D_2(k - 1)}{\sqrt{h}}\right)^n \right\}} \quad (26)$$

$$\frac{[\eta](t)}{[\eta](\infty)} = \exp\left[\frac{-\epsilon D_2^2}{2h} \right] \Big/ \left\{ 1 + \sum_{n=3}^{\infty} \left(\frac{A_n}{n!}\right)\left(\frac{-\epsilon D_2}{\sqrt{h}}\right)^n \right\} \quad (27)$$

Truncation of the Gram-Charlier series in Eq. (21) leads to the Edgeworth series (24), where $A_5 = 0$, $A_6 = 10A_3^2$ and $A_n = 0$ for $n > 7$. Expressions for $\bar{M}_n(t)$, $\bar{M}_w(t)$, and $[\eta](t)$ may be written in terms of the fundamental parameters h, μ_3, and μ_4 as follows:

$$\bar{M}_n(t) = \bar{M}_n(\infty) \exp(D_2^2/2h) \left\{ 1 + \frac{D_2^3 \mu_3}{6} + \frac{D_2^4}{24}\left(\mu_4 - \frac{3}{h^2}\right) + \frac{1}{2}\left(\frac{D_2^3 \mu_3}{6}\right)^2 \right\} \quad (28)$$

$$\bar{M}_w(t) = \bar{M}_w(\infty) \exp(-D_2^2/2h) \Big/ \left\{ 1 - \frac{D_2^3 \mu_3}{6} + \frac{D_2^4}{24}\left(\mu_4 - \frac{3}{h^2}\right) \right.$$

$$\left. + \frac{1}{2}\left(\frac{D_2^3 \mu_3}{6}\right)^2 \right\} \quad (29)$$

$$[\eta](t) = [\eta](\infty) \exp(-\epsilon^2 D_2^2/2h) \Big/ \left\{ 1 - \frac{\epsilon^3 D_2^3 \mu_3}{6} \right.$$

$$+ \frac{\epsilon^4 D_2^4}{24}\left(\mu_4 - \frac{3}{h^2}\right) + \frac{1}{2}\left(\frac{\epsilon^3 D_2^3 \mu_3}{6}\right)^2\Big\} \qquad (30)$$

DETERMINATION OF h, μ_3, AND μ_4 FROM STANDARDS

$\bar{M}_n$, $\bar{M}_w$ Supplied

If the coefficients of the Mark-Houwink intrinsic viscosity-molecular weight relationship are unavailable, then only $\bar{M}_n(t)$ and $\bar{M}_w(t)$ can be used to determine h and μ_3 in Eqs. (28) and (29). If we denote the ratios $\bar{M}_n(t)/\bar{M}_n(\infty)$ and $\bar{M}_w(\infty)/\bar{M}_w(t)$ as $R_n(t,\infty)$ and $R_w(\infty,t)$, respectively, Eqs. (28) and (29) can be combined to yield

$$[R_n(t,\infty) - R_w(\infty,t)] \exp(-D_2^2/2h) = 2[D_2^3\mu_3/6] \qquad (31)$$

$$[R_n(t,\infty) + R_w(\infty,t)] \exp(-D_2^2/2h) = 2\left[1 + \frac{D_2^4}{24}\left(\mu_4 - \frac{3}{h^2}\right)\right.$$
$$\left. + \frac{1}{2}\left(\frac{D_2^3\mu_3}{6}\right)^2\right] \qquad (32)$$

If

$$\frac{D_2^4}{24}\left(\mu_4 - \frac{3}{h^2}\right) + \frac{1}{2}\left[\frac{D_2^3\mu_3}{6}\right]^2 \ll 1 \qquad (33)$$

Eqs. (31) and (32) can be combined to give simple algebraic expressions for h and μ_3 as follows:

$$h = \left(\frac{D_2^2}{2}\right)\Big/ \log_e\left\{\frac{[R_n(t,\infty) + R_w(\infty,t)]}{2}\right\} \qquad (34)$$

$$\mu_3 = \left(\frac{6}{D_2^3}\right)\left\{\frac{R_n(t,\infty) - R_w(\infty,t)}{R_n(t,\infty) + R_w(\infty,t)}\right\} \qquad (35)$$

A calibration curve for h and μ_3 as functions of PEV can be obtained from standards via Eqs. (34) and (35). The corrected number- and weight-average molecular weights and intrinsic viscosity of other samples (provided the Mark-Houwink constants are known), which shall be designated as $\bar{M}_n(h,\mu_3)$, $\bar{M}_w(h,\mu_3)$, and $[\eta](h,\mu_3)$, respectively, can be obtained from

$$\bar{M}_n(h,\mu_3) = \bar{M}_n(\infty) \exp(D_2^2/2h)\{1 + D_2^3\mu_3/6\} \qquad (36)$$
$$\bar{M}_w(h,\mu_3) = \bar{M}_w(\infty) \exp(-D_2^2/2h)\{1 - D_2^3\mu_3/6\}^{-1} \qquad (37)$$
$$[\eta](h,\mu_3) = [\eta](\infty) \exp(-\epsilon^2 D_2^2/2h)\{1 - \epsilon_2^3 D_2^3\mu_3/6\}^{-1} \qquad (38)$$

In using Eqs. (34) through (38), it is important to be certain that Eq. (33) holds so that terms in D_2^4 or higher can be neglected.

When $D_2^3\mu_3/6 \ll 1$, the term in braces in Eq. (37) can be expanded to give the same term that appears in braces in Eq. (36). Under this condition, Eqs. (36) and (37) then become equivalent to the empirical correction equations obtained by Balke and Hamielec (11). Their empirical skewing operator sk and resolution factor h' are then related to μ_3 and h, respectively, according to the relations

$$sk = D_2^3\mu_3/3$$
$$h' = h/2 \tag{39}$$

$\bar{M}_n$, [η] Supplied

The measurement of the weight-average molecular weight by light-scattering techniques is a time consuming and often experimentally difficult task compared to the determination of the number-average molecular weight by membrane or vapor pressure osmometry. When $\bar{M}_n$ and [η] are relatively easy to obtain, compared to $\bar{M}_w$, and the Mark-Houwink coefficients are available, Eqs. (28) and (30) can be used to obtain h and μ_3.

Upon denoting the ratio [η](∞)/[η](t) as $R_v(\infty,t)$, Eqs. (28) and (30) can be combined to yield

$$R_n(t,\infty) \exp(-D_2^2/2h) - R_v(\infty,t) \exp(-\epsilon^2 D_2^2/2h) = \mu_3 D_2^3(1 + \epsilon^3)/6 \tag{40}$$

$$R_n(t,\infty) \exp(-D_2^2/2h) + R_v(\infty,t) \exp(-\epsilon^2 D_2^2/2h) - 2 = \mu_3 D_2^3(1 - \epsilon^3)/6 \tag{41}$$

where again terms in D_2^4 and L_2^6 have been neglected. Manipulation of Eqs. (40) and (41) leads to the following expression for h

$$h = \left(\frac{D_2^2}{2}\right) \Big/ \log_e\left[\frac{\epsilon^3 R_n(t,\infty)}{(1 + \epsilon^3) + R_v(\infty,t)[1 - (\epsilon^2 D_2^2/2h)]}\right] \tag{42}$$

A value for h can be rapidly obtained from Eq. (42) by an iterative procedure with a desk calculator. A good initial guess is obtained by setting the term ($\epsilon^2 D_2^2/4h$) to zero. Then, the guessed value is inserted into Eq. (42) and a new value computed. This procedure is continued until there is satisfactory agreement between two successively calculated values of h. Once h is obtained, μ_3 can be readily obtained from either Eq. (40) or (41). Then, the corrected number- and weight-average molecular weights and intrinsic viscosity of other samples can be obtained from Eqs. (36) through (38).

$\bar{M}_n$, $\bar{M}_w$, $[\eta]$ Supplied

When values of the absolute number- and weight-average molecular weights and intrinsic viscosity are available, h, μ_3, and μ_4 can be obtained from Eqs. (28), (29), and (30) by solving for them with the aid of an "algorithm for least squares estimation of nonlinear parameters" and a computer. When the values of h obtained in this manner are large, $h >$ 2.0, the exponential factor in $D_2^2/2h$ approaches unity and can be neglected. Then, the corrected number- and weight-average molecular weights and intrinsic viscosities can be obtained from the following equations

$$\bar{M}_n(\mu_3,\mu_4) = \bar{M}_n(\infty)\left\{1 + \frac{D_2^3\mu_3}{6} + \frac{D_2^4\mu_4}{24} + \frac{1}{2}\left(\frac{D_2^2\mu_3}{6}\right)^2\right\} \tag{43}$$

$$\bar{M}_w(\mu_3,\mu_4) = \bar{M}_w(\infty)\left\{1 - \frac{D_2^3\mu_3}{6} + \frac{D_2^4\mu_4}{24} + \frac{1}{2}\left(\frac{D_2^3\mu_3}{6}\right)^2\right\}^{-1} \tag{44}$$

$$[\eta](\mu_3,\mu_4) = [\eta](\infty)\left\{1 - \frac{\epsilon^3 D_2^3\mu_3}{6} + \frac{\epsilon^4 D_2^4\mu_4}{24} + \frac{1}{2}\left(\frac{\epsilon^3 D_2^3\mu_3}{6}\right)^2\right\}^{-1} \tag{45}$$

DISCUSSION OF THE EFFECTS OF h, μ_3, AND μ_4 ON $\bar{M}_n(\infty)$, $\bar{M}_w(\infty)$, AND $[\eta](\infty)$

The effect that skewed GPC chromatograms have upon calculated number- and weight-average molecular weights and intrinsic viscosity is graphically illustrated in Fig. 4. The correction Eqs. (28), (29), and (30) have been plotted as a function of μ_3 for the situation where axial dispersion is absent, $h = \infty$ and $\mu_4 = 0$. Typical experimental values have been assigned to D_2 and ϵ, $D_2 = 0.5$ and $\epsilon = 0.7$.

Positive values of μ_3 indicate that the chromatogram is skewed to high elution volumes (low molecular weight). As a result, $\bar{M}_n(\infty)$ always will be less than the true value, $\bar{M}_n(\mu_3)$. As μ_3 increases, the correction necessary to raise $\bar{M}_n(\infty)$ to $\bar{M}_n(\mu_3)$ monotonically increases. $\bar{M}_w(\infty)$ is less than the true value, $\bar{M}_w(\mu_3)$, in the range $0 < \mu_3 < 12/D_2^3$. As μ_3 increases, the correction necessary to raise $\bar{M}_w(\infty)$ to $\bar{M}_w(\mu_3)$ at first increases monotonically. At $\mu_3 = 6/D_2^3$ the correction reaches a maximum value $[\bar{M}_w(\mu_3)/\bar{M}_w(\infty) = 2]$ and then decreases monotonically. When $\mu_3 > 12D/_2^3$, $\bar{M}_w(\infty) > \bar{M}_w(\mu_3)$ and the correction necessary to lower $\bar{M}_w(\infty)$ to $\bar{M}_w(\mu_3)$ increases in a monotonic fashion. This unusual behavior is due to the effect of the quadratic term in μ_3 in Eq. (29). Large values of μ_3, $\mu_3 > 15$, are due to a loss of resolution at high molecular weights. The high molecular weight species are not adequately frac-

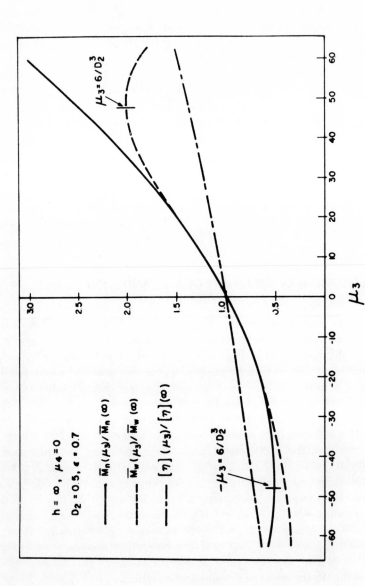

FIG. 4. Correction factor curves for $\overline{M}_n$, $\overline{M}_w$, and $[\eta]$ vs μ_3 for $h = \infty$ and $\mu_4 = 0$.

tionated by the gel and tend to elute over a very narrow range of elution volumes. This effect causes the chromatogram to be extremely skewed toward higher elution volumes and causes the calibration curve to tail up sharply at high molecular weights.

Negative values of μ_3 indicate that the chromatogram is skewed to low elution volumes (high molecular weight). This is due to a loss of resolution at low molecular weights and is associated with a sharp downturn in the calibration curve at high elution volumes. Thus, for all negative values of μ_3, $\bar{M}_w(\infty) > \bar{M}_w(\mu_3)$ and the correction necessary to lower $\bar{M}_w(\infty)$ to $\bar{M}_w(\mu_3)$ increases in a monotonic fashion. Because of the quadratic term in μ_3 in Eq. (28), $\bar{M}_n(\infty) > \bar{M}_n(\mu_3)$ for $-12/D_2^3 < \mu_3 < 0$ and $\bar{M}_n(\infty) < \bar{M}_n(\mu_3)$ for $\mu_3 < -12/D_2^3$. As μ_3 decreases, the correction necessary to lower $\bar{M}_n(\infty)$ to $\bar{M}_n(\mu_3)$ at first increases monotonically. At $\mu_3 = -6/D_2^3$ the correction reaches a maximum value $[\bar{M}_n(\mu_3)/\bar{M}_n(\infty) = 0.5]$ and then decreases monotonically. For $\mu_3 < -12/D_2^3$ the correction to raise $\bar{M}_n(\infty)$ to $\bar{M}_n(\mu_3)$ increases in a monotonic fashion.

The empirically defined skewing operator, sk, of Balke and Hamielec (11) has the effect of raising $\bar{M}_n(\infty)$ and lowering $\bar{M}_w(\infty)$ equally through their correction equations, which as previously mentioned are limiting cases of Eqs. (36) and (37),

$$\bar{M}_n(t) = \bar{M}_n(\infty) \exp{(D_2^2/4h')}[1 + \tfrac{1}{2}sk] \qquad (46)$$

$$\bar{M}_w(t) = \bar{M}_w(\infty) \exp{(-D_2^2/4h')}[1 + \tfrac{1}{2}sk] \qquad (47)$$

These equations are based on the assumption that the main effect of the skewing correction is to shift the intercept of the molecular weight calibration curve. Figure 4 shows that this assumption is only valid for a narrow range of μ_3 values, and thus a narrow range of elution volumes. For $D_2 = 0.5$, $\bar{M}_n(t)/\bar{M}_n(\infty) \simeq \bar{M}_w(t)/\bar{M}_w(\infty)$ only in the range $-15 \leq \mu_3 \leq 15$, $-0.625 \leq sk \leq 0.625$.

The behavior of $[\eta](\infty)$ as a function of μ_3 parallels that of $\bar{M}_w(\infty)$, except that the ϵ-parameter in Eq. (30) damps out the effect of the quadratic term in μ_3 over the range of μ_3 which is of practical interest.

The separate effect of h ($\mu_3 = \mu_4 = 0$) and μ_4 ($h = \infty$, $\mu_3 = 0$), upon $\bar{M}_n(\infty)$, $\bar{M}_w(\infty)$, and $[\eta](\infty)$ are shown in Fig. 5 for $D_2 = 0.5$ and $\epsilon = 0.7$. It can be seen that h raises $\bar{M}_n(\infty)$ and lowers $\bar{M}_w(\infty)$ and $[\eta](\infty)$ to their respective values $\bar{M}_n(h)$, $\bar{M}_w(h)$, and $[\eta](h)$. $\bar{M}_n(\infty)$ is raised and $\bar{M}_w(\infty)$ is lowered by the same amount for $h > 1.00$. For $h < 1.00$ and $D_2 = 0.5$, the effects of axial dispersion although symmetric in elution volume space do not shift the DMWD curve of narrow MWD

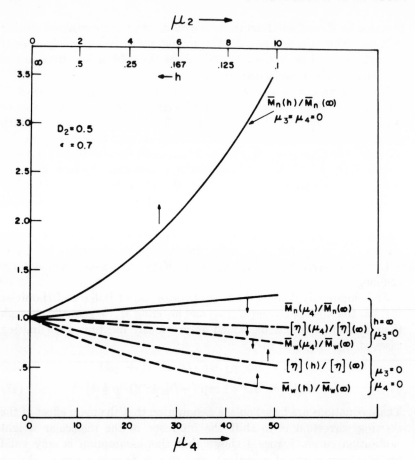

FIG. 5. Correction factor curves for $\bar{M}_n$, $\bar{M}_w$, and $[\eta]$ vs h and μ_2 for $\mu_3 = \mu_4 = 0$ and vs μ_4 for $h = \infty$ and $\mu_3 = 0$.

standards in a symmetric manner in molecular weight space. The exponential term containing h in Eqs. (28), (29), and (30) will serve to enhance the effect of μ_3 on $\bar{M}_n(\infty)$ by dilating the $\bar{M}_n(t)/\bar{M}_n(\infty)$ correction factor and will serve to reduce the effect of μ_3 on $\bar{M}_w(\infty)$ by contracting the $\bar{M}_w(t)/\bar{M}_w(\infty)$ correction factor.

The coefficient μ_4 effects $\bar{M}_n(\infty)$, $\bar{M}_w(\infty)$, and $[\eta](\infty)$ in the same way as does the resolution factor h. For $\mu_4 < 10$ and $D_2 = 0.5$, $\bar{M}_n(\infty)$ is raised and $\bar{M}_w(\infty)$ is lowered by the same amount. The effect of the interaction of μ_4 with μ_3 in Eqs. (43), (44), and (45) is to increase the correction factor for $\bar{M}_n(\infty)$ when $\mu_3 < 0$, and to raise the minimum in the $\bar{M}_n(\mu_3, \mu_4)/\bar{M}_n(\infty)$ curve when $\mu_3 < 0$. The interaction effect of μ_4 with μ_3 upon $\bar{M}_w(\infty)$ when $\mu_3 < 0$ is to decrease the correction factor for $\bar{M}_w(\infty)$, and when $\mu_3 > 0$ to initially increase the correction factor for $\bar{M}_w(\infty)$ and then decrease it by lowering the maximum in the $\bar{M}_w(\mu_3,\mu_4)/\bar{M}_n(\infty)$ curve. The effect of the interaction of μ_4 with μ_3 on $[\eta](\infty)$ is similar to the effect upon $\bar{M}_w(\infty)$ except that the ϵ-parameter damps out the effect of the quadratic term in μ_3 over the range of μ_3 which is of practical interest.

RESULTS

The values of the absolute and infinite resolution number- and weight-average molecular weights and intrinsic viscosities shown in Table 1 were used in conjunction with Eqs. (28), (29), and (30) to generate values of h, μ_3, and μ_4 which are shown in Tables 3 and 4. These parameters were obtained, with the aid of a CDC-6400 computer,* by a least squares method for estimating nonlinear parameters (25) which utilizes a combination of gradient and Newton-Raphson methods. Values of h and μ_3 were obtained by fitting to Eqs. (28) and (29) with $\mu_4 = 0$ ($\{\bar{M}_n, \bar{M}_w\}$ supplied), and fitting to Eqs. (28) and (30) with $\mu_4 = 0$ ($\{\bar{M}_n, [\eta]\}$ supplied). Where the Mark-Houwink coefficients were available, values of h, μ_3, and μ_4 were obtained by fitting to Eqs. (28), (29), and (30) ($\{\bar{M}_n, \bar{M}_w, [\eta]\}$ supplied).

The correlations obtained between μ_3 and PEV for PS, PBD, and PVC standards at a flow rate of 1 ml/min are shown in Fig. 9. The values of μ_3 obtained for each standard by fitting to the parameter sets $\{\bar{M}_n, \bar{M}_w\}$, $\{\bar{M}_n, [\eta]\}$ and $\{\bar{M}_n, \bar{M}_w, [\eta]\}$ agree well within the experimental errors associated with absolute and infinite resolution values of $\bar{M}_n$, $\bar{M}_w$, and $[\eta]$. In fact the values of μ_3 for the polystyrene standards obtained by fitting to $\{\bar{M}_n, \bar{M}_w, [\eta]\}$ are not shown in Fig. 9 because these values would be indistinguishable from the μ_3-values obtained by fitting to $\{\bar{M}_n, \bar{M}_w\}$. The differences in μ_3-values for PS, PVC, and PBD standards over comparable elution volume ranges are very slight. It is interesting to note that the ultranarrow recycle-PS-standards [$P(t) < 1.009$] have slightly lower values of μ_3 (dotted curve) than the narrow PS standards [$1.05 \leq P(t) \leq 1.25$]. The linear portion of the μ_3 vs PEV curve for PS corresponds to the PEV range of optimum

---

* A Fortran IV computer program which makes use of Marquardt's algorithm to solve a set of nonlinear equations can be found in E. J. Henley and E. M. Rosen, *Material and Energy Balance Computations*, Wiley, New York, 1969, pp. 547f and 560–566.

TABLE 3

Values of h, μ_3, and μ_4 for Polystyrene Standards

Run	PEV	D_2	Fit to $\bar{M}_n$ and $\bar{M}_w$		Fit to $\bar{M}_n$ and $[\eta]$		Fit to $\bar{M}_n$, $\bar{M}_w$, and $[\eta]$		
			h	μ_3	h	μ_3	h	μ_3	μ_4
					Flow Rate = 1 ml/min				
180-176	38.20	0.4896	1.95	-2.53	—	—	—	—	—
180-177	37.60	0.4896	1.88	-1.76	—	—	—	—	—
180-178	36.60	0.4896	2.69	3.06	2.28	2.62	1.67	2.97	0
180-179	34.55	0.4896	1.51	7.17	2.35	8.85	∞	7.78	29.6
180-180	33.25	0.4896	1.49	6.76	3.58	9.56	∞	7.50	29.7
180-182	29.70	0.4580	2.88	7.76	∞	10.7	∞	8.42	13.1
181-183	28.50	0.4580	1.95	7.38	4.69	9.60	4.23	7.91	25.0
181-184	28.30	0.4580	1.22	5.56	1.57	6.86	1.15	5.91	32.1
181-185	27.85	0.4580	0.895	1.26	0.957	17.5	2.28	1.29	16.4
181-186	26.25	0.4580	0.531	11.7	0.706	15.4	0.346	13.6	76.8
181-187	25.60	0.4580	1.06	3.21	0.632	-1.03	1.285	2.62	0
181-188	24.60	0.4580	0.631	12.6	0.768	14.8	∞	13.6	37.0
181-189	23.70	0.4580	0.241	37.8	0.475	61.8		58.7	55.2
181-190	23.65	0.4580	0.213	42.1	0.439	72.2		69.2	11.2
184-207	34.50	0.4896	1.61	3.18	1.74	3.49	2.09	3.23	8.24

184-208	33.45	0.4896	2.39	5.84	71.1	8.80	∞	6.36	20.1
184-209	31.39	0.4580	2.33	4.06	8.08	6.22	∞	4.41	22.7
184-210	29.65	0.4580	2.83	5.09	∞	7.93	∞	5.74	13.0
184-211	28.50	0.4580	1.68	4.76	3.32	6.77	1.58	5.21	28.2
184-212	34.51	0.4896	1.59	2.52	1.59	2.52		2.52	0.390
184-213	33.48	0.4896	2.32	7.86	∞	11.3	∞	8.60	17.9
184-214	31.42	0.4580	2.58	3.97	12.4	6.13	∞	4.31	20.2

Flow Rate = 2 ml/min

193-280	37.00	0.4859	1.79	-3.84	1.01	-6.25	0.506	-3.25	0
193-281	35.90	0.4859	2.20	1.52	1.55	3.09	0.925	1.41	0
193-282	33.90	0.4859	1.30	5.92	1.59	6.88	2.94	6.23	1.96
193-283	32.45	0.4409	1.77	3.11	1.93	3.44	1.76	3.10	0.0838
193-285	28.95	0.4409	2.39	1.56	1.93	0.863	1.33	1.51	0
194-286	27.80	0.4409	6.88	5.61	∞	6.81	∞	5.85	5.83
194-287	27.40	0.4409	2.52	-2.80	1.50	-4.55	0.767	-2.57	0
194-288	27.25	0.4409	0.767	-2.39	0.640	-4.06	0.462	-2.20	0
194-289	25.40	0.4409	0.851	2.83	0.774	2.00	0.640	2.73	0
194-290	24.60	0.4409	1.70	-4.67	0.535	-1.23	0.428	-4.49	0
194-291	23.80	0.4409	∞	-3.78	11.3	-7.08	0.818	-3.35	0
194-292	23.10	0.4409	0.413	36.81	1.80	58.0	∞	47.6	16.3
194-293	23.00	0.4409	0.459	39.13	8.28	63.3	∞	49.7	0

TABLE 4

Values of h, μ_3, and μ_4 for Polybutadiene and Polyvinyl Chloride Standards

			Fit to $\bar{M}_n$ and $\bar{M}_w$		Fit to $\bar{M}_n$ and $[\eta]$		Fit to $\bar{M}_n$, $\bar{M}_w$ and $[\eta]$		
Run	PEV	D_2	h	μ_3	h	μ_3	h	μ_3	μ_4
			Polybutadiene, 1 ml/min						
159-045	32.01	0.4785	0.654	5.17	—	—	—	—	—
159-046	27.16	0.4785	0.637	14.6	—	—	—	—	—
159-047	26.10	0.4785	0.485	6.61	—	—	—	—	—
159-048	25.80	0.4785	0.303	9.38	—	—	—	—	—
159-049	25.22	0.5693	0.302	10.5	—	—	—	—	—
			Polyvinyl Chloride, 1 ml/min						
158-042	30.80	0.4750	∞	5.35	1.52	−0.653	0.339	3.83	0
158-043	29.70	0.4750	∞	6.98	∞	3.97	1.52	6.48	0
158-044	29.31	0.4750	5.51	8.49	2.53	6.95	0.969	7.71	0

resolution in the molecular weight calibration curve. The upturn at low PEV (high molecular weight) to large μ_3-values and the downturn at high PEV (low molecular weight) to negative μ_3-values is due to the loss of resolution at the high and low molecular weight ends of the calibration curve. The increase in μ_3 to large positive values can be expected to be greater than the decrease in μ_3 to large negative values because of increased viscosity effects at the high molecular weight end of the calibration curve.

The correlations obtained between h and PEV for PS and PBD standards at a flow rate of 1 ml/min by fitting to the parameter sets $\{\bar{M}_n, \bar{M}_w\}$ and $\{\bar{M}_n, [\eta]\}$ are shown in Figs. 6 and 8. A scatter envelope (dotted curves) corresponding to reasonable experimental errors is shown in Fig. 6 for the polystyrene standards. The observed scatter in the data can be attributed to uncertainties in the experimental number- and weight-average molecular weights (particularly at low molecular weights), and the Mark-Houwink coefficients of the narrow MWD standards.

The recycle-PS-standards scatter to the same extent as the broader PS standards. The values of h obtained for the PVC standards were extremely scattered. The experimental polydispersity of these samples was greater than the GPC infinite resolution values and was reproducible. This can be attributed to possible experimental errors in the determination of $\bar{M}_n(t)$ and $\bar{M}_w(t)$, and the nature of the molecular weight

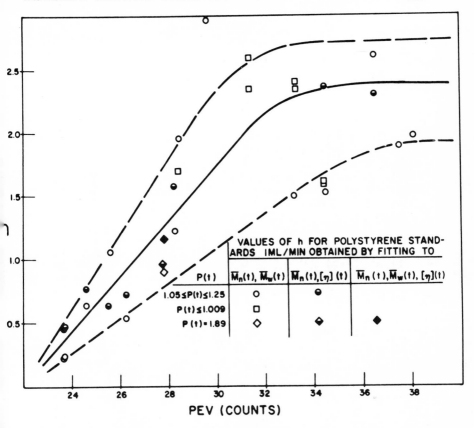

FIG. 6. Resolution factor h vs PEV for polystyrene, 1 ml/min flow rate.

calibration curve. At first the weight-average and viscosity-average molecular weights were associated with PEV to obtain a calibration curve. This resulted in $\bar{M}_w(\infty) > \bar{M}_w(t)$, $\bar{M}_n(\infty) > \bar{M}_n(t)$, and $P(\infty) < P(t)$. Then, $[\bar{M}_n(t) \cdot \bar{M}_w(t)]^{1/2}$ was associated with PEV. This resulted in the more usual GPC conditions, $\bar{M}_n(\infty) < \bar{M}_n(t)$ and $\bar{M}_w(\infty) \leq \bar{M}_w(t)$. However, $P(\infty) < P(t)$. This condition could result from the limited elution volume-molecular weight range obtained from experiment and the necessary extrapolation of the molecular weight calibration curve outside this range for the calculation of $\bar{M}_n(\infty)$, $\bar{M}_w(\infty)$, and $[\eta](\infty)$. However, a rather large increase in the slope of the calibration curve $(D_2/2.303)$ would be necessary for $P(\infty) \geq P(t)$. Meyerhoff (26) also has observed this phenomena with broad standards. He attributes

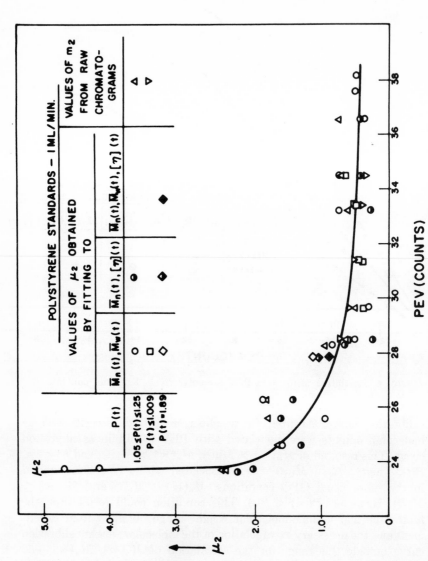

FIG. 7. Variance μ_2 vs PEV for polystyrene, 1 ml/min flow rate.

this behavior to extremely high resolving power of GPC columns over the elution volume range of the samples.

Since the magnitude of the correction for axial dispersion is inversely proportional to h, it is more useful to look at a plot of μ_2 vs PEV when assessing the effects of experimental scatter in h on the magnitude of the correction. Such a plot for PS standards at a flow rate of 1 ml/min is shown in Fig. 7. For typical experimental uncertainty ranges in h of $1.4 \leq h \leq 2.8$, $0.8 \leq h \leq 1.6$, and $0.28 \leq h \leq 0.6$ (corresponding to ranges in μ_2 of $0.36 \leq \mu_2 \leq 0.71$, $0.63 \leq \mu_2 \leq 1.25$, and $1.7 \leq \mu_2 \leq 3.6$), the corresponding ranges in the correction factors obtainable from Fig. 3 are, respectively: $1.06 \leq \bar{M}_n(t)/\bar{M}_n(\infty)$, $\bar{M}_w(t)/\bar{M}_w(\infty) \leq 1.10$; $1.09 \leq \bar{M}_n(t)/\bar{M}_n(\infty)$, $\bar{M}_w(t)/\bar{M}_w(\infty) \leq 1.16$; and $1.22 \leq \bar{M}_w(t)/\bar{M}_w(\infty) \leq 1.56$, $1.16 \leq \bar{M}_n(t)/\bar{M}_n(\infty) \leq 1.37$. For $h > 0.6$, the magnitude of the errors in the correction factors produced by the experimental uncertainties in h are not too severe and can be tolerated.

A similar analysis of the effect of experimental errors in μ_3 upon the correction factors obtainable from Fig. 4 can be done. For $6 \leq \mu_3 \leq 12$, the range in the correction factor is $1.12 \leq \bar{M}_n(t)/\bar{M}_n(\infty)$, $\bar{M}_n(t)/\bar{M}_w(\infty) \leq 1.26$. For $25 \leq \mu_3 \leq 40$, the range in correction factors is $1.64 \leq \bar{M}_n(t)/\bar{M}_n(\infty) \leq 2.18$ and $1.8 \leq \bar{M}_w(t)/\bar{M}_w(\infty) \leq 1.96$. Thus at large values of μ_3, where loss of resolution in the calibration curve is observed, extreme errors in the correction factors are obtained.

When values of h, μ_3, and μ_4 at a flow rate of 1 ml/min are obtained by fitting to the parameter set $\{\bar{M}_n, \bar{M}_w, [\eta]\}$, the values of h are generally much larger than those obtained by fitting to the parameter sets $\{\bar{M}_n, \bar{M}_w\}$ or $\{\bar{M}_n, [\eta]\}$. In fact, for most cases, h is sufficiently large that the exponential factors $\exp(\pm D_2^2/2h) \simeq 1$ and the correction equations in μ_3 and μ_4 (Eqs. 43, 44, and 45) can be used to compute the corrected values $\bar{M}_n(\mu_3, \mu_4)$, $\bar{M}_w(\mu_3, \mu_4)$ and $[\eta](\mu_3, \mu_4)$. Both h and μ_4 are measures of dispersion in elution volume space and thus interact. The large values of h (small values of μ_2) obtained in this manner indicate that at a flow rate of 1 ml/min the parameter of μ_4 is a stronger function of the dispersion than is h. A plot of μ_4 vs PEV is shown in Fig. 10.

Earlier in this paper it was pointed out that the computed second, third, and fourth moments, m_2, m_3, and m_4, about the mean elution volume of the normalized observed chromatogram would be equivalent to the moments μ_2, μ_3, and μ_4 of the general shape function in the limit of the ideal monodisperse standard, $P(t) = 1$. The computed moments m_2, m_3, and m_4 of some of the polystyrene standards at 1 ml/min. flow rate are shown in Table 5. Values of m_2 are directly compared with μ_2 in

TABLE 5

Values of m_2, m_3, and m_4 for Polystyrene Standards, 1 ml/min Obtained from the Normalized Observed Chromatograms

Run	m_2	m_3	m_4
180-178	0.767	0.276	18.6
180-179	0.663	0.284	17.0
180-180	0.600	0.135	12.1
180-182	0.540	0.335	13.9
181-183	0.642	0.473	21.2
181-184	0.933	0.457	42.2
181-185	4.04	3.49	51.2
181-186	1.83	3.39	19.1
181-187	1.80	3.17	17.4
181-188	1.69	3.46	17.3
181-189	2.49	5.83	33.0
181-190	2.46	5.69	31.6
184-207	0.318	0.0471	0.317
184-208	0.406	0.128	0.572
184-209	0.524	0.450	1.60
184-210	0.556	0.350	1.27
184-211	0.650	0.479	1.93

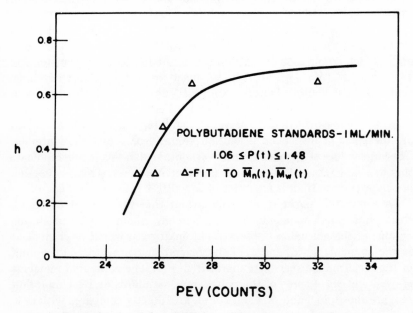

FIG. 8. Resolution factor h vs PEV for polybutadiene, 1 ml/min flow rate.

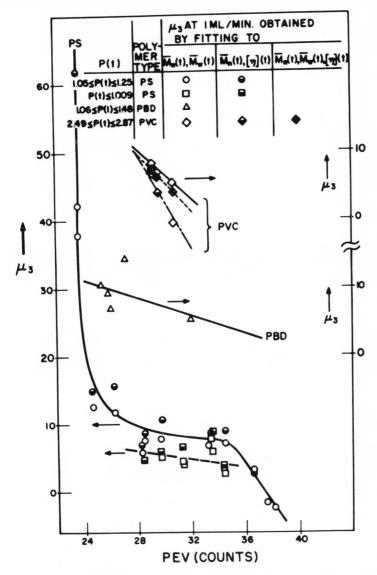

FIG. 9. Skewing coefficient μ_3 vs PEV for polystyrene, polybutadiene, and polyvinyl chloride, 1 ml/min flow rate.

Fig. 5. In general $\mu_2 \approx m_2$ for these standards. Comparison of m_3 in Table 5 with μ_3 in Table 3 shows that $\mu_3 \gg m_3$. Similarly $\mu_4 > m_4$ as shown in Fig. 8. The large differences between m_3 and μ_3 and m_4 and μ_4 are attributable to the natural skewness and flatness associated with

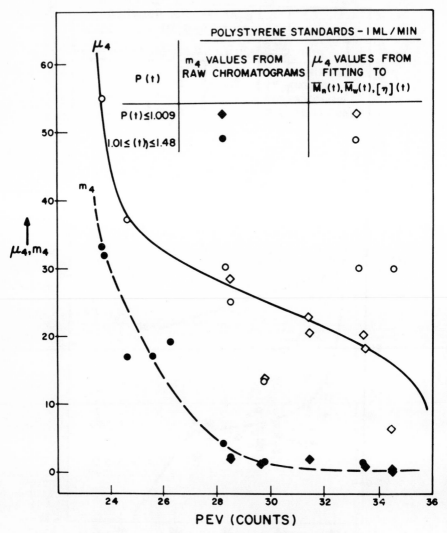

FIG. 10. μ_4, m_4 vs PEV for polystyrene, 1 ml/min flow rate.

the MWD of the PS-calibration standards, the approximation of the molecular weight calibration curve by two linear segments, and the experimental errors associated with determination of $\bar{M}_n(t)$, $\bar{M}_w(t)$, and $[\eta](t)$.

$\bar{M}_n(\infty)$, $\bar{M}_w(\infty)$, and $[\eta](\infty)$ (where applicable) for PS and PBD-

TABLE 6

Corrected Values of $\bar{M}_n$, $\bar{M}_w$, and $[\eta]$ for Polystyrene Standards, 1 ml/min

Run	$10^{-3}\bar{M}_n(h,\mu_3)$	$10^{-3}\bar{M}_w(h,\mu_3)$	$[\eta](h,\mu_3)$	$P(h,\mu_3)$	$10^{-3}\bar{M}_n(\mu_3,\mu_4)$	$10^{-3}\bar{M}_w(\mu_3,\mu_4)$	$[\eta](\mu_3,\mu_4)$	$P(\mu_3,\mu_4)$
180-176	1.20	1.42	—	1.18	—	—	—	—
180-177	1.75	1.92	—	1.10	—	—	—	—
180-178	3.16	3.46	0.0492	1.10	3.04	3.60	0.0503	1.18
180-179	9.44	10.6	0.106	1.12	9.35	10.6	0.107	1.13
180-180	19.4	20.8	0.167	1.07	19.3	20.6	0.170	1.07
180-182	103	100	0.504	<1.0	103	98.8	0.514	<1.0
181-183	172	175	0.759	1.02	168	176	0.777	1.05
181-184	178	193	0.808	1.08	174	195	0.829	1.12
181-185	149	309	1.067	2.07	145	312	1.095	2.15
181-186	360	446	1.433	1.23	342	457	1.489	1.34
181-187	482	588	1.731	1.22	450	609	1.815	1.35
181-188	834	928	2.351	1.11	748	980	2.522	1.31
181-189	1420	2176	3.231	1.53	1229	1714	3.554	1.39
181-190	1480	2573	3.303	1.73	1303	1738	3.600	1.33
184-207	10.0	10.4	0.108	1.04	9.90	10.5	0.110	1.06
184-208	16.3	16.2	0.145	~1.0	16.2	16.2	0.147	1.00
184-209	43.0	42.9	0.288	~1.0	42.5	43.2	0.292	1.02
184-210	97.2	92.2	0.493	<1.0	95.7	92.6	0.503	<1.0
184-211	161	162	0.736	1.006	157	165	0.754	1.05
184-212	10.1	10.6	0.109	1.05	9.99	10.6	0.111	1.06
184-213	15.9	15.7	0.142	~1.0	15.8	15.7	0.144	~1.0
184-214	43.2	42.8	0.288	~1.0	42.7	43.0	0.292	1.007

TABLE 7

Corrected Values of $\bar{M}_n$ and $\bar{M}_w$ for Polybutadiene Standards, 1 ml/min

Run	$10^{-3}\bar{M}_n(h,\mu_3)$	$10^{-3}\bar{M}_w(h,\mu_3)$	$P(h,\mu_3)$
159-045	16.2	17.5	1.08
159-046	126	147	1.17
159-047	221	287	1.30
159-048	208	307	1.78
159-049	288	426	1.48

calibration standards at a flow rate of 1 ml/min have been corrected for skewing and axial dispersion with the aid of Eqs. (28), (29), and (30) using values of h, μ_3, and μ_4 obtained from the average smooth curves drawn in Figs. 6, 8, 9, and 10. The corrected data is shown in Tables 6 and 7. $\bar{M}_n(\mu_3,\mu_4)$, $\bar{M}_w(\mu_3,\mu_4)$, and $[\eta](\mu_3,\mu_4)$ have been calculated from Eqs. (43), (44), and (45) under the assumption that h is very large ($\mu_2 \simeq 0$). There is good agreement between the corrected values $\bar{M}_n(\mu_3,\mu_4)$, $\bar{M}_w(\mu_3,\mu_4)$, and $[\eta](\mu_3,\mu_4)$ and the corrected values $\bar{M}_n(h,\mu_3)$, $\bar{M}_w(h,\mu_3)$, and $[\eta](h,\mu_3)$ when the h referred to in the data set $\{h,\mu_3,\mu_4\}$ of Table 3 is greater than 2. Also, the corrected value of $\bar{M}_n$, $\bar{M}_w$, and $[\eta]$ compare favorably with the true values, $\bar{M}_n(t)$, $\bar{M}_w(t)$, and $[\eta](t)$ of Table 1. Most of the corrected values lie well within the range of the experimental errors associated with the determination of $\bar{M}_n(t)$, $\bar{M}_w(t)$, $[\eta](t)$, and $P(t)$. Deviations occur for the very broad standard NBS-706, and for the ultrahigh molecular weight standards ($M \geq 10^6$) where the uncertainties in h, μ_3, and μ_4 are very large.

The PS-calibration standards run at 2 ml/min yielded values of h slightly smaller than those at 1 ml/min, while the values of μ_3 and μ_4 at 2 ml/min were considerably smaller than those at 1 ml/min. The values of μ_3 at 2 ml/min scattered considerably more than at 1 ml/min. This scatter can be attributed to baseline instabilities at the higher flow rate and a smaller data sampling frequency. Values of h and μ_3 at 1 ml/min obtained by fitting to the parameter sets $\{\bar{M}_n, \bar{M}_w\}$ and $\{\bar{M}_n, [\eta]\}$, and values of h, μ_3, and μ_4 at 2 ml/min obtained by fitting to the parameter set $\{\bar{M}_n, \bar{M}_w, [\eta]\}$ are shown in Table 3. When the flow rate is increased from 1 to 2 ml/min, there is an increase in instrument spreading due to axial dispersion and a reduction in instrument spreading due to skewing and flattening.

Some polystyrene blends run at a flow rate of 2 ml/min having complex multimodal shapes shown in Figs. 2 and 3, and having a poly-

TABLE 8

Corrected Values of $\bar{M}_n$, $\bar{M}_w$, $[\eta]$, and P for Polystyrene
Blends, 2 ml/min for Ranges of h and μ_3 Values

Run	h	μ_3	10^{-3} $\bar{M}_n(h,\mu_3)$	10^{-3} $\bar{M}_w(h,\mu_3)$	$[\eta](h,\mu_3)$	$P(h,\mu_3)$
196-302-2	1.7	1.5	66.9	71.3	0.419	1.06
	1.7	3.0	68.3	72.7	0.422	1.06
	2.4	1.5	65.8	72.3	0.423	1.10
	2.4	3.0	67.2	74.0	0.426	1.10
196-303-2	1.7	1.5	29.1	35.5	0.253	1.22
	1.7	3.0	29.7	36.3	0.255	1.22
	2.4	1.5	28.6	36.1	0.255	1.26
	2.4	3.0	29.2	36.9	0.257	1.26
196-304-2	1.7	1.5	33.1	59.7	0.356	1.80
	1.7	3.0	33.7	61.0	0.358	1.81
	2.4	1.5	32.6	60.6	0.358	1.86
	2.4	3.0	33.2	62.0	0.361	1.86
196-305-2	1.7	1.5	23.6	28.2	0.215	1.20
	1.7	3.0	24.1	28.8	0.216	1.20
	2.4	1.5	23.2	28.7	0.216	1.24
	2.4	3.0	23.7	29.3	0.218	1.24
196-306-2	1.7	1.5	37.1	41.1	0.283	1.11
	1.7	3.0	37.9	42.0	0.285	1.11
	2.4	1.5	36.5	41.8	0.286	1.15
	2.4	3.0	37.2	42.7	0.288	1.15
196-307-2	1.7	1.5	37.4	54.9	0.340	1.47
	1.7	3.0	38.2	56.1	0.342	1.48
	2.4	1.5	36.8	55.8	0.343	1.52
	2.4	3.0	37.6	57.1	0.345	1.52
196-308-2	1.7	1.5	50.9	68.0	0.401	1.34
	1.7	3.0	52.0	69.6	0.404	1.34
	2.4	1.5	50.1	69.2	0.404	1.38
	2.4	3.0	51.1	70.7	0.407	1.38
196-309-2	1.7	1.5	30.1	42.9	0.284	1.43
	1.7	3.0	30.7	43.8	0.286	1.43
	2.4	1.5	29.6	43.6	0.287	1.47
	2.4	3.0	30.2	44.5	0.289	1.48

dispersity range $1.11 \leq P(t) \leq 1.79$, were corrected for axial dispersion and skewing. Values of h and μ_3, obtained from calibration standards covering the elution volume range of these blends, were in the ranges $1.7 \leq h \leq 2.4$ and $1.5 \leq \mu_3 \leq 3.0$, respectively. Corrected values of $\bar{M}_n(h,\mu_3)$, $\bar{M}_w(h,\mu_3)$, $[\eta](h,\mu_3)$, and $P(h,\mu_3)$ are shown in Table 8 for $h = 1.7$ and 2.4, and $\mu_3 = 1.5$ and 3.0. These corrected values compare favorably with the absolute values in Table 2 over a reasonably wide range of experimental uncertainties in h and μ_3 values.

More experimental work is needed to ascertain whether broad calibration standards, $P(t) > 2$, can be used to obtain valid values of h, μ_3, and μ_4. The main problem with using broad MWD standards is that average values of h, μ_3, and μ_4 will be obtained over wide elution volume ranges. These values will be unreliable if h, μ_3, and μ_4 are changing rapidly over the elution volume range. For these reasons, values of h, μ_3, and μ_4 obtained from narrow MWD standards may tend to overcorrect very broad MWD standards. A method will be proposed in the next section that will circumvent this problem.

CORRECTION METHODS FOR THE DMWD CURVE

Fitting for an Effective Linear Calibration

If the calibration curve is truly linear over the molecular weight range of interest, the corrected values $\bar{M}_n(t)$ and $\bar{M}_w(t)$ can be used to fit for the slope and intercept of the corrected calibration curve. Then, the corrected DMWD curve can be obtained from this corrected calibration curve. Hamielec and co-workers (11, 27) have demonstrated the feasibility of this approach. If the corrected molecular weight calibration constants are denoted by $D_1(t)$ and $D_2(t)$, and the infinite resolution values by $D_1(\infty)$ and $D_2(\infty)$, then

$$\frac{\bar{M}_n(t)}{\bar{M}_n(\infty)} = \left(\frac{D_1(t)}{D_1(\infty)}\right)\frac{\bar{F}[-D_2(\infty)]}{\bar{F}[-D_2(t)]} \tag{48}$$

$$\frac{\bar{M}_w(t)}{\bar{M}_w(\infty)} = \left(\frac{D_1(t)}{D_1(\infty)}\right)\frac{\bar{F}[D_2(t)]}{\bar{F}[D_2(\infty)]} \tag{49}$$

Use of Eqs. (48) and (49) with (36) and (37) yields the following useful relations

$$\frac{\bar{F}[-D_2(\infty)]\bar{F}[D_2(\infty)]}{\bar{F}[-D_2(t)]\bar{F}[D_2(t)]} = \exp[D_2^2(\infty)/h]\{1 - [D_2^3(\infty)\mu_3/6]^2\} \tag{50}$$

$$D_1(t) = D_1(\infty)\left\{\left(\frac{1 + D_2^3(\infty)\mu_3/6}{1 - D_2^3(\infty)\mu_3/6}\right)\left(\frac{\bar{F}[-D_2(t)]\bar{F}[D_2(\infty)]}{\bar{F}[-D_2(\infty)]\bar{F}[D_2(t)]}\right)\right\}^{1/2} \tag{51}$$

where $\bar{F}$ is the previously defined bilateral Laplace transform function. Equation (50) shows that the slope of the corrected calibration curve $[D_2(t)/2.303]$ is a function of $D_2(\infty)$, h, and μ_3. Equation (51) shows that the intercept of the corrected calibration curve $[\log_{10} D_1(t)]$ is a function of $D_1(\infty)$, $D_2(\infty)$, h, and μ_3. Thus, in order to compensate for the effects of skewing and axial dispersion, the calibration curve must be both translated and rotated.

The Hydrodynamic Volume Concept Approach

Another method which is useful for correcting the infinite resolution DMWD for skewing and axial dispersion involves the use of the hydrodynamic volume concept. Benoit et al. (28) have shown that when the product $\{[\eta]\bar{M}_w\}$ is plotted against PEV, a common calibration curve results for linear and branched homopolymers and grafted copolymers. In particular, they have shown that linear PS, PBD, and PVC fall on a common $\{[\eta]\bar{M}_w\}$ vs PEV plot. Pickett et al. (14) have obtained expressions for $\bar{M}_n$ and $\bar{M}_w$ in terms of the DMWD, (da/dM), as follows:

$$\bar{M}_n = \left[\int_{M_L}^{M_H} \frac{1}{M} \left(\frac{da}{dM} \right) dM \right]^{-1} \tag{52}$$

$$\bar{M}_w = \int_{M_L}^{M_H} M \left(\frac{da}{dM} \right) dM \tag{53}$$

where

$$\frac{da}{dM} = \frac{C(v_M)}{\int_{v_H}^{v_L} C(v)\,dv} \cdot \frac{1}{\left(\dfrac{df}{dv} \right)_{v_M}} \cdot \frac{\log_{10} e}{M} \tag{54}$$

The first term on the right-hand side of Eq. (54) is the normalized chromatogram height at elution volume v_M and the second term is the reciprocal of the slope of the calibration curve, $f(v) = \log_{10} M$. Expressing the hydrodynamic volume as

$$Z = \{[\eta]M\} = KM^{\epsilon+1} \tag{55}$$

where K and ϵ are the standard Mark-Houwink parameters, and substituting this expression into Eqs. (52), (53), and (54) leads to the expressions

$$\bar{M}_n = \left[\int_{Z_L}^{Z_H} \left(\frac{Z}{K} \right)^{1/(\epsilon+1)} \left(\frac{da}{dZ} \right) dZ \right]^{-1} \tag{56}$$

$$\bar{M}_w = \int_{Z_L}^{Z_H} \left(\frac{Z}{K} \right)^{1/(\epsilon+1)} \left(\frac{da}{dZ} \right) dZ \tag{57}$$

$$\frac{da}{dZ} = \frac{C(v_Z)}{\displaystyle\int_{v_H}^{v_L} C(v)\,dv} \cdot \frac{1}{\left(\dfrac{df'}{dv}\right)_{v_Z}} \cdot \frac{\log_{10} e}{Z} \tag{58}$$

where now the hydrodynamic volume calibration curve is expressed as $f'(v) = \log_{10} Z$. By using the corrected values of $\bar{M}_n(t)$ and $\bar{M}_w(t)$ obtained from Eqs. (28) and (29) in conjunction with the hydrodynamic volume calibration curve, Eqs. (56) and (57) can be fit for effective values of ϵ and K that correct the infinite resolution values, $\bar{M}_n(\infty)$ and $\bar{M}_w(\infty)$ to the true values $\bar{M}_n(t)$ and $\bar{M}_w(t)$. Once ϵ and K are obtained, the corrected molecular weight calibration curve can be obtained from Eq. (55) and, subsequently, the corrected DMWD can be obtained from Eq. (54). The relationship between the corrected calibration curve, $f_t(v) = \log_{10} M_t$, and the uncorrected curve $f(v)$ can be expressed, according to the formalism of Coll and Prusinowski (29), as

$$\log_{10} M_t = \left(\frac{1}{1 + \epsilon_t}\right) \log_{10}\left(\frac{K}{K_t}\right) + \left(\frac{1 + \epsilon}{1 + \epsilon_t}\right) f(v) \tag{59}$$

where ϵ and K are the Mark-Houwink parameters used in the construction of the hydrodynamic volume calibration curve, and ϵ_t and K_t are the effective Mark-Houwink parameters obtained upon fitting Eqs. (56) and (57) to $\bar{M}_n(t)$ and $\bar{M}_w(t)$. This hydrodynamic volume approach for obtaining the corrected DMWD curves is quite general and does not require a linear molecular weight calibration curve over the entire elution volume range of interest. For the special case of a linear calibration curve, the coefficients $D_1(t)$ and $D_2(t)$ of the corrected calibration curve can be related to the coefficients $D_1(\infty)$ and $D_2(\infty)$ of the uncorrected curve with the aid of Eqs. (6) and (59).

$$D_1(t) = \left(\frac{K}{K_t}\right)^{1/(1+\epsilon_t)} (D_1(\infty))^{(1+\epsilon)/(1+\epsilon_t)} \tag{60}$$

$$D_2(t) = \left(\frac{1 + \epsilon}{1 + \epsilon_t}\right) D_2(\infty) \tag{61}$$

From the previous analysis made with Eqs. (50) and (51), it can be seen that ϵ_t is a function of $D_2(\infty)$, ϵ, h, and μ_3 while K_t is a function of $D_1(\infty)$, ϵ, K, $D_2(\infty)$, h, and μ_3. Thus, both K_t and ϵ_t are affected by GPC chromatogram spreading due to axial dispersion and skewing.

The DMWD curves of several of the standards have been corrected by this hydrodynamic volume fitting procedure. The mathematical method (25) used in the fitting procedure has been discussed previously.

TABLE 9

Fitted Values of ϵ_t and K_t and Calculated Values of $\bar{M}_n(t)$, $\bar{M}_w(t)$, $[\eta](t)$, and $P(t)$

Run	ϵ_t	$10^3 K_t$	$10^{-3}\bar{M}_n(t)$	$10^{-3}\bar{M}_w(t)$	$[\eta](t)$	$P(t)$
				Calculated values		
			Polyvinyl Chloride, 1 ml/min			
158-042	0.474	2.39	25.5	68.7	0.799	2.70
158-043	0.283	20.6	41.1	118	1.202	2.87
158-044	0.437	3.78	54.9	127	1.253	2.30
			Polybutadiene, 1 ml/min			
159-047	0.865	0.0335	200	282	—	1.41
159-049	1.10	0.0010	265	443	—	1.67
			Polystyrene, 1 ml/min			
180-177	0.727	0.145	1.69	1.89	—	1.12
180-178	0.548	0.498	3.18	3.53	0.0489	1.11
180-179	0.749	0.0862	9.71	10.3	1.053	1.06
181-185	0.871	0.0188	135	261	0.944	1.94
181-186	0.603	0.337	392	412	1.277	1.05
			Polystyrene Blends, 2 ml/min			
196-302-2	0.799	0.0541	64.3	75.4	0.435	1.17
196-303-2	0.865	0.0309	27.7	35.9	0.255	1.30
196-304-2	0.871	0.0269	32.6	59.0	0.355	1.81
196-305-2	0.907	0.0208	22.5	28.3	0.215	1.26
196-306-2	0.782	0.0684	35.5	43.3	0.293	1.22
196-307-2	0.843	0.0352	36.7	56.3	0.346	1.54
196-308-2	0.814	0.0485	48.7	69.6	0.408	1.48
196-309-2	0.872	0.0271	29.4	43.3	0.287	1.47

The hydrodynamic volume calibration curve was obtained from the $[\eta](t)$–$\bar{M}_w(t)$ data for polystyrene in Table 1. The values of $\bar{M}_n(t)$ and $\bar{M}_w(t)$ used in the fitting of Eqs. (56) and (57) are listed in Table 1. The values of ϵ_t and K_t obtained from the fit along with values of $\bar{M}_n(t)$, $\bar{M}_w(t)$, $[\eta](t)$, and $P(t)$ calculated from the fitted ϵ_t and K_t values are shown in Table 9. Some plots of the uncorrected and corrected DMWD curves $[da/dM](\infty)$ and $[da/dM](t)$, respectively, are shown in Figs. 11–16. The values of $\bar{M}_n(t)$, $\bar{M}_w(t)$, $[\eta](t)$, and $P(t)$ in Table 9 are in excellent agreement with the corresponding values in Table 1. The range of values in ϵ_t and K_t shown in Table 9 for PS samples reflect the dispersion and skewing corrections, experimental errors in $\bar{M}_n(t)$ and $\bar{M}_w(t)$, and chromatogram baseline errors.

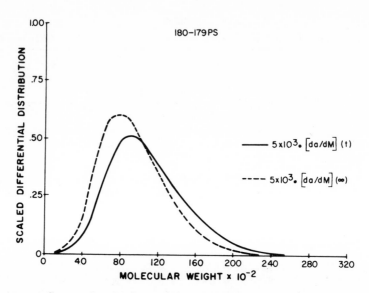

FIG. 11. Corrected and infinite resolution DMWD vs M curves for polystyrene sample 180-179.

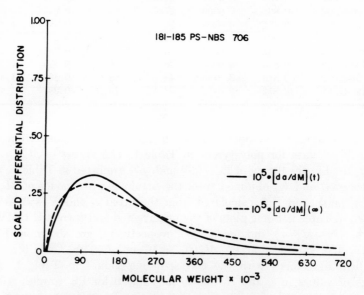

FIG. 12. Corrected and infinite resolution DMWD vs M curves for polystyrene sample 181-185.

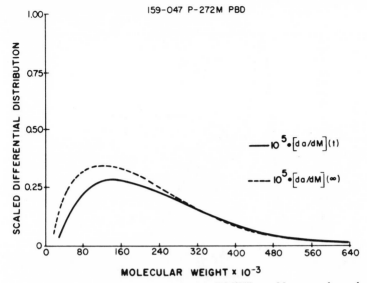

FIG. 13. Corrected and infinite resolution DMWD vs M curves for polybutadiene sample 159-047.

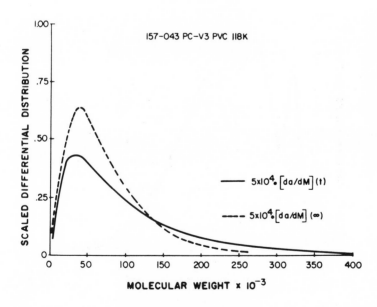

FIG. 14. Corrected and infinite resolution DMWD vs M curves for polyvinyl chloride sample 157-043.

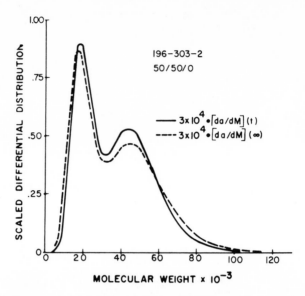

FIG. 15. Corrected and infinite resolution DMWD vs M curves for polystyrene blend 196-303-2.

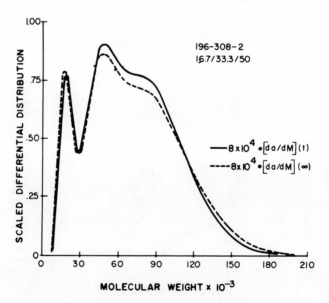

FIG. 16. Corrected and infinite resolution DMWD vs M curves for polystyrene blend 196-308-2.

In principle, the corrected chromatogram $W(v)$ can be obtained from the DMWD curve corrected by the hydrodynamic volume fitting procedure. However, by use of Eq. (54), $W(v)$ can be expressed in terms of $F(v)$ and the uncorrected and corrected molecular weight calibration curves. The corrected DMWD curve obtained from $W(v)$ is given by

$$\left[\frac{da}{dM}\right](t) = W(v) \cdot \frac{1}{\left(\frac{\partial f}{\partial v}\right)_{v_{M(\infty)}}} \cdot \frac{\log_{10} e}{M(\infty)} \tag{62}$$

and the corrected DMWD curve obtained from the hydrodynamic volume fitting procedure is given by

$$\left[\frac{da}{dM}\right](t) = F(v) \cdot \frac{1}{\left(\frac{df_t}{dv}\right)_{v_{M_t}}} \cdot \frac{\log_{10} e}{M_t} \tag{63}$$

Combining Eqs. (62) and (63) yields

$$W(v) = F(v) \left[\left(\frac{df}{dv}\right)_{v_{M(\infty)}} \bigg/ \left(\frac{df_t}{dv}\right)_{v_{M_t}}\right]\left[\frac{M(\infty)}{M_t}\right] \tag{64}$$

For the special case of a linear molecular weight calibration curve over the elution volume range of interest,

$$W(v) = F(v) \frac{D_2(\infty)D_1(\infty)}{D_2(t)D_1(t)} \exp\{-[D_2(\infty) - D_2(t)]v\} \tag{65}$$

From Eqs. (60) and (61), it can be seen that $W(v)$ is then an explicit function of the effective Mark-Houwink parameters ϵ_t and K_t obtained from the hydrodynamic volume fitting procedure.

The Fourier Transform Method

Recently, Tung (*12*) has used a numerical Fourier transform method for correcting observed chromatograms with a Gaussian instrument spreading shape function. The general shape function of Eq. (14) is readily adaptable to the Fourier transform method. Since the statistical coefficients h, μ_3, and μ_4 can be determined as a function of elution volume with narrow MWD standards, corrected chromatograms and subsequently corrected DMWD curves can be determined for broad as well as for narrow MWD samples. Following Tung's notation, the corrected chromatogram is given by

$$W(v) = (1/\sqrt{2\pi}) \int_{-\infty}^{\infty} [W_r(k) \cos (kv) + W_i(k) \sin (kv)] \, dk \quad (66)$$

where

$$W_r(k) = \frac{[F_r(k)G_r(k) + F_i(k)G_i(k)]}{\sqrt{2\pi}[G_r^2(k) + G_i^2(k)]} \quad (67)$$

$$W_i(k) = \frac{[F_i(k)G_r(k) - F_r(k)G_i(k)]}{\sqrt{2\pi}[G_r^2(k) + G_i^2(k)]} \quad (68)$$

The functions $F_r(k)$ and $F_i(k)$, and $G_r(k)$ and $G_i(k)$ refer to the real and imaginary parts of the Fourier transforms of the normalized observed chromatogram, $F(v)$, and the shape function $G(v)$, respectively. The functions $F_r(k)$ and $F_i(k)$ are evaluated numerically. The functions $G_r(k)$ and $G_i(k)$ can be obtained analytically. The Fourier transforms of $G(v - y)$ expressed in terms of the reduced variable $x = (v - y)h^{\frac{1}{2}}$ is

$$G(k) = (1/\sqrt{2\pi}) \int_{-\infty}^{\infty} G(x) \exp (ikx/\sqrt{h}) \, dx \quad (69)$$

Through the use of the generating function (20) for Hermite polynomials, the exponential term in the integrand of Eq. (69) becomes

$$\exp (ikx/\sqrt{h}) = \exp (-k^2/2h) \sum_{n=0}^{\infty} \frac{H_n(x)}{n!} \left(\frac{ik}{\sqrt{h}}\right)^n \quad (70)$$

Substitution of Eqs. (70) and (21) into Eq. (69), along with the use of the ortho-normality conditions for Hermite functions as expressed by Eq. (24), leads to the following general expression for $G(k)$,

$$G(k) = (1/\sqrt{2\pi}) \exp (-k^2/2h) \left[1 + \sum_{n=3}^{\infty} \{(A_n/n!)(ik/h)^n\}\right] \quad (71)$$

Using the Edgeworth series (24) form of Eq. (14), where $A_5 = 0$ and $A_6 = 10A_3^2$ and $A_n = 0$ for $n \geq 7$, the expressions for $G_r(k)$ and $G_i(k)$ expressed in terms of h, μ_3, and μ_4 are

$$G_r(k) = (1/\sqrt{2\pi}) \exp (-k^2/2h) \left[1 + \left(\frac{\mu_4}{24} - \frac{1}{8h^2}\right)k^4 - \frac{1}{2}(\mu_3/6)^2k^6\right] \quad (72)$$

$$G_i(k) = (1/\sqrt{2\pi}) \exp (-k^2/2h)[-(\mu_3/6)k^3] \quad (73)$$

Use of Eqs. (72) and (73) in Eqs. (66), (67), and (68) will lead to the corrected chromatogram $W(v)$. The use of the Fourier transform method with the general shape function can be used for nonlinear as well as linear calibration curves. The application of this formalism to nonlinear calibration curves will be extensively discussed in Part II of this series.

CONCLUSIONS

A general analytical instrument spreading shape function containing statistical coefficients h, μ_3, and μ_4 has been proposed to correct GPC chromatograms for axial dispersion, skewing, and flattening. This shape function has been used to develop simple algebraic equations which correct $\bar{M}_n(\infty)$, $\bar{M}_w(\infty)$, and $[\eta](\infty)$ to their true values. Application of these correction equations to infinite resolution GPC data has yielded corrected values in reasonable agreement with the experimental absolute values. The parameters h and μ_3 may be determined from the observed chromatograms of characterized standards from either knowing $\bar{M}_n$ and $\bar{M}_w$ or knowing $\bar{M}_n$ and $[\eta]$. Determination of h, μ_3, and μ_4 requires knowledge of $\bar{M}_n$, $\bar{M}_w$, and $[\eta]$. The use of $\bar{M}_n$ and $[\eta]$ to determine h and μ_3 is of particular advantage for situations where the determination of $\bar{M}_w$ is not experimentally feasible.

The use of the hydrodynamic volume concept to fit for calibration and DMWD curves corrected for skewing and axial dispersion has been demonstrated. The potential use of the general instrument spreading shape function with the numerical Fourier transform method for correcting DMWD curves has been outlined. The general instrument spreading shape function should be readily adaptable to the correction methods of Smith (*3*); Pierce and Armonas (*6*); Pickett, Cantow, and Johnson (*5*); and the method of Chang and Huang (*10*) for correcting DMWD curves. This shape function also should have applicability to other areas of chromatography and to many fields of spectroscopy where it is important to accurately and analytically describe deviations from the Gaussian peak shape and separate overlapping skewed peaks (*30, 31*).

Acknowledgments

The authors would like to acknowledge the contribution of Mr. Wilfred J. Renaudette in assisting the experimental GPC work, Mr. James H. Clark for his assistance in the GPC data reduction, and Mrs. C. M. Fraser and Miss L. Freeman for helpful assistance in the programming of some of the computer techniques in this work.

List of Symbols

A_n the coefficients in the general shape function in Eq. (14)

A_3 coefficient of general shape function which is a statistical measure of skewness and defined by Eq. (16)

A_4 coefficient of general shape function which is a statistical measure of flatness and defined by Eq. (17)

$C(v_M)$ normalized chromatogram height at elution volume v_M

$D_1, D_1(\infty), D_2, D_2(\infty)$ infinite resolution molecular weight vs counts calibration curve constants and defined by Eq. (6)

$D_1(t), D_2(t)$ corrected molecular weight vs counts calibration curve constants

(da/dM) DMWD function of molecular weight as defined by Eq. (54)

$[da/dM](\infty), [da/dM](t)$ DMWD functions of molecular weight based on the infinite resolution and corrected molecular weight calibration curves, respectively

(da/dZ) differential distribution of hydrodynamic volume as defined by Eq. (58)

$(df/dv)_{v_{M(\infty)}}, (df_t/dv)_{v_{M_t}}, (df'/dv)_{v_Z}$ slope of the infinite resolution molecular weight, corrected molecular weight and hydrodynamic volume calibration curves, respectively

DMWD differential molecular weight distribution

$F(v)$ function describing normalized observed chromatogram heights at elution volume v

$\bar{F}$ bilateral Laplace transform of $F(v)$

$F_r(k), F_i(k)$ real and imaginary parts of the Fourier transform of $F(v)$

$f(v)$ function describing infinite resolution molecular weight calibration curve at elution volume v

$f'(v)$	function describing infinite resolution hydrodynamic volume calibration curve at elution volume v
$f_t(v)$	function describing molecular weight calibration curve corrected for instrument spreading
$G(v-y)$, $G(x)$	general shape functions defined by Eqs. (14) and (21), respectively
$\bar{G}$	bilateral Laplace transform of $G(v-y)$
$G_T(v-y)$	normal form of the Gaussian instrument spreading function
$G_T^n(v-y)$	nth order derivatives of $G_T(v-y)$ with respect to v
$G_r(k)$, $G_i(k)$	real and imaginary parts of the Fourier transform of $G(v-y)$
$H_n(x)$	Hermite polynomials of order n
h, h'	resolution factors defined by Eqs. (18) and (39), respectively
k	Fourier transform variable
K, K_t	empirical and effective Mark-Houwink pre-exponential parameters, respectively
M, $M(v)$	molecular weight in calibration curve corresponding to ideal monodisperse species at elution volume v as defined by Eq. (6)
M_L, M_H	lowest and highest molecular weights of the sample, respectively
$M_k(\infty)$, $M_k(t)$	infinite resolution and absolute or corrected kth molecular weight average as defined by Eq. (4)
$\bar{M}_n$, $\bar{M}_n(t)$	true or absolute value of the number-average molecular weight
$\bar{M}_n(\infty)$	infinite resolution value of the number-average molecular weight
$\bar{M}_n(h)$, $\bar{M}_n(\mu_3)$, $\bar{M}_n(h,\mu_3)$, $\bar{M}_n(\mu_4)$, $\bar{M}_n(\mu_3,\mu_4)$	number-average molecular weights corrected for axial dispersion, skewing, axial dispersion and skewing, flattening, and skewing and flattening, respectively
M_t	molecular weight in corrected calibration curve as defined by Eq. (59)
$\bar{M}_v$	viscosity-average molecular weight

$\bar{M}_w, \bar{M}_w(t)$	true or absolute value of the weight-average molecular weight
$\bar{M}_w(\infty)$	infinite resolution value of the weight-average molecular weight
$\bar{M}_w(h), \bar{M}_w(\mu_3), \bar{M}_w(h,\mu_3),$ $\bar{M}_w(\mu_4), \bar{M}_w(\mu_3,\mu_4)$	weight-average molecular weights corrected for axial dispersion, skewing, axial dispersion and skewing, flattening, and skewing and flattening, respectively
$M(\infty)$	infinite resolution molecular weight as defined by Eq. (6)
MWD	molecular weight distribution
m_2, m_3, m_4	calculated second, third, and fourth moments about the mean, respectively, of normalized observed chromatogram
$P, P(t)$	true or absolute value of polydispersity ratio $[\bar{M}_w(t)/\bar{M}_n(t)]$
$P(\infty)$	infinite resolution value of polydispersity ratio
$P(h,\mu_3), P(\mu_3,\mu_4)$	polydispersity ratios corrected for axial dispersion and skewing, and skewing and flattening, respectively
PEV	peak elution volume
$R_n(t,\infty), R_w(\infty,t), R_v(\infty,t)$	the ratios $\bar{M}_n(t)/M_n(\infty)$, $M_w(\infty)/\bar{M}_w(t)$, and $[\eta](\infty)/[\eta](t)$, respectively
s	bilateral Laplace transform variable
sk	empirical skewing operator of Balke and Hamielec
v	elution volume
v_L, v_H	elution volumes corresponding to lowest and highest molecular weight of the sample, respectively
v_M	elution volume at molecular weight M
v_Z	elution volume at hydrodynamic volume Z
$W(v), W(y)$	functions describing normalized corrected chromatogram height at elution volume v and y, respectively
$\bar{W}$	bilateral Laplace transform of $W(v)$
$W_r(k), W_i(k)$	real and imaginary parts of the Fourier transform of $W(v)$

x	reduced dimensionless variable defined by equation (19)
y	elution volume y
Z	hydrodynamic volume as defined by Eq. (55)
Z_L, Z_H	lowest and highest hydrodynamic volume of the sample, respectively

Greek Letters

ϵ, ϵ_t	empirical and effective Mark-Houwink exponential parameters, respectively
$[\eta], [\eta](t)$	true or absolute value of the intrinsic viscosity
$[\eta](\infty)$	infinite resolution value of the intrinsic viscosity
$[\eta](h), [\eta](\mu_3), [\eta](h,\mu_3),$ $[\eta](\mu_4), [\eta](\mu_3,\mu_4)$	intrinsic viscosities corrected for axial dispersion, skewing, axial dispersion and skewing, flattening, and skewing and flattening, respectively
$\mu_2, \mu_3, \mu_4, \mu_n$	second, third, fourth, and nth order moments about elution volume of the instrument spreading function
$\phi(x)$	normal form of the Gaussian instrument spreading function in reduced variable notation
$\phi^n(x)$	nth order derivative of $\phi(x)$ with respect to x

REFERENCES

1. L. H. Tung, *J. Appl. Polym. Sci.*, **10**, 375 (1966).
2. J. H. Duerksen and A. E. Hamielec, *J. Polym. Sci., Part C*, **21**, 83 (1968).
3. W. N. Smith, *J. Appl. Polym. Sci.*, **11**, 639 (1967).
4. M. Hess and R. F. Kratz, *J. Polym. Sci., Part A-2*, **4**, 731 (1966).
5. H. E. Pickett, M. J. R. Cantow, and J. F. Johnson, *J. Polym. Sci., Part C*, **21**, 67 (1968).
6. P. E. Pierce and J. E. Armonas, *J. Polym. Sci., Part C*, **21**, 23 (1968).
7. L. H. Tung, J. C. Moore, and G. W. Knight, *J. Appl. Polym. Sci.*, **10**, 1261 (1966).
8. L. H. Tung, *J. Appl. Polym. Sci.*, **10**, 1271 (1966).
9. S. T. E. Aldhouse and D. M. Stanford, Paper presented at the 5th International GPC Seminar, London, May, 1968.

10. K. S. Chang and R. Y. M. Huang, *J. Appl. Polym. Sci.*, **13**, 1459 (1969).
11. S. T. Balke and A. E. Hamielec, *J. Appl. Polym. Sci.*, **13**, 1381 (1969).
12. L. H. Tung, *J. Appl. Polym. Sci.*, **13**, 775 (1969).
13. W. W. Yau, H. L. Suchan, and C. P. Malone, *J. Polym. Sci., Part A-2*, **6**, 1349 (1968).
14. H. E. Pickett, M. J. R. Cantow, and J. F. Johnson, *J. Appl. Polym. Sci.*, **10**, 917 (1966).
15. M. Freeman and P. P. Manning, *J. Polym. Sci., Part A*, **2**, 2017 (1964).
16. ArRo Laboratories, Inc., Private communication.
17. A. C. Aitken, *Statistical Mathematics*, Oliver and Boyd, Edinburg, 1962, pp. 64–66.
18. M. Abramowitz and I. A. Stegun, *Handbook of Mathematical Functions*, National Bureau of Standards Applied Mathematics Series 55, Washington, D.C. 1964, pp. 927–935.
19. R. S. Burington and D. C. May, *Handbook of Probability and Statistics*, Handbook Publishers, Sandusky, Ohio, 1953, pp. 91f.
20. H. Cramer, *Mathematical Methods of Statistics*, Princeton Univ. Press, Princeton, New Jersey, 1954, p. 133.
21. V. S. Pugachev, *Theory of Random Functions*, Addison-Wesley, New York, New York, 1962, pp. 44f.
22. H. Freeman, *Introduction to Statistical Inference*, Addison-Wesley, New York, New York, 1963, Chapter 18.
23. J. A. Greenwood and H. O. Hartley, *Guide to Tables in Mathematical Statistics*, Princeton Univ. Press, Princeton, New Jersey, 1962, pp. 407f.
24. D. E. Barton and K. E. Dennis, *Biometrika*, **39**, 425 (1952).
25. D. W. Marquardt, *J. Soc. Ind. Appl. Math.*, **2**, 431 (1963).
26. G. Meyerhoff, *J. Polymer Sci., Part C*, **21**, 31 (1968).
27. A. E. Hamielec, Published Lecture Notes from Two-Day Short Course in Gel Permeation Chromatography at Washington University, St. Louis, April 25–26, 1969.
28. Z. Grubisic, P. Rempp, and H. Benoit, *J. Polym. Sci., Part B*, **5**, 753 (1967).
29. H. Coll and L. R. Prusinowski, *J. Polym. Sci., Part B*, **5**, 1153 (1967).
30. E. Grushka, M. N. Myers, P. D. Schettler, and J. C. Giddings, *Anal. Chem.*, **41**, 889 (1969).
31. H. M. Gladney, B. F. Dowden, and J. D. Swalen, *Anal. Chem.*, **41**, 883 (1969).
32. A. H. Wachter and W. Simon, *Anal. Chem.*, **41**, 90 (1969).

The Instrument Spreading Correction in GPC.
II. The General Shape Function Using the Fourier
Transform Method with a Nonlinear Calibration Curve*

EDWARD M. ROSEN and THEODORE PROVDER†

MONSANTO COMPANY

ST. LOUIS, MISSOURI 63166

Summary

The instrument spreading function suggested in Part I of this series is investigated for use with the Fourier transform method for generating corrected elution volume chromatograms. The instrument spreading parameters are obtained using linear theory on narrow molecular weight distribution standards, as indicated in Part I. The corrected chromatogram is then combined with a nonlinear molecular weight calibration curve which was fit with a function suggested by Yau and Malone to generate true values of the number- and weight-average molecular weights.

The instrument spreading function is shown to qualitatively and quantitatively describe the dispersion, skewing, and flattening effects ordinarily found in GPC chromatograms due to imperfect resolution by the GPC columns. The Yau-Malone function is shown to be a very useful function for fitting nonlinear molecular weight vs elution volume calibration data. Although the Fourier transform method is shown to work well with analytically generated data, it is shown that a number of numerical problems must be overcome before it can quantitatively produce corrected elution volume chromatograms. Some of these numerical problems are discussed.

*Presented at the ACS Symposium on Gel Permeation Chromatography, sponsored by the Division of Petroleum Chemistry at the 159th National Meeting of the American Chemical Society, Houston, Texas, February, 1970.

†Author to whom all correspondence should be addressed at: SCM Corporation, Glidden-Durkee Division, Dwight P. Joyce Research Center, 16651 Sprague Road, Strongsville, Ohio 44136.

291

1. INTRODUCTION

In the first paper of this series (1) [denoted as (I.) throughout the present paper], a general instrument spreading function was suggested which, together with a linear molecular weight calibration curve, was used to correct the infinite resolution gel permeation chromatography (GPC) number- and weight-average molecular weights and intrinsic viscosities, denoted respectively by $\bar{M}_n(\infty)$, $\bar{M}_w(\infty)$, and $[\eta](\infty)$, to their true values, denoted respectively by $\bar{M}_n(t)$, $\bar{M}_w(t)$ and $[\eta](t)$. In addition, it was shown that the hydrodynamic volume concept could be used to evaluate a reasonable approximation to the corrected differential molecular weight distribution (DMWD) curves.

This approach had three major limitations. (a) The $\log_{10} M$ vs elution volume calibration curve had to be linear over the elution volume range of interest. (b) The instrument spreading correction factors were assumed to be constant over the elution volume range of interest. (c) The corrected elution volume chromatogram could not be calculated directly.

In order to overcome the first difficulty, use was made of a nonlinear calibration curve, the functional form of which was suggested by Yau and Malone (2). To circumvent the second and third limitations, use was made of the Fourier transform method (3, 4).

2. THE INSTRUMENT SPREADING FUNCTION

Tung (3) has shown that the normalized (i.e., area under the curve is unity) observed GPC chromatogram $F(v)$ at elution volume v is related to the instrument spreading function $G(v - y)$ and the normalized corrected chromatogram $W(v)$ by means of the equation

$$F(v) = \int_{-\infty}^{\infty} G(v - y)W(y)\, dy \qquad (1)$$

where $F(v)$ is known, $G(v - y)$ is postulated, and $W(y)$ is to be evaluated. Physically, $G(v - y)$ can be thought of as the distribution function of elution volume about an elution volume y describing the shape of the chromatogram resulting from an ideal monodisperse species passing through a set of GPC columns, and $W(y)$ as the weighting factor for the species such that Eq. (1) is satisfied. Whatever form is chosen for $G(v - y)$, it should be capable of describing the behavior of the observed chromatogram of a nearly monodisperse standard.

As indicated in (I.), the instrument spreading function chosen has been widely used in the statistical literature (5, 6) to describe perturbed Gaussian density functions and recently has been used to describe chromatogram shapes in gas–liquid chromatography (7). The instrument spreading function is given by

$$G(x) = \phi(x)\left[1 + \sum_{n=3}^{\infty} (A_n/n!)H_n(x)\right] \qquad (2)$$

where

$$\phi(x) = (1/2\pi)^{1/2} \exp(-x^2/2) \qquad (3)$$

and

$$x = (v - y)/\sqrt{\mu_2} \qquad (4)$$

$H_n(x)$ are the Hermite polynomials defined in Table 1 and μ_2 is the

TABLE 1

Properties of Hermite Polynomials

Hermite Polynomials

$$H_0(x) = 1$$
$$H_1(x) = x$$
$$H_2(x) = x^2 - 1$$
$$H_3(x) = x^3 - 3x$$
$$H_4(x) = x^4 - 6x^2 + 3$$

Generating Function

$$\sum_{m=0}^{\infty} \frac{H_m(x)}{m!} t^m = \exp[t^2/2 + tx]$$

Orthonormality Relationship

$$(1/2\pi)^{1/2} \int_{-\infty}^{\infty} H_n(x)H_m(x) \exp(-x^2/2)\,dx = \begin{array}{ll} n! & n = m \\ 0 & n \neq m \end{array}$$

second moment of the instrument spreading function about an elution volume y. When $A_n = 0$ for $n \geq 5$, Eq. (2) becomes the Gram-Charlier (A) series which shall be used throughout this paper and is given by

$$G(v - y) = (1/2\pi\mu_2)^{1/2} \exp[-(v - y)^2/2\mu_2]$$
$$\times \{1 + (A_3/3!)H_3[(v - y)/\sqrt{\mu_2}] + (A_4/4!)H_4[(v - y)/\sqrt{\mu_2}]\} \qquad (5)$$

where A_3 and A_4 are skewing and flatness parameters defined by the equations

$$A_3 = \mu_3/\mu_2^{3/2} = \sqrt{\beta_1} \qquad (6)$$
$$A_4 = \mu_4/\mu_2^2 - 3 = \beta_2 - 3 \qquad (7)$$

The parameters μ_2, μ_3, and μ_4 are the second, third, and fourth moments of $G(v - y)$ about elution volume y. In general,

$$\mu_n = \int_{-\infty}^{\infty} (v - y)^n G(v - y)\, d(v - y) \qquad n = 2, 3, 4 \qquad (8)$$

Since the instrument spreading parameters μ_n cannot be measured directly, they will be determined by an indirect procedure. Barton and Dennis (8) examined Eq. (5) for regions in which it remained positive definite and unimodal. Their results are shown in Fig. 1 in terms of the parameters β_1 and β_2 defined by Eqs. (6) and (7). The corresponding regions for the Edgeworth series (Eq. 2 with $A_6 = 10A_3^2$ and $A_n = 0$ for $n \geq 7$) used in (I.) are also shown. Figure 1 shall be referred to in the discussion of the determination of the instrument spreading parameters in Section 6.

A number of relationships between the moments of $F(v)$ and the moments of $W(y)$ can be derived directly from Eqs. (1) and (2) by using the properties of the Hermite polynomials. These relationships together with the notations used are summarized in Table 2.

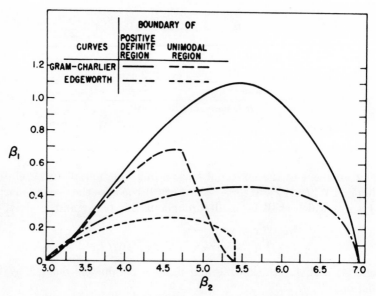

FIG. 1. Unimodal and positive definite regions in the β_1-β_2 plane for Gram-Charlier and Edgeworth Curves.

TABLE 2

Nomenclature and Relationships Among the Moments of the Raw and Corrected
Chromatograms and the Instrument Spreading Function

	Nomenclature		
Parameter	Raw chromatogram $F(v)$ (moments about μ)	Instrument spreading function $G(v\text{-}y)$ (moments about y)	Corrected chromatogram $W(v)$ (moments about μ)
Unnormalized area	A	1	A
Normalized area	1	1	1
Mean	μ	0	μ
Second moment	m_2	μ_2	m_2^*
Third moment	m_3	μ_3	m_3^*
Fourth moment	m_4	μ_4	m_4^*

Moment	Relation
Zero	Area of unnormalized $W(v)$ = Area of unnormalized $F(v)$
First	Mean of $W(v)$ = Mean of $F(v)$
Second	$m_2^* = m_2 - \mu_2$
Third	$m_3^* = m_3 - \mu_3$
Fourth	$m_4^* = m_4 - 6\mu_2 m_2 + 6\mu_2^2 - \mu_4$

To illustrate the general method used to derive these relationships, consider the third moment calculations. The third moment about zero elution volume is given by

$$\langle v^3 \rangle = \int_{-\infty}^{\infty} v^3 F(v)\, dv = \int_{-\infty}^{\infty} \int_{-\infty}^{\infty} v^3 G(v - y) W(y)\, dy\, dv \qquad (9)$$

Substituting Eqs. (2), (3), and (4) into (9) and reversing the order of integration yields

$$\langle v^3 \rangle = \int_{-\infty}^{\infty} W(y) \int_{-\infty}^{\infty} \phi(x)[H_0(x) + (A_3/3!)H_3(x) + (A_4/4!)H_4(x)]$$
$$\cdot\, [\sqrt{\mu_2}x + y]^3\, dx\, dy \qquad (10)$$

Expanding and integrating term by term yields

$$\langle v^3 \rangle = \int_{-\infty}^{\infty} W(y)[A_3\mu_2^{3/2} + 3\mu_2 y + y^3]\, dy \qquad (11)$$

Then

$$\langle v^3 \rangle = \langle y^3 \rangle + A_3\mu_2^{3/2} + 3\mu_2\mu \qquad (12)$$

Expressing Eq. (12) in terms of moments about the mean results in the following relation

$$m_3^* = m_3 - \mu_3 \tag{13}$$

3. THE FOURIER TRANSFORM METHOD

Equation (1) is an integral equation of the convolution type and classically has been solved by means of Fourier transforms. This transformation is performed because the relationship can be reduced to a simple algebraic expression in the transformed k-space. Following Tung (3), the Fourier transform of both sides of Eq. (1) yields the relation

$$W(k) = (1/2\pi)^{1/2}[F(k)/G(k)] \tag{14}$$

where

$$F(k) = (1/2\pi)^{1/2} \int_{-\infty}^{\infty} F(v) \exp (ikv) \, dv \tag{15}$$

$$G(k) = (1/2\pi)^{1/2} \int_{-\infty}^{\infty} G(v - y) \exp [ik(v - y)] \, d(v - y) \tag{16}$$

$$W(k) = (1/2\pi)^{1/2} \int_{-\infty}^{\infty} W(v) \exp (ikv) \, dv \tag{17}$$

and

$$F(k) = F_r(k) + iF_i(k) \tag{18}$$
$$G(k) = G_r(k) + iG_i(k) \tag{19}$$
$$W(k) = W_r(k) + iW_i(k) \tag{20}$$

Use of Eqs. (18), (19), and (20) in Eq. (14) leads to the following relations for $W_r(k)$ and $W_i(k)$

$$W_r(k) = \frac{F_r(k)G_r(k) + F_i(k)G_i(k)}{\sqrt{2\pi} \, [G_r^2(k) + G_i^2(k)]} \tag{21}$$

$$W_i(k) = \frac{F_i(k)G_r(k) - F_r(k)G_i(k)}{\sqrt{2\pi} \, [G_r^2(k) + G_i^2(k)]} \tag{22}$$

Once $W(k)$ is determined, $W(v)$ can be obtained from

$$W(v) = (1/2\pi)^{1/2} \int_{-\infty}^{\infty} W(k) \exp (-ikv) \, dv \tag{23}$$

Upon noting that $W_r(k)$ and $W_i(k)$ are even and odd functions, respectively, in k-space, $W(v)$ can be expressed in terms of $W_r(k)$ and $W_i(k)$ by the equation

$$W(v) = (1/2\pi)^{1/2} \int_{-\infty}^{\infty} [W_r(k) \cos (kv) + W_i(k) \sin (kv)] \, dk \qquad (24)$$

The operational procedure used for the determination of $W(v)$ consists of the following steps. First, $F_r(k)$ and $F_i(k)$ are numerically evaluated from the normalized observed chromatogram $F(v)$.

$$F_r(k) = (1/2\pi)^{1/2} \int_{-\infty}^{\infty} F(v) \cos (kv) \, dv \qquad (25)$$

$$F_i(k) = (1/2\pi)^{1/2} \int_{-\infty}^{\infty} F(v) \sin (kv) \, dv \qquad (26)$$

Using the generating function of Table 1 and Eqs. (5), (6), and (7) in Eqs. (16) and (19), analytical expressions are obtained for $G_r(k)$ and $G_i(k)$.

$$G_r(k) = (1/2\pi)^{1/2} \exp (-k^2\mu_2/2)[1 + (k^4/24)(\mu_4 - 3\mu_2^2)] \qquad (27)$$

$$G_i(k) = (1/2\pi)^{1/2} \exp (-k^2\mu_2/2)[-k^3\mu_3/6] \qquad (28)$$

Upon knowing or assuming values for μ_2, μ_3, and μ_4, $G_r(k)$ and $G_i(k)$ are evaluated and then $W_r(k)$ and $W_i(k)$ are obtained from Eqs. (21) and (22). Finally, $W(v)$ is evaluated from Eq. (24).

Although the parameters μ_2, μ_3, and μ_4 are assumed to be constants or slowly varying functions of the elution volume v during the integrations, in general they shall be considered to be strong functions of the elution volume v in the evaluation of $W(v)$. In general, $G(k)$ and $W(k)$ then will both be functions of v. This inconsistency will remain a limitation of the method, particularly when μ_2, μ_3, and μ_4 are no longer slowly varying functions of v. Note, however, regardless of the nature of $G(k)$, $F(k)$ needs to be evaluated only once because it is only a function of the observed chromatogram.

4. NUMERICAL PROCEDURES

The evaluation of the integrals in Eqs. (25) and (26) have been the subject of considerable numerical investigations (9, 10). Tung (3) used the discrete Fourier transform (DFT), which considers the integrals as infinite sums.* The fast Fourier transform (FFT) is an efficient method for the evaluation of the DFT when the number of data points tends to be very large. However, for GPC data the sampling interval is not excessively small nor is the range of elution volumes very large. Hence, the integral form can be used without excessive computational

---

* Examination of the Fortran computer program supplied by Dr. Tung indicated the use of the DFT.

work. In order to evaluate Eqs. (25) and (26), the range of integration must be transformed to some convenient point about the center of the chromatogram, v_0, which is an even multiple of the increment size used in k-space. Upon letting

$$z = v - v_0 \tag{29}$$

Eqs. (25) and (26) become

$$F_r(k) = \cos (kv_0)F_1(k) - \sin (kv_0)F_2(k) \tag{30}$$

and

$$F_i(k) = \cos (kv_0)F_2(k) + \sin (kv_0)F_1(k) \tag{31}$$

where

$$F_1(k) = (1/2\pi)^{1/2} \int_{-\infty}^{\infty} F(z) \cos (kz) \, dz \tag{32}$$

and

$$F_2(k) = (1/2\pi)^{1/2} \int_{-\infty}^{\infty} F(z) \sin (kz) \tag{33}$$

Under this transformation Eq. (24) becomes

$$W(v) = (1/2\pi)^{1/2} \int_{-\infty}^{\infty} [W_1(k) \cos (kz) + W_2(k) \sin (kz)] \, dk \tag{34}$$

where

$$W_1(k) = \frac{F_1(k)G_r(k) + F_2(k)G_i(k)}{\sqrt{2\pi} \, [G_r^2(k) + G_i^2(k)]} \tag{35}$$

$$W_2(k) = \frac{F_2(k)G_r(k) - F_1(k)G_i(k)}{\sqrt{2\pi} \, [G_r^2(k) + G_i^2(k)]} \tag{36}$$

Equations (32) and (33) are evaluated by breaking $F(z)$ into small straight-line segments and analytically integrating over the segments. This technique is known generally as the Filon Quadrature (11). For example, for the jth interval let $F_j(z) = a_j + b_j z$. For N data points (not necessarily equally spaced), Eq. (32) becomes

$$F_1(k) = (1/2\pi)^{1/2} \Bigg\{ \sum_{j=1}^{N-1} (a_j/k)[\sin (kz_{j+1}) - \sin (kz_j)]$$

$$+ \sum_{j=1}^{N-1} (b_j/k^2)[\cos (kz_{j+1}) + kz_{j+1} \sin (kz_{j+1})$$

$$- \cos (kz_j) - kz_j \sin (kz_j)] \Bigg\} \qquad k \neq 0 \tag{37}$$

In practice, both $F_1(k)$ and $F_2(k)$ are evaluated at equal intervals of k. By noting that $F_1(k) = F_1(-k)$ and $F_2(k) = -F_2(-k)$, the evaluation starts at $k = 0$ and simultaneously proceeds out in the positive and negative directions in k-space. The actual interval used in k-space varies because it is dependent on the value of k_{max} required for the evaluation of $F_1(k)$ and $F_2(k)$. Generally, the intervals in k were in the range $\pi/600 \leq \Delta k \leq \pi/20$. Since equal increments in k-space were selected, it was convenient to avoid multiple evaluations of the sine and cosine functions by means of expressions such as

$$\sin (k_{i+1}z_j) = \sin (k_i z_j + \Delta k z_j)$$
$$= \sin (k_i z_j) \cos (\Delta k z_j) + \cos (k_i z_i) \sin (\Delta k z_j) \qquad (38)$$

One of the major difficulties of the entire method, however, was determining the value of k_{max} out to which $F(k)$ should be evaluated. When $F(v)$ is a smooth analytical function, the magnitude of $F(k)$ goes to zero very smoothly as $|k|$ increases. Therefore $|F(k)|$ can be monitored and the generation terminated when $|F(k)|$ becomes suitably small. However, when $F(v)$ is actual experimental data, no such smooth behavior is encountered. In fact, rather slowly undulating oscillations are encountered as $|k|$ increases. This is due to two factors: (a) Noise and oscillations actually are present in $F(v)$; and (b) whenever a discontinuity appears in $F(v)$, due to the fact that the experimental data is chopped off at some arbitrary point before having a chance to approach zero asymptotically, oscillations will appear in the transform. Taking the transform of any simple function which is nonzero within a finite interval and zero outside of the interval will illustrate this point (*12*).

It is very difficult indeed to separate these two effects and determine the value of $|k|$ at which to terminate the evaluation of $F(k)$. If the numerical evaluation of $F(k)$ is performed out to values of $|k|$ that are too large, round-off errors will predominate because $F(k)$ is divided by $G(k)$, which itself goes to zero, resulting in a $0 \cdot \infty$ calculation. Termination of the evaluation of $F(k)$ at values of $|k|$ that are much too small will give poor results because much of the information is lost. Tung (*3*) used an empirical rule to determine the value of k_{max} and indicated that the results could be improved by varying this value for the particular problem.

A theoretical upper limit on k_{max} as given by the sampling theorem is

$$k_{max} = \pi/\Delta v \qquad (39)$$

where Δv is the sampling interval. Some studies (*13*) have indicated that, in practice, this should be closer to $0.1(\pi/\Delta v)$. However, a hard and fast rule appears difficult to ascertain.

A number of approaches has been suggested to overcome this problem (10). One approach is the use of a "spectral window" (10) to provide a weighting factor to force the tails of the chromatogram to approach zero smoothly. The use of a spectral window will be discussed further in Section 8 as well as the method used in practice to determine k_{max}.

The evaluation of $G(k)$ is made over the same range of k as used to determine $F(k)$. After $W(k)$ is evaluated, $W(v)$ is calculated using the same quadrature method that was used to evaluate $F(k)$. The spacing of the $W(v)$ function in v-space is determined by the spacing that was used in k-space. Generally, $F(v)$ curves covering narrow ranges of elution volume require evaluations in k-space out to large values of k (a coarse k-spacing) which in turn require very fine spacing in v-space for the $W(v)$ calculation. In this work, spacing was based on fractions of 600. Thus if the interval in k-space was $\pi/75$, then spacing in the corrected v-space was at count increments of $1/12$.

5. THE CALIBRATION CURVE

Yau and Malone (2) derived a theoretical expression describing a relationship between molecular weight and elution volume. Their equation is of the form

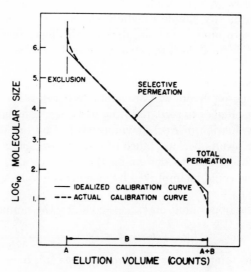

FIG. 2. Illustrative GPC calibration curve.

$$v = A + B\{(1/\sqrt{\pi}\psi)[1 - \exp(-\psi^2)] + \text{erfc}(\psi)\} \qquad (40)$$

where

$$\psi = M^D/C \qquad (41)$$

The form of Eq. (40) has the characteristics of experimental GPC molecular weight calibration curves and has a number of convenient mathematical properties. If A, B, C, and D are treated as parameters, then Eq. (40) can be used to fit experimental molecular weight vs count data.

The value of A, once determined, is the effective exclusion volume and $A + B$ is the total volume as indicated in Fig. 2 taken from Ref. *14*. If sufficient data are not present in the steep portions of the curve at very high and very low elution volumes to determine A and B adequately, then this physical significance is lost.

Calibration curves such as shown in Fig. 2 are arbitrary to the extent that a particular molecular weight average such as $\bar{M}_v$, $\bar{M}_w$ or $(\bar{M}_w \cdot \bar{M}_n)^{\frac{1}{2}}$

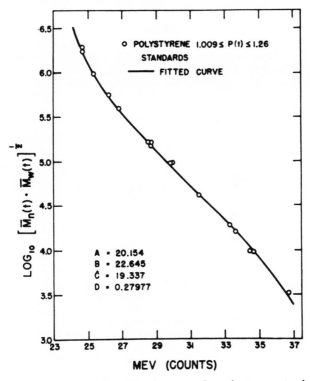

FIG. 3. Molecular weight calibration curve for polystyrene standards.

is associated with the peak elution volume (PEV) or mean elution volume (MEV).

Whatever is used, nevertheless, appears to be satisfactorily fit with Eq. (40). Figure 3 is a plot of the data given in Table 3 in which $(\bar{M}_n \cdot \bar{M}_w)^{\frac{1}{2}}$ is plotted vs MEV. Values* of A, B, C, and D were determined from a nonlinear least squares fit (15, 16) of the data and are shown in Fig. 3.

Pickett, Cantow, and Johnson (17) have shown that $\bar{M}_n$ and $\bar{M}_w$ can be evaluated from

$$\bar{M}_n = \left[\int_{M_L}^{M_H} (1/M)(da/dM) \, dM \right]^{-1} \tag{42}$$

$$\bar{M}_w = \int_{M_L}^{M_H} M(da/dM) \, dM \tag{43}$$

where

$$(da/dM) = - \frac{C(v_M)}{\int_{V_L}^{V_H} C(v) \, dv} \cdot \frac{1}{(df/dv)_{v_M}} \cdot \frac{\log_{10} e}{M} \tag{44}$$

TABLE 3

Statistical and Molecular Weight Data for Polystyrene
Standards Based on Normalized Chromatograms at 1ml/min Flow Rate

Run[a]	$10^{-3}\bar{M}_n(t)$	$10^{-3}\bar{M}_w(t)$	$P(t)$	MEV	m_2	m_3	m_4
181-189	1610	1900	1.18	24.65	2.487	5.833	33.05
181-190	1780	2145	1.21	24.67	2.463	5.690	31.61
181-188	773	867	1.12	25.31	1.686	3.458	17.25
181-187	404	507	1.26	26.27	1.801	3.164	17.36
181-186	392	411	1.05	26.84	1.833	3.390	19.06
181-184	170.9	179.3	1.05	28.52	0.933	0.457	4.224
181-183	164	173	1.05	28.68	0.642	0.473	2.119
184-211	160.6	162	≤1.009	28.71	0.649	0.479	1.927
184-210	95.1	96	1.02	29.88	0.556	0.350	1.268
180-182	96.2	98.2	≤1.009	29.90	0.540	0.335	1.394
184-209	42.12	42.5	≤1.009	31.56	0.524	0.450	1.603
180-180	19.65	19.85	1.01	33.32	0.600	0.135	1.212
184-208	16.35	16.5	≤1.009	33.61	0.406	0.128	0.572
184-207	9.91	10.0	≤1.009	34.57	0.318	0.0471	0.316
180-179	9.70	10.3	1.06	34.73	0.662	0.283	1.701
180-178	3.18	3.53	≤1.11	36.78	0.766	0.276	1.863

[a] Refer to (I.) for the commercial sources of the polystyrene standards.

* A Fortran IV computer program which makes use of Marquardt's algorithm to solve a set of nonlinear equations can be found in Ref. 16.

The first term on the right-hand side of Eq. (44) is the normalized baseline-adjusted chromatogram height at elution volume v_M, where $C(v_M)$ is the raw chromatogram height at v_M and $\int_{V_L}^{V_H} C(v)\ dv$ is the area of the raw chromatogram. The second term on the right-hand side of Eq. (44) is the reciprocal of the slope of the calibration curve, $f(v) = \log_{10} M$. Equation (44) can be expressed as

$$(da/dM) = -[C(v_M)/\text{area}](dv/dM)v_M \tag{45}$$

From Eqs. (40) and (41)

$$(dv/dM)v_M = -(DB/\psi M \sqrt{\pi})[1 - \exp(-\psi^2)] \tag{46}$$

Since B, C, and D are always positive, $(dv/dM)v_M$ is always negative. Throughout this paper, $\bar{M}_n$ and $\bar{M}_w$ are determined by means of Eqs. (42), (43), (45), and (46).

6. EVALUATION OF THE INSTRUMENT SPREADING PARAMETERS

Before the Fourier transformed instrument spreading function defined by Eqs. (27) and (28) can be evaluated, it is necessary to determine the values of μ_2, μ_3, and μ_4 as a function of elution volume. In principle, these parameters can be determined from a knowledge of $\bar{M}_n(t)$, $\bar{M}_w(t)$, and $[\eta](t)$ or $\bar{M}_v(t)$. However, in this paper it shall be assumed that only $\bar{M}_n(t)$ and $\bar{M}_w(t)$ are available. In addition an assumption shall be made regarding the nature of the corrected chromatogram for narrow MWD standards.

If the chromatograms of the characterized standard polymer samples cover narrow ranges of elution volume, it can be assumed that (a) μ_2, μ_3, and μ_4 are slowly varying functions of elution volume and will be reasonably constant over this elution volume range, and (b) the calibration curve will be reasonably linear over this elution volume range. Under these assumptions, it has been shown in (I.) that the following correction equations are applicable.

$$\frac{\bar{M}_n(t)}{\bar{M}_n(\infty)} = \exp(\alpha^2/2)\{1 + \sqrt{\beta_1}\alpha^3/6 + (\alpha^4/24)(\beta_2 - 3)\} \tag{47}$$

$$\frac{\bar{M}_w(\infty)}{\bar{M}_w(t)} = \exp(\alpha^2/2)\{1 - \sqrt{\beta_1}\alpha^3/6 + (\alpha^4/24)(\beta_2 - 3)\} \tag{48}$$

where

$$\alpha = D_2 \sqrt{\mu_2} \tag{49}$$

$$\beta_1 = \mu_3^2/\mu_2^3 \tag{50}$$

$$\beta_2 = \mu_4/\mu_2^2 \tag{51}$$

D_2 is the slope of the $\log_e M$ vs elution volume calibration curve at a specified elution volume v_s and is given by

$$D_2 = (d \log_e M/dv)_{v_s} = [M(dv/dM)_{v_s}]^{-1} \tag{52}$$

In terms of the parameters of the Yau-Malone function, D_2 is expressed as

$$D_2 = -(\sqrt{\pi}\,\psi/DB)[1 - \exp(-\psi^2)]^{-1} \tag{53}$$

Addition of Eqs. (47) and (48) results in an expression for the fundamental parameter α.

$$\alpha^2 = 2 \log_e \left\{ \frac{[\bar{M}_n(t)/\bar{M}_n(\infty)] + [\bar{M}_w(\infty)/\bar{M}_w(t)]}{2[1 + (\alpha^4/24)(\beta_2 - 3)]} \right\} \tag{54}$$

Subtracting Eq. (48) from Eq. (47) and combining the result with Eq. (54) yields an expression for the fundamental parameter β_1.

$$\sqrt{\beta_1} = (6/\alpha^3)[1 + (\alpha^4/24)(\beta_2 - 3)]$$
$$\times \left\{ \frac{[\bar{M}_n(t)/\bar{M}_n(\infty)] - [\bar{M}_w(\infty)/\bar{M}_w(t)]}{[\bar{M}_n(t)/\bar{M}_n(\infty)] + [\bar{M}_w(\infty)/\bar{M}_w(t)]} \right\} \tag{55}$$

Equations (54) and (55) are two equations in the three unknowns α, β_1, and β_2. For the third equation, consider the relationship between the fourth moment of the observed and corrected chromatogram as shown in Table 1.

$$m_4^* = m_4 - 6\mu_2 m_2 + 6\mu_2^2 - \mu_4 \tag{56}$$

If the corrected chromatogram for the standard is assumed to be nearly Gaussian in shape, then $m_4^* = 3m_2^{*2}$. Using this assumption along with the relation $m_2^* = m_2 - \mu_2$ in Eq. (56) results in

$$\mu_4 - 3\mu_2^2 = m_4 - 3m_2^2 \tag{57}$$

Combining Eq. (57) with Eqs. (49) and (51) leads to the necessary third equation

$$(\alpha^4/24)(\beta_2 - 3) = (D_2^4/24)(m_4 - 3m_2^2) \tag{58}$$

Use of Eq. (58) in (54) and (55) leads to expressions for α and β_1 in terms of quantities that can be determined by measurement or calculation.

$$\alpha^2 = 2 \log_e \left\{ \frac{[\bar{M}_n(t)/\bar{M}_n(\infty)] + [\bar{M}_w(\infty)/\bar{M}_w(t)]}{2[1 + (D_2^4/24)(m_4 - 3m_2^2)]} \right\} \tag{59}$$

$$\alpha^3 \sqrt{\beta_1} = 6[1 + (D_2^4/24)(m_4 - 3m_2^2)] \\ \times \left\{ \frac{[\bar{M}_n(t)/\bar{M}_n(\infty)] - [\bar{M}_w(\infty)/\bar{M}_w(t)]}{[\bar{M}_n(t)/\bar{M}_n(\infty)] + [\bar{M}_w(\infty)/\bar{M}_w(t)]} \right\} \tag{60}$$

It is important to note that α is negative due to the negativity of D_2 in Eq. (49).

The calibration procedure to obtain μ_2, μ_3, and μ_4 as a function of elution volume is described below. The raw data of the standards used in this procedure are the data shown in Table 3 along with the corresponding raw elution volume curves. The experimental GPC conditions and procedures used for the generation of the raw chromatograms have been described previously in (I.).

(a) The raw $\bar{M}_n(t)$ vs MEV and $\bar{M}_w(t)$ vs MEV data sets in Table 3 are each fit to the functional form of the Yau-Malone Eq. (40). The smooth values $\bar{M}_n(t)_s$ and $\bar{M}_w(t)_s$ are calculated from the fitted functions. The results are shown in Table 4 with the subscript s denoting smooth values.

(b) The infinite resolution molecular weight calibration curve is obtained by using Eq. (40) to fit $[\bar{M}_n(t) \cdot \bar{M}_w(t)]^{1/2}$ vs MEV data. Values

TABLE 4

Smooth Values of Molecular Weight for Polystyrene Standards

Run	MEV	$10^{-3}\bar{M}_n(t)_s$	$10^{-3}\bar{M}_w(t)_s$	$P(t)_s$	$10^{-3}\bar{M}_n(\infty)_s$	$10^{-3}\bar{M}_w(\infty)_s$	$P(\infty)_s$
181-189	24.65	1510	1780	1.18	925	3090	3.34
181-190	24.67	1490	1740	1.17	913	3020	3.31
181-188	25.31	934	1100	1.18	661	1580	2.39
181-187	26.27	513	576	1.12	408	725	1.78
181-186	26.84	372	418	1.12	317	490	1.55
181-184	28.52	166	174	1.05	152	186	1.22
181-183	28.68	155	162	1.05	142	173	1.22
184-211	28.71	152	158	1.04	141	170	1.21
184-210	29.88	89.2	92.7	1.04	85.1	96.8	1.14
180-182	29.90	89.2	92.5	1.04	85.0	96.2	1.13
184-209	31.56	43.7	43.8	1.002	41.7	44.7	1.07
180-180	33.32	19.5	19.6	1.005	18.6	20.4	1.10
184-208	33.61	17.0	17.1	1.006	15.9	17.8	1.12
184-207	34.57	10.5	11.0	1.05	9.55	11.2	1.17
180-179	34.73	9.78	10.0	1.02	8.91	10.5	1.18
180-178	36.78	2.89	3.24	1.12	2.40	3.39	1.41

of the Yau-Malone parameters are given together with the fitted curve in Fig. 3.

(c) The infinite resolution calibration curve and the raw chromatograms of the standards are used to calculate $\bar{M}_n(\infty)$ and $\bar{M}_w(\infty)$. These values are again smoothed by fitting to Eq. (40). Values of $\bar{M}_n(\infty)_s$ and $\bar{M}_w(\infty)_s$ are shown in Table 4.

(d) The values of $\bar{M}_n(t)_s$, $\bar{M}_w(t)_s$, $\bar{M}_n(\infty)_s$, and $\bar{M}_w(\infty)_s$ together with the statistics of the standards given in Table 3 are used to find the value of α^2 from Eq. (59) at each MEV. Equation (53) is used to evaluate D_2 at each MEV. Plots of α and α^2 vs MEV are shown in Fig. 4 and a plot of D_2 vs MEV is shown in Fig. 5.

(e) Values of μ_2 are calculated at equal increments of elution volume using Eq. (49) and Fig. 5. These results are given in Table 5 and plotted in Fig. 6.

(f) Values of $\sqrt{\beta_1}$ are calculated at each MEV using Eqs. (60) and

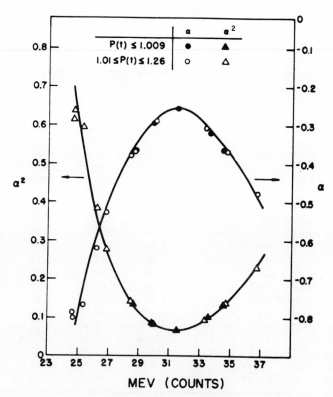

FIG. 4. α and α^2 vs MEV.

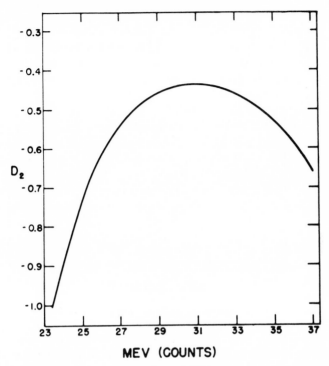

FIG. 5. D_2 vs MEV.

TABLE 5

Interpolated Values of μ_2, μ_3, and μ_4 at Equal Increments of the Mean Elution Volume

MEV	μ_2	μ_3	μ_4
23	0.825	0.970	19.1
24	0.950	0.695	19.0
25	1.05	0.270	17.1
26	1.05	0.160	13.2
27	0.927	0.047	8.15
28	0.750	0.00	4.32
29	0.593	0.0145	2.22
30	0.465	0.0346	1.15
31	0.397	−0.0750	0.706
32	0.370	−0.213	0.550
33	0.425	−0.500	0.650
34	0.510	−0.890	0.859
35	0.550	−1.18	0.966
36	0.595	−1.52	1.10
37	0.580	−1.63	1.01

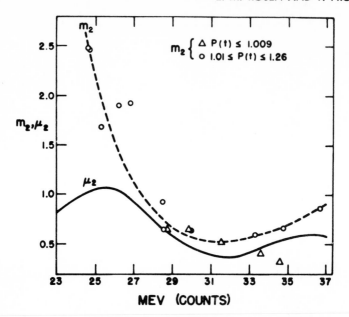

FIG. 6. m_2 and μ_2 vs MEV.

(53) and Fig. 4. Then values of μ_3 at equal increments of elution volume are obtained from a plot of $\sqrt{\beta_1}$ vs MEV in Fig. 7, Eq. (50), and Fig. 6.

(f) Values of μ_3 at equal increments of elution volume are given in Table 5 and plotted in Fig. 8.

(g) Values of β_2 are calculated at each MEV from Eqs. (58) and (53) and Fig. 4. Then values of μ_4 at equal increments of elution volume are obtained from a plot of β_2 vs MEV in Fig. 9, Eq. (51), and Fig. 6. Values of μ_4 at equal increments of elution volume are given in Table 5 and plotted in Fig. 10.

Several points in the above procedure are worthy of note. (a) The smoothing operations on $\bar{M}_n(t)$, $\bar{M}_w(t)$, $\bar{M}_n(\infty)$, and $\bar{M}_w(\infty)$ eliminated much of the scatter in the plots of μ_2, μ_3, and μ_4 vs MEV. (b) The entire calibration procedure was rather arbitrary because, in principle, $\bar{M}_w(t)$ vs PEV could have been used instead of $[\bar{M}_n(t) \cdot \bar{M}_w(t)]^{\frac{1}{2}}$ vs MEV. However, use of $\bar{M}_w(t)$ vs PEV gave less satisfactory numerical values of β_1.

The rather surprising shape of the μ_2 vs MEV curve in Fig. 6 can perhaps be justified by plotting values of m_2 on the same plot. Since m_2^* must always be positive, this implies that $m_2 > \mu_2$. With the exception of two points of the ultranarrow recycle polystyrene standards, the

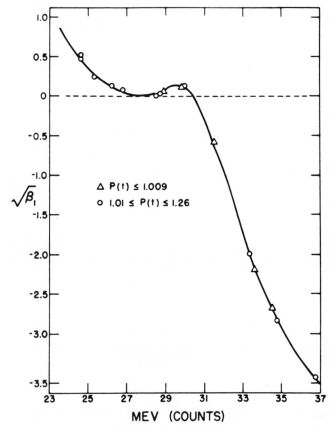

FIG. 7. $\sqrt{\beta_1}$ vs MEV.

values of m_2 are, in fact, greater than μ_2 within experimental errors.

Similar reasoning can be applied to Fig. 10. Since $m_4^* > 0$, it follows from Eq. (56) that

$$m_4^* + \mu_4 = m_4 - 6\mu_2 m_2 + 6\mu_2^2 > \mu_4 \qquad (61)$$

Generally, $m_4^* \geq 0.03$ over most of the elution volume range. Therefore, Eq. (61) holds. Figure 10 illustrates that, generally, $m_4^* \ll \mu_4$ because many of the experimental data points are almost indistinguishable from the μ_4 curve.

It is difficult to assess the effect of the use of the approximation given by Eq. (58) except to note that, generally, $D_2^4(m_4 - 3m_2^2)/24 \ll 1$.

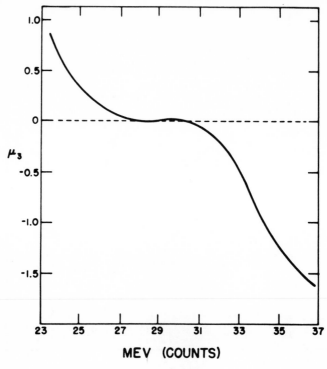

FIG. 8. μ_3 vs MEV.

However, when $|D_2|$ begins to climb rapidly in the low MEV region, the above inequality is no longer valid and the approximation made in Eq. (58) may introduce serious errors in the determination of μ_2, μ_3, and μ_4.

Over some regions of elution volume the values of β_1 and β_2 fall outside the unimodal and positive definite regions of the Gram-Charlier curves as indicated in Fig. 1. However, this does not necessarily mean that the instrument spreading function is predicting unrealistic values over the elution volume range of the sample. Unrealistic values of $G(v-y)$ that do occur over the elution volume range of the sample may be so small that they have no practical significance with respect to the values of $W(v)$ generated by the numerical Fourier transform method. Consider, for example, that the normalized observed chromatogram can be described by the skewed Gaussian function

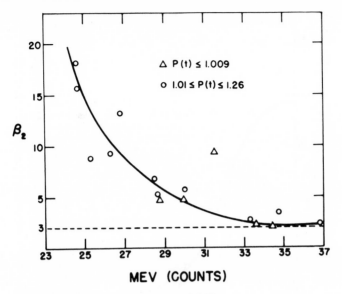

FIG. 9. β_2 vs MEV.

$$F(v) = (1/2\pi m_2)^{1/2} \exp\left[-(v-\mu)^2/2m_2\right]$$
$$\times \{1 + (C_3/3!)H_3[(v-\mu)/\sqrt{m_2}] + (C_4/4!)H_4[(v-\mu)/\sqrt{m_2}]\} \quad (62)$$

where

$$C_3 = m_3/m_2^{3/2} \quad (63)$$
$$C_4 = m_4/m_2^2 - 3 \quad (64)$$

Then

$$F(k) = (1/\sqrt{2\pi}) \exp(-k^2 m_2/2)\{1 + (C_3/3!)(ik\sqrt{m_2})^3 + (C_4/4!)k^4 m_2^2\} \quad (65)$$

From Eqs. (27), (28), and (14)

$$W(k) = (1/\sqrt{2\pi}) \exp\left[\frac{-k^2(m_2-\mu_2)}{2}\right] \left\{\frac{1 + \dfrac{C_3}{3!}(ik\sqrt{m_2})^3 + \dfrac{C_4}{4!}k^4 m_2^2}{1 + \dfrac{A_3}{3!}(ik\sqrt{\mu_2})^3 + \dfrac{A_4}{4!}k^4 \mu_2^2}\right\} \quad (66)$$

Equation (66) contains an exponential in m_2^* multiplied by a ratio of polynomials in the parameters β_1 and β_2 through the coefficients A_3 and

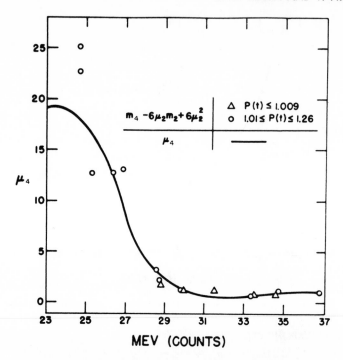

MEV (COUNTS)

FIG. 10. μ_4 and $m_4 - 6\mu_2 m_2 + 6\mu_2^2$ vs MEV.

A_4, and in the parameters m_3 and m_4 through the coefficients C_3 and C_4. If m_2^* ($m_2^* = m_2 - \mu_2$) is large, it is not necessary to go out too far in k-space in order for the exponential term to wash out the effect of the bracketed term in Eq. (66) and force $|W(k)|$ to go to zero. On the other hand, when m_2^* is small, it is necessary to go out farther in k-space. Then, the effect of noise in the numerical $F(v)$ data and round-off errors tend to be greatly magnified at large values of k and prevents $|W(k)|$ from going to zero. If $|W(k)|$ is poorly bounded, $W(v)$ will tend to have unrealistic values manifested by large positive and negative oscillations.

7. TEST RESULTS—ANALYTICAL DATA

In order to check the numerical procedures used in the Fourier transform method, two analytical functions were chosen for study. The first test function was designed to check the functioning of the dispersion, skewing, and flattening parameters. The observed chromatogram was described by a skewed Gaussian function which transformed to a corrected chromatogram that was described by a pure Gaussian function. The pure Gaussian function in k-space is given by

$$W(k) = (1/2\pi)^{1/2} \exp{(-k^2 m_2^{*}/2)} \qquad (67)$$

Using Eqs. (27), (28), and (14), the following expression can be derived for $F(v)$

$$F(v) = (1/2\pi m_2)^{1/2} \exp{[-(v - \mu)^2/2m_2]}\{1 + (\mu_3/6m_2^{3/2})H_3[(v - \mu)/\sqrt{m_2}]$$
$$+ \{(\mu_4 - 3\mu_2^2)/24m_2^2\}H_4[(v - \mu)/\sqrt{m_2}\} \qquad (68)$$

where

$$m_2 = m_2^{*} + \mu_2 \qquad (69)$$

$F(v)$ was generated over the range $18 \le v \le 42$ about $\mu = 30$ with $\Delta v = 0.1$.

Equation (67) immediately suggests a method for determining a means of estimating the value of $k_{\max}$ in k-space out to which $F(k)$ should be evaluated. If $k_{\max}$ is such that $\exp{(-k^2 m^{*}/2)} = 5 \times 10^{-6}$, then

$$k_{\max} = [-2 \log_e{(5 \times 10^{-6})/(m_2 - \mu_2)}]^{1/2} \qquad (70)$$

This was the method used to estimate $k_{\max}$ throughout this paper. If $|F(k)|$ reached a minimum value of 5×10^{-5} before $k_{\max}$ was encountered, then that value of k was the largest value of k used.

Table 6 illustrates the performance of the numerical procedures used in the computer program to transform the test $F(v)$ function in Eq. (68) to the corrected Gaussian function for varying values of m_2^{*}. As m^{*} decreased the value of $k_{\max}$ increased, and larger values of Δk were required to proceed to the desired value of $k_{\max}$ because a fixed number of computer storage locations were allotted for Δk-increments in k-space. In all cases the minimum value of $|F(k)|$ was encountered before $k_{\max}$ was reached. When m_2^{*} was small, this resulted in $|W(k)|$, the critical quantity, being too large at the termination of the generation. Consequently, there was a loss in the accuracy of the computation as measured by the statistics.

It can be argued that a better procedure would be to always allow the computation of $F(k)$ to proceed until $k_{\max}$ is encountered. However, very small values of $|F(k)|$ would be used and roundoff and integration errors would begin to predominate. With the above method there is a

TABLE 6

Results for Analytical Test Functions

Statistic	Test function, Eq. (68), $\mu_3 = -0.7$, $\mu_4 = 4.5$, fixed values of m_2^*				Tung's function Eq. (71) $H = 0.4$
	1.0	0.5	0.3	0.1	
Raw[a] area	1.0000	1.0000	1.0000	0.99999	0.99999
Corrected[b] area	0.99977	1.0003	1.0025	1.0266	0.99959
Raw MEV	30.000	30.000	30.000	30.000	27.40
Corrected MEV	30.000	30.000	30.000	30.000	27.39
m_2	2.0000	1.5000	1.3000	1.1000	16.49
μ_2	1.0000	1.0000	1.0000	1.0000	3.125
Predicted[c] m_2^*	1.0000	0.50000	0.30001	0.10003	13.37
Corrected m_2^*	1.0031	0.50442	0.31519	0.11245	13.60
m_3	-0.69952	-0.69980	-0.69972	-0.69984	10.37
μ_3	-0.7	-0.7	-0.7	-0.7	0
Predicted m_3^*	0.00048	0.00019	0.00028	0.00016	10.37
Corrected m_3^*	0.00017	0.000058	0.000016	0.000003	10.32
m_4	13.500	8.2510	6.5696	5.1298	679.64
μ_4	4.5000	4.5000	4.5000	4.5000	29.29
Predicted m_4^*	2.9998	0.75098	0.26960	0.029583	399.60
Corrected m_4^*	3.0555	0.76979	0.34160	0.029843	397.28
k_{max} from Eq. (70)	4.94	6.98	9.02	15.62	1.35
Δk	$\pi/50$	$\pi/40$	$\pi/30$	$\pi/20$	$\pi/20$
$|F(k)|_{min}$[d]	5×10^{-6}	5×10^{-6}	5×10^{-6}	5×10^{-6}	5×10^{-6}
k_{max} used[e]	3.39	4.00	4.39	4.86	1.35
$|W(k)|$ at k_{max} used	0.00125	0.00713	0.0215	0.119	0.00334

[a] All the statistics preceded by the word "raw" are those calculated from $F(v)$, the normalized observed chromatogram.

[b] All the statistics preceded by the word "corrected" are those calculated from the $W(v)$ obtained from $F(v)$ by the numerical Fourier transformation procedures.

[c] All the statistics preceded by the word "predicted" are those calculated from the relations in Table 2.

[d] $|F(k)|_{min}$ was set equal to 5×10^{-6}.

[e] k_{max} used refers to the actual largest value of k used to evaluate $F(k)$.

limitation on the value of m_2^* that can be handled with accuracy. With experimental data the value of k_{max} is always used because noise and oscillations prevent $|F(k)|$ from ever reaching any reasonably small minimum value of $|F(k)|$. Figure 11 is a plot of $F(v)$, the true $W(v)$, and the calculated $W(v)$ obtained from $F(v)$ for the case where $m_2^* = 0.3$, $\mu_2 = 1.0$, $\mu_3 = -0.7$, and $\mu_4 = 4.5$.

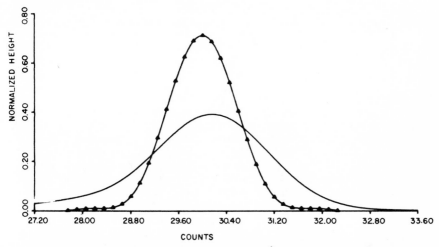

FIG. 11. Test function-skewed Gaussian for $m_2^* = 0.3$, $\mu_2 = 1$, $\mu_3 = -0.7$, $\mu_4 = 4.5$: $-F(v)$ from Eq. (68), ▲ corrected $W(v)$. Note true values of $W(v)$ calculated from Gaussian function are not shown because they are coincident with the corrected values.

The second function studied was suggested by Tung (*3*) and is

$$F(v) = \frac{0.325H}{\sqrt{\pi[(0.325)^2 + H^2]}} \left\{ 0.6 \exp\left[-\frac{(0.325)^2 H^2 (v - 25)^2}{(0.325)^2 + H^2} \right] \right.$$
$$\left. + 0.4 \exp\left[-\frac{(0.325)^2 H^2 (v - 31)^2}{(0.325)^2 + H^2} \right] \right\} \quad (71)$$

with $H = 0.4$. Note that

$$H = 1/\sqrt{2\mu_2} \quad (72)$$

The corrected chromatogram is

$$W(v) = \frac{0.325}{\sqrt{\pi}}$$
$$\times \{0.6 \exp[-(0.325)^2(v - 25)^2] + 0.4 \exp[-(0.325)^2(v - 31)^2]\} \quad (73)$$

$F(v)$ was generated over the range $12 \le v \le 44$ with $\Delta v = 0.1$. The numerical results for this case are also shown in Table 6. A plot of $F(v)$, the true $W(v)$, and the calculated $W(v)$ obtained from $F(v)$ are shown in Fig. 12.

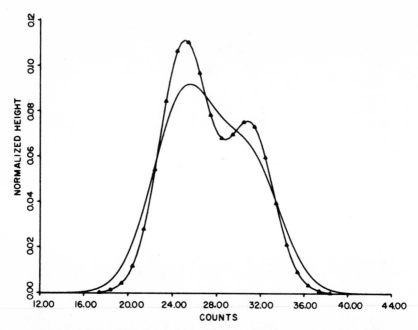

FIG. 12. Tung's analytical test function for $H = 0.4$, $\mu_3 = 0$, $\mu_4 = 29.29$: $-F(v)$ from Eq. (71), ▲ corrected $W(v)$. Note true values of $W(v)$ calculated from Eq. (73) are not shown because they are coincident with the corrected values.

8. TEST RESULTS—EXPERIMENTAL DATA

In order to test the method on actual experimental data, five samples were selected. The first was a broad MWD standard. The other four were mixtures prepared by mixing three of the calibration standards together in various proportions by weight. The resulting polyblends were narrow in MWD and distinctly multimodal. These five samples are described in Table 7.

Since the raw data tended to be noisy at the tails, the "spectral window" shown in Fig. 13 was used to smooth the tails of the chromatogram. The window used was an extended cosine-bell data window (10). This window generated a "factor" between 0 and 1 by which the raw data was multiplied over the first and last 10% of the total number of data points. The middle 80% of the total number of data points was unchanged.

TABLE 7
Composition and Molecular Weight Data for Experimental Test Samples

Run[a]	Wt%[b] composition	$10^{-3}\bar{M}_n(t)$[c]	$10^{-3}\bar{M}_w(t)$	$P(t)$
181-185	NBS-706	136.5	257.8	1.889
196-302-1	0/50/50	64.9	74.6	1.15
196-305-1	75/25/0	23.1	27.6	1.20
196-306-1	25/75/0	35.6	43.2	1.21
196-308-1	16.7/33.3/50	48.8	69.4	1.42

[a] All samples were run at a flow rate of 1 ml/min.

[b] The wt% of each polystyrene standard in the mixture refers to the standards with the run numbers in the following order: 180-180, 180-181, and 180-182.

[c] The true values, $\bar{M}_n(t)$, $\bar{M}_w(t)$, and $P(t)$ of the mixtures were calculated from the true values of the individual components and compositions of the mixtures.

The distance in k-space, out to which $F(k)$ was evaluated, was determined from Eq. (70) with μ_2 evaluated at the MEV of the sample. Values of $W(v)$, which were generated out in each direction from v_0, generally became negative at the ends of the chromatogram. If more than five negative values were encountered, this was used as an indication to terminate the evaluation of $W(v)$. However, this criterion allowed

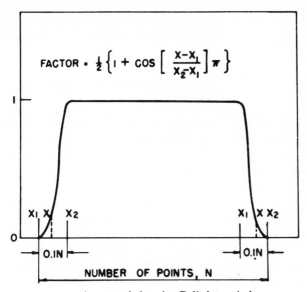

FIG. 13. An extended cosine-Bell data window.

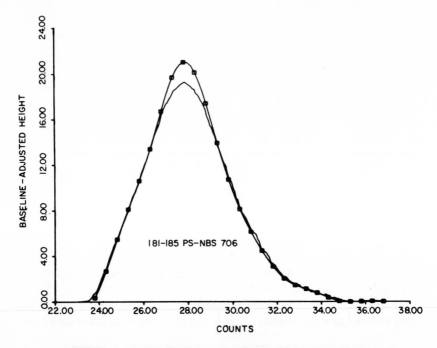

FIG. 14. Observed $\{-F(v)\}$ and corrected $\{\boxplus, W(v)\}$ chromatograms for sample 181-185.

the generation of completely separated peaks. All negative values of $W(v)$ were set equal to zero.

Figures 14–18 are plots of $F(v)$ and $W(v)$ for the samples shown in Table 7. The variations of the μ_2, μ_3, and μ_4 with elution volume that were used in the calculations are given in Table 5. Linear interpolation was used between the tabulated points.

Values of $\bar{M}_n(\infty)$ and $\bar{M}_w(\infty)$ obtained from the calibration curve of Fig. 3 and the observed chromatogram $F(v)$, and values of $\bar{M}_n(t)_w$ and $\bar{M}_w(t)_w$ obtained from the corrected chromatogram $W(v)$ and the calibration curve of Fig. 3 are shown in Table 8. For comparison purposes the values of μ_2, μ_3, and μ_4 were evaluated at the MEV's and were subsequently used to evaluate $\bar{M}_n(t)_E$ and $\bar{M}_w(t)_E$ from Eqs. (47) and (48).

As can be seen from a study of Table 8, the areas of $W(v)$ are almost always consistently higher than those of $F(v)$. When the parameters μ_2, μ_3, and μ_4 are functions of elution volume, it is difficult to prove mathematically that the area of the raw chromatogram is equal to the area of the corrected chromatogram. From Eq. (1)

$$\int_{-\infty}^{\infty} F(v)\, dv = \int_{-\infty}^{\infty} \int_{-\infty}^{\infty} G(v-y)W(y)\, dy\, dv \qquad (74)$$

The use of the convolution theorem or a transformation of variables to interchange the order of integration with respect to y and v to prove the area equality requires the constancy of μ_2 with v. The same argument leads to the conclusion that none of the relationships of Table 2 strictly hold. However, it should be noted from Table 8 and Fig. 16 that in the case of sample 196-305-1, where the sample was in the range of reasonable constancy of the μ_n's, the areas indeed gave a reasonable check. From physical arguments the areas must check within experimental errors because the area under the chromatogram represents a certain number of milligrams of material that has passed through the GPC columns and the differential refractometer detector.

Table 8 also shows that the evaluation of $F(k)$ always proceeded out to the value of $k_{\max}$. The minimum value of $|F(k)|$ was not encountered first because of the noise and oscillations in the data. Since there was

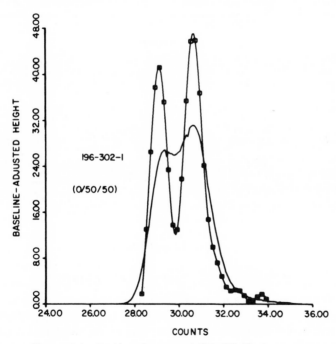

FIG. 15. Observed $\{-F(v)\}$ and corrected $\{\boxminus,\ W(v)\}$ chromatograms for sample 196-302-1.

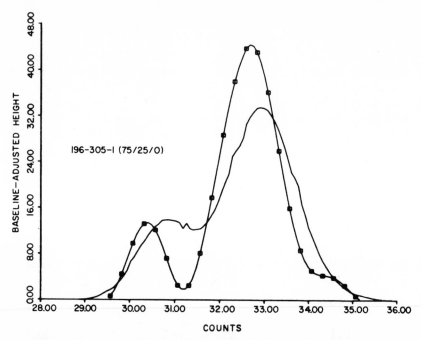

FIG. 16. Observed $\{-F(v)\}$ and corrected $\{\boxminus,\ W(v)\}$ chromatograms for sample 196-305-1.

considerable doubt that the value of k_{max} used was adequate, the value of k_{max} was increased for each of the samples to $1\frac{1}{2}$ times the value indicated in Table 8. Not much effect was observed in the broad MWD sample. However, greater resolution of the components in the 196-series of samples (small m_2^*'s) was achieved. How realistic this increased resolution was is open to question. The "correct" value of k_{max} to use is unresolved and remains a major area for further investigation.

9. CONCLUSIONS

Based on this study several conclusions can be made.

(a) The instrument spreading function given by Eq. (5) is useful for characterizing the dispersion, skewing, and flattening effects encountered in raw GPC chromatograms due to the imperfect resolution by the GPC columns. It is limited, however, to applications where

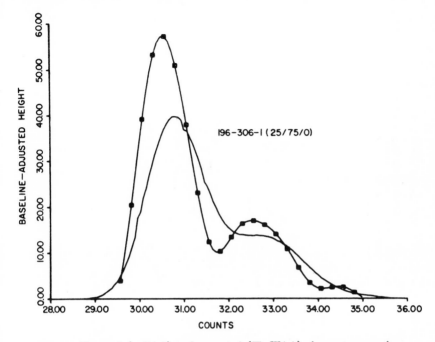

FIG. 17. Observed $\{-F(v)\}$ and corrected $\{\boxminus,\ W(v)\}$ chromatograms for sample 196-306-1.

m_2^* is large enough to "wash out" large oscillations caused by large values of β_1 and β_2 relative to the values of m_2, m_3, and m_4.

(b) The Yau-Malone function given by Eq. (40) is a flexible and useful equation for the fitting of molecular weight vs elution volume data.

(c) The Fourier transform method, while it can be demonstrated to work well on analytically generated data, has a number of problems associated with it when applied to actual experimental data. At this time these problems appear to be associated with two major considerations: (1) It is not clear how far out in k-space to go in generating $F(k)$ and hence $W(k)$. Equation (70), while probably qualitatively useful, does not appear to be adequate. (2) It is not clear how serious are the assumptions of the constancy of the parameters μ_2, μ_3, and μ_4 with elution volume in situations where they are reasonably strong functions of elution volume and/or the sample covers a broad range of elution volume.

(d) The Fourier transform method is useful in qualitatively resolving

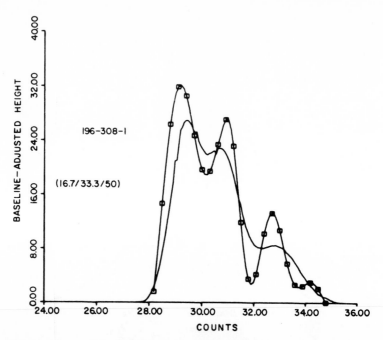

FIG. 18. Observed $\{-F(v)\}$ and corrected $\{\boxplus,\ W(v)\}$ chromatograms for sample 196-308-1.

peaks.

(e) The calibration procedure used, though arbitrary, gives values of the parameters μ_2, μ_3, and μ_4 which are physically reasonable. When the sample is a narrow MWD sample, then Eqs. (47) and (48) can be used, with parameters evaluated at the MEV of the sample, to evaluate $\bar{M}_n(t)$ and $\bar{M}_w(t)$. When the samples are broad in MWD, the adequacy of Eqs. (47) and (48) are in doubt because of the variability of the μ_n's with elution volume.

10. SUGGESTIONS FOR IMPROVEMENTS IN THE CALIBRATION AND NUMERICAL PROCEDURES*

The use of Eqs. (47) and (48) tied together the determination of the instrument spreading parameters with errors in the determination of

Note added in proof. Since this manuscript was written the improvements suggested in this section have been tried and have not been successful. A new numerical technique for correcting observed chromatograms, which eliminates some of the limitations of the Fourier Transform method, has been developed and will be published elsewhere (*18*).

TABLE 8

Results for Experimental Test Samples

Statistic	Run				
	181-185	196-302-1	196-305-1	196-306-1	196-308-1
$10^{-3}\bar{M}_n(\infty)$	133.3	69.0	25.7	38.8	51.5
$10^{-3}\bar{M}_w(\infty)$	364.5	87.8	34.2	51.4	78.3
$P(\infty)$	2.72	1.28	1.33	1.32	1.52
$10^{-3}\bar{M}_n(t)_w{}^a$	137.2	74.0	26.8	43.7	56.2
$10^{-3}\bar{M}_w(t)_w$	352.7	91.1	33.8	56.5	85.1
$P(t)_w$	2.57	1.23	1.31	1.29	1.51
$10^{-3}\bar{M}_n(t)_E{}^b$	144.0	71.8	26.6	40.0	53.5
$10^{-3}\bar{M}_w(t)_E$	328.4	83.5	32.6	49.2	74.7
$P(t)_E$	2.26	1.27	1.22	1.22	1.40
Raw area (smoothed)^c	92.57	88.22	85.71	91.42	87.22
Corrected area^d	94.35	93.81	84.28	104.1	94.43
Raw MEV	28.12	30.24	32.44	31.46	30.65
m_2	3.928	1.192	1.400	1.40	2.11
μ_2 at MEV	0.7311	0.4487	0.3940	0.3840	0.4200
Predicted $m_2^*{}^e$	3.197	0.7435	1.005	1.017	1.680
Corrected m_2^*	3.670	1.045	1.133	1.278	2.090
k_{max} from Eq. (70)	2.76	5.73	4.92	4.89	3.80
Δk	$\pi/100$	$\pi/40$	$\pi/50$	$\pi/50$	$\pi/60$
k_{max} used^f	2.73	5.65	4.90	4.83	3.76

^{a} All the statistics with a subscript w are those calculated from the corrected chromatogram $W(v)$.

^{b} All the statistics with a subscripted E are those calculated from Eqs. (47) and (48).

^{c} All the statistics preceded by the word "raw" are those calculated from $F(v)$, the normalized observed chromatogram.

^{d} All the statistics preceded by the word "corrected" are those calculated from the $W(v)$ obtained from $F(v)$ by the numerical Fourier transformation procedures.

^{e} All the statistics preceded by the word "predicted" are those calculated from the relations in Table 2.

^{f} k_{max} used refers to the actual largest value of k used to evaluate $F(k)$.

the experimental values of $\bar{M}_n$ and $\bar{M}_w$. It may be better to separate the determination of the instrument spreading parameters from the experimental values of $\bar{M}_n$ and $\bar{M}_w$ because these parameters should be dependent only on the flow behavior of the polymer molecules through the GPC columns. This may be done by assuming that Eqs. (62), (63), and (64) adequately describe the observed chromatogram of a nearly monodisperse standard.

Then μ_2, μ_3, and μ_4 are directly obtained from the moments of the $F(v)$ curve.

$$\mu_2 = m_2 \tag{75}$$

$$\mu_3 = m_3 \tag{76}$$

$$\mu_4 = m_4$$

Plots of μ_2, μ_3, and μ_4 vs MEV would then result. The plot of μ_3 vs MEV would be quite different from that obtained by directly using $\bar{M}_n(t)$ and $\bar{M}_w(t)$ data. This difference would be reflected in the molecular weight calibration curve to be determined by the method described below.

Since $F(v) \approx W(v)$ for each MEV, then the Yau-Malone parameters can be sought such that $\bar{M}_n(t)_w$ and $\bar{M}_w(t)_w$ are fit as closely as possible to the experimentally determined values of $\bar{M}_n(t)$ and $\bar{M}_w(t)$. In this way experimental errors associated with $\bar{M}_n(t)$ and $\bar{M}_w(t)$ would be smeared out over the entire calibration curve.

A major source of difficulty in the numerical Fourier transform method is the determination of the proper value of $k_{\max}$ at which to stop the evaluation of $W(k)$ before the effects of noise in the numerical $F(v)$ data and roundoff errors prevent $|F(k)|$ and thus $|W(k)|$ from smoothly approaching zero. If $F(k)$ can be fit by a function having a generalized form of the type shown in Eq. (65), then $W(k)$ may be expressed by an analytical function of the form similar to Eq. (66), which appears to have better numerical properties in that $|W(k)|$ will be forced to smoothly approach zero. As was shown previously, the Fourier transform method, indeed, does work well on analytically generated data.

Acknowledgments

The authors would like to acknowledge the contributions of Mr. Wilfred J. Renaudette and Mr. James H. Clark for assistance in the experimental GPC work and data reduction calculations, respectively.

List of Symbols*

A, B, C, D	parameters of the Yau-Malone nonlinear molecular weight calibration curve as defined by Eq. (40) and illustrated by Fig. 3
a_j, b_j	intercept and slope parameters of a straight line segment approximating $F(v)$ over jth interval for the Filon Quadrature method

* Symbols and notation found in the text and not defined here have been previously described in (I.).

C_3, C_4 skewness and flatness coefficients of the analytical skewed Gaussian function described by Eq. (62)

DFT   discrete Fourier transform

FFT   fast Fourier transform

$|F(k)|_{min}$ minimum value of $|F(k)|$ chosen as the point in k-space to terminate the evaluation of $F(k)$

$F_j(z)$ value of $F(v)$ over jth interval approximated by straight line segment

$F_1(k), F_2(k)$ real and imaginary parts of the Fourier transformation of the observed chromatogram $F(v)$ about v_0 as defined by Eqs. (29), (32), and (33)

H resolution factor as defined by Tung (*3*) and described by Eq. (72)

(I.)   Paper I of this series described in Ref. *1*

k_{max} the maximum value of k in k-space out to which $F(k)$ should be evaluated

MEV   mean elution volume

$\bar{M}_n(t)_E$ value of number-average molecular weight obtained from values of μ_2, μ_3, and μ_4 at the MEV used in Eq. (47)

$\bar{M}_n(\infty)_s, \bar{M}_n(t)_s$ smoothed infinite resolution and true number-average molecular weights, respectively, by use of the Yau-Malone function, Eq. (40)

$\bar{M}_n(t)_w$ number-average molecular weight calculated from corrected chromatogram, $W(v)$

$\bar{M}_w(t)_E$ value of weight-average molecular weight obtained from values of μ_2, μ_3, and μ_4 at the MEV used in Eq. (48)

$\bar{M}_w(\infty)_s, \bar{M}_w(t)_s$ smoothed infinite resolution and true weight-average molecular weights, respectively, obtained by use of the Yau-Malone function, Eq. (40)

$\bar{M}_w(t)_w$ weight-average molecular weight calculated from corrected chromatogram, $W(v)$

m_2^*, m_3^*, m_4^* second, third, and fourth moments about the mean of the corrected chromatogram, $W(v)$

$P(t)_E$ value of the polydispersity ratio obtained from the values of μ_2, μ_3, and μ_4 at the MEV used in Eqs. (47) and (48)

$P(\infty)_s, P(t)_s$ smoothed infinite resolution and true polydispersity ratio, respectively

$P(t)_w$ polydispersity ratio obtained from the corrected chromatogram, $W(v)$

v_0 elution volume at center of the chromatogram about which to transform the range of integration in Eqs. (25) and (26)

v_s specified elution volume at which $(dv/dM)_{v_s}$ is to be evaluated

$\langle v^3 \rangle$ third moment about the origin of the normalized observed chromatogram

$\langle y^3 \rangle$ third moment about the origin of the normalized corrected chromatogram

$W_1(k), W_2(k)$ real and imaginary parts of the Fourier transformation of the corrected chromatogram $W(v)$ about v as defined by Eqs. (29), (35), and (36)

z transformed variable defined by Eq. (29)

Greek Letters

α reduced parameter defined by Eq. (49)

β_1, β_2 reduced parameters of $G(v - y)$ function defined by Eqs. (6) and (7)

μ mean elution volume

ψ reduced parameter of Yau-Malone function defined by Eq. (41)

REFERENCES

1. T. Provder and E. M. Rosen, *Separ. Sci.,* **5**, 437 (1970).
2. W. W. Yau and C. P. Malone, *J. Polym. Sci., Part B,* **5**, 663 (1967).
3. L. H. Tung, *J. Appl. Polym. Sci.,* **13**, 775 (1969).
4. H. T. Davis, *Introduction to Nonlinear Differential and Integral Equations,* U.S. Atomic Energy Commission, Supt. of Documents, Washington, D.C., 1960, pp. 434ff.
5. V. S. Pugachev, *Theory of Random Functions,* Addison-Wesley, New York, 1962, pp. 44ff.
6. H. Freeman, *Introduction to Statistical Inference,* Addison-Wesley, New York, 1963, Chapter 18.
7. E. Grushka, M. N. Myers, P. D. Schettler, and J. C. Giddings, *Anal. Chem.,* **41**, 889 (1969).
8. D. E. Barton and K. E. Dennis, *Biometrika,* **39**, 425 (1952).
9. G-AE Subcommittee on Measurement Concepts, *IEEE Trans. Audio Electroacoustics,* **AU-15(2)**, 45 (1967).
10. G. D. Bergland, *IEEE Spectrum,* **41**, (July 1969).
11. S. M. Chase and L. D. Fosdick, *Commun. Ass. Comput. Mach.,* **12**, 457 (1969).

12. A. Papoulis, *The Fourier Integral and Its Applications,* McGraw-Hill, New York, 1962, pp. 20f.

13. P. W. Murrill, R. W. Pike, and C. L. Smith, *Chem. Eng.,* **76,** 105 (February 1969).

14. K. J. Bombaugh, W. A. Dark, and R. F. Levangie, *Separ. Sci.,* **3,** 375 (1968).

15. D. W. Marquardt, *J. Soc. Ind. Appl. Math.,* **2,** 431 (1963).

16. E. J. Henley and E. M. Rosen, *Material and Energy Balance Computations,* Wiley, New York, 1969, pp. 547f and 560-566.

17. H. E. Pickett, M. J. R. Cantow, and J. F. Johnson, *J. Appl. Polym. Sci.,* **10,** 917 (1966).

18. E. M. Rosen and T. Provder, Paper to be presented at the Ninth International GPC Seminar, Miami Beach, Florida, October, 1970.

Behavior of Micellar Solutions in Gel Permeation Chromatography. A Theory Based on a Simple Model*

HANS COLL

SHELL DEVELOPMENT COMPANY
EMERYVILLE, CALIFORNIA

Summary

Gel permeation chromatograms are interpreted in terms of V, the peak elution volume, which is characteristic for the size of the solute species. If solute association takes place, V may become extensively concentration dependent. To further the understanding of this effect a simple theory, which ignores axial diffusion, has been developed, the model for molecular association being a micellar surfactant in equilibrium with its monomer.

INTRODUCTION

Gel permeation chromatography (GPC) separates molecules according to size. The method has therefore found wide application in the study of molecular weight distribution of polymers and oligomers. A special situation arises if two or more species, being in a dynamic equilibrium with each other, enter into the GPC process. An example with certain unique properties is micellar surfactant (A_n) in equilibrium with its monomer (A), $nA \rightleftharpoons A_n$. Micelles, being larger than the monomer, tend to move faster down the GPC column than the monomer, but in this event they dissociate to regenerate the equilibrium concentration of monomer. As a result, the surfactant front will emerge from the column at times which should be distinctly dependent on the concentration of the sample that had

* Presented at the ACS Symposium on Gel Permeation Chromatography sponsored by the Division of Petroleum Chemistry at the 159th National Meeting of the American Chemical Society, Houston, Texas, February, 1970.

329

been injected into the column. In the following an attempt will be made to give a quantitative description of the process, based on a very simple model. This model ignores band spreading, a simplification that appears justified if a very broad band of solute is introduced in the column. This has been the experimental approach of Tokiwa et al. (1) The complications arising from band spreading will be discussed although rigorous treatment cannot be given at the present time.

THEORY

It is usually assumed that the association of surfactant molecules is approximately governed by the law of mass action,

$$K \simeq c_2/c_1^n \tag{1}$$

where K is an equilibrium constant and c_2 and c_1 are the mass concentrations of the micelles and monomer molecules, respectively. The association number, n, is usually larger than 10. It then follows from Eq. (1) that up to the so-called critical micelle concentration (cmc) the surfactant is almost exclusively present as the monomer, while at a total concentration larger than the cmc the monomer concentration remains essentially constant ($c_1 \simeq$ cmc).

We shall assume the micelles to be monodisperse, and further, that equilibration between micelles and monomer is virtually instantaneous. On the other hand, we tentatively assume that longitudinal diffusion in the column which gives rise to the familiar broadening of peaks to be insignificant. Thus, a solution introduced into the column as a band of discrete width is assumed to give rise to a square-wave signal in the detector monitoring the column effluent. The width of the signal is to be the same as that of the sample band (expressed as cubic centimeters of effluent). Although this assumption appears to be rather unrealistic, we believe that some of the essential features of the GPC process are still represented despite this simplification.

Figure 1A represents the concentration profile of a sample band just introduced in the column. As this band moves along, three velocities can be distinguished: the monomer velocity, u_1; the velocity of the undissociated micelle, u_2; and a velocity, u_2^*, of the dissociating micelles. Generally, $u_1 < u_2^* < u_2$. The velocities are expressed as centimeters per second. They can be immediately converted into

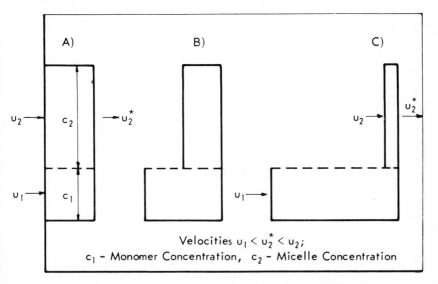

FIG. 1. Concentration profile of micellar solute in the GPC column.

flow rates (cubic centimeters per second) by multiplying with the (constant) cross sectional area, A, of the mobile phase in the column.

If V_0 is the interstitial (void) volume of the column packing, V_1 the emergence volume of the monomer, V_2 the emergence volume of the undissociating micelles, and u_0 the velocity of the solvent in the interstices, then $u_1 = u_0 V_0 / V_1$ and $u_2 = u_0 V_0 / V_2$. Furthermore, it is easy to show that the surfactant front moves with a velocity

$$u_2^* = u_1 + (u_2 - u_1)c_2/(c_1 + c_2) \qquad (2)$$

The factor $c_2/(c_1 + c_2)$ accounts for the slowing down of the micellar front as a result of dissociation. The concentrations, c_1 and c_2, of monomer and micelles are related to the total surfactant concentration, c_0, in the sampling loop by

$$c_0 V_0 = c_1 V_1 + c_2 V_2 \qquad (3)$$

Equation (3) accounts for the dilution effect once the surfactant enters the column and disperses into the accessible pores in the gel: $c_1 \simeq$ cmc is a constant for a given system.

As the surfactant band travels down the column, the micellar "hump" in Figs. 1B and 1C becomes increasingly narrower. We define a time t^* when the back of the "hump" has caught up with the front, and micelles disappear altogether from the system. At times greater than t^* only monomer at concentration c_1 is present in the column. Micelles cannot reform since no concentrating process occurs in the column. If τ_0 is the time interval during which sample is injected into the column, we can equate distances of travel

$$u_2^* t^* = u_2(t^* - \tau_0) \tag{4}$$

Substitution for u_2 and u_2^* yields

$$t^* = \tau_0 a(k - a + b)/b(a - b) \tag{5}$$

introducing the dimensionless parameters $a = V_1/V_0$, $b = V_2/V_0$, and $k = c_0/c_1$.

Next we must compare t^* with the emergence time of the micellar front from the column, $t_e = V_0/u_2^* A$. Writing $t_0 = V_0/Au_0$ for the time required by the solvent to traverse the column, we obtain again by substitution

$$t_e = t_0 ab(k + b - a)/(ak + b^2 - a^2) \tag{6}$$

We may then discern three cases with respect to the emergence of the solute front in the column effluent:

(a) Monomer and micelles elute from the column. The condition is $t_e < t^*$. The relative elution volume ($=$ elution volume divided by V_0), v_I, can be calculated from

$$v_I = u_0 A t_e/V_0 = ab(k + b - a)/(ak + b^2 - a^2) \tag{7}$$

(b) Although micelles were originally present in the column, only monomer emerges. The condition for this is $t_e \geq t^*$. The relative elution volume for this case, v_{II}, can be found from the following consideration: up to time t^* the solute front travels at a velocity u_2^*, thereafter, for a period t_1, until leaving the column, the velocity is that of the monomer, u_1. Hence,

$$v_{II} = u_0 A (t^* + t_1)/V_0 \tag{8}$$

t_1 can be expressed by means of the relationship $u_1 t_1 + u_2^* t_2^* = u_0 t_0$, and t^* by means of Eq. (5). One obtains

$$v_{II} = a - (ak - a^2)w/b^2 \tag{9}$$

where $w = \tau_0 u_0 A/V_0$, the volume of solution injected divided by the void volume of the column.

(c) The original sample concentration was $c_0 \leq a \times$ cmc, in which case monomer only is transported in the column. Therefore,

$$v_{III} = a \qquad (10)$$

These equations show that, depending on the case, the emergence volume of the sample front may not only be a function of the emergence volumes of the particle species (a, b), but also of the total sample concentration relative to the cmc (k), and of the sample volume relative to the void volume of the column (w).

It follows from Eq. (7) that in the case of the cmc being very small compared to the concentration of the sample $(k \gg 1)$,

$$v_{I} \simeq b$$

since a, b are of the order of unity. The parameter b, by definition, is the relative elution volume of undissociating micelles, and in this case the surfactant should behave like an ordinary polymer. This situation may, for instance, be expected in the case of ethylene oxide alcohols which were found to be very stable by measurements of osmotic pressure (2).

The peak width can readily be calculated (disregarding diffusional peak spreading, as stipulated earlier). In Case I the relative width of the micellar "hump," h_{I}, as it emerges from the column is given by

$$h_{I} = w - b^{2}(a - b)/(ak - a^{2} + b^{2}) \qquad (11)$$

the total width by

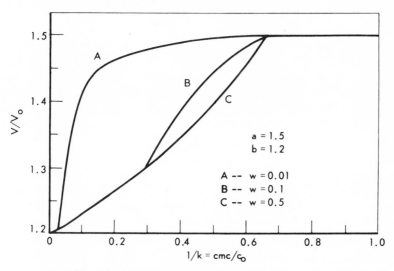

FIG. 2. Plot of relative elution volume against reciprocal reduced concentration.

$$H_I = w + a(a - b)(k - a)/(ak - a^2 + b^2) \tag{12}$$

The total width in Case II is

$$H_{II} = w + (ak - a^2)w/b^2 \tag{13}$$

In Case III the width is simply equal to w.

Figure 2 shows a plot of the reduced emergence volume of the surfactant front, calculated by means of Eq. (8) and (9). Curves A and B exhibit three segments, corresponding to the three cases discussed above. In Curve C, where $w = 0.5$, only two segments are apparent.

SPREADING OF SOLUTION BANDS

Considerable complications arise if band spreading (axial dispersion) has to be taken into account. Since chromatographic elution curves of monodisperse solutes are usually Gaussian, it appears justified to treat broadening as a diffusion problem. If the reduced width of the solution band (width divided by length of column) is denoted as $2\bar{h}$, one may consider as a starting point for a discussion the equation (3).

$$2c(\bar{x},t) = c(0)\{\operatorname{erf}[(\bar{h} - \bar{x})/2(\bar{D}t)^{\frac{1}{2}}] + \operatorname{erf}[(\bar{h} + \bar{x})/2(\bar{D}t)^{\frac{1}{2}}]\} \tag{14}$$

$c(0)$ is the initial concentration at $t = 0$, $\bar{x}$ and $\bar{D}$ (sec^{-1}) are the reduced space coordinate and the reduced diffusion coefficient, respectively. The latter may be obtained from calibration with some reference solute if one assumes that $\bar{D}$ is the same for all solutes in a given column system. Equation (14) expresses lowering of concentration as a consequence of diffusion, an effect particularly pronounced if the solution band is very narrow (Fig. 3). In this case micelles in the column will evidently disappear sooner than follows from Eq. (5). The frontal velocity of the surfactant, u_2^*, will not only be affected by the lowering of concentration but also by the concentration profile of the band, which is no longer square. A complete treatment of the combined processes of diffusion and chromatographic transport seems to be out of question. The most important effect of axial dispersion of a narrow band appears to be the lowering of the micelle concentration, and an approximation by numerical calculation may be feasible.

An experimental alternative suggests itself if one introduces a very broad band into the column. As shown in Fig. 3, the peak concentration shows only relatively little lowering in this case. However, if the micellar "hump" considerably narrows during its passage down the column, diffusion effects should again become significant.

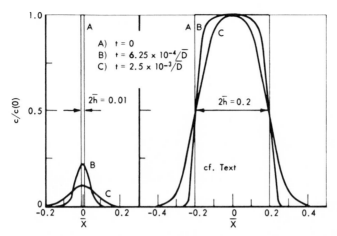

FIG. 3. Diffusive spreading of a narrow and a wide solution band.

Hence, Case II of the previous section will be most affected by band spreading. An evaluation of these effects must be left to experimental tests.

EXPERIMENTAL

A few preliminary experimental results obtained with narrow bands of sodium dodecyl sulfate (NaDDS) in aqueous sodium nitrate solution (0.03 M) are shown in Fig. 4.

These results were obtained with a column 120-cm long having an inside diameter of 0.33 cm. The column was packed with porous glass beads Corning CPG 10-240 (Water Associates, Framingham, Mass.), the particle size being 36–75 μ. The void volume, V_0, was calculated as 5.4 cc. The solvent flow rate was approximately 0.08 cc/min. A sensitive differential refractometer (4) served as the detector (temperature, 24°C).

In all cases the stock solution contained 5 g/liter of highly purified NaDDS (the critical micelle concentration is approximately 0.8 g/liter (5). The stock solution was charged for 10, 30, 60, and 120 sec, respectively, corresponding to relative band widths, w, of 0.0025, 0.0076, 0.0149, and 0.0304.

DISCUSSION

The curves in Fig. 4 have, at least qualitatively, the expected appearance. While in cases of A and B the concentration in the effluent

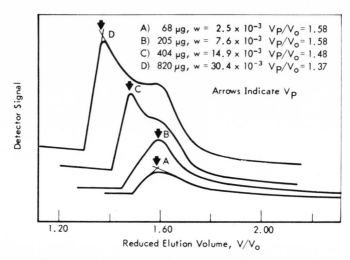

FIG. 4. Chromatographic peaks of NaDDS at different sample charges.

is below the cmc and the peak elution volumes are the same, micellar peaks are apparent in cases of C and D. Correspondingly, the peak elution volumes decrease with increasing concentration. The concentration differences are primarily brought about by band spreading, and a quantitative interpretation in terms of the present theory was not attempted. As mentioned before, the spreading effect can be minimized by injecting broad bands, for instance, with $w > 0.2$.

This latter approach was chosen by Tokiwa et al. (1), who studied the behavior of alkyl sulfates in aqueous solutions by means of Sephadex columns. The band width was larger than $w = 0.5$, and Case II, i.e., disappearance of micelles in the column, probably did not occur. The experimental results, expressed in terms of appearance volumes (V) of the surfactant front are, however, not in complete accord with the present theory: the ascending branch of the plot of V/V_0 versus reciprocal concentration was found to be linear, while the plot in Fig. 2 shows curvature. But more seriously, the break in the experimental curves of Tokiwa et al. occurred at the cmc, while according to our reasoning it should correspond to $a \times$ cmc.

CONCLUSION

The present treatment was restricted to the special case of an associating system where one could assume a constant equilibrium concentration of monomer. An extension of the theory to include other types of association (e.g., dimer formation) would require a different mathematical approach.

Experimental data obtained under well-defined conditions are needed to verify the present theory and to assess the influence of band spreading on the peak elution volumes.

Acknowledgment

The author wishes to thank Dr. A. G. Polgar for his help in the chromatographic experiments.

REFERENCES

1. F. Tokiwa, K. Ohki, and I. Kokubo, *Bull. Chem. Soc. Jap.* **41**, 2285, 2845 (1968).
2. H. Coll, *J. Amer. Oil Chem. Soc.*, **46**, 593 (1969).
3. J. Crank, *Mathematics of Diffusion*, Clarendon Press, Oxford. 1964.
4. H. W. Johnson, Jr., V. A. Campanile, and H. A. LeFebre, *Anal. Chem.*, **39**, 32 (1967).
5. H. F. Huisman, *Konink. Ned. Akad. Wetenschapp. Proc.*, **B67**, 367 (1964).

Gel Permeation Analysis of Macromolecular Association by an Equilibrium Method*

BRUCE F. CAMERON, LARRY SKLAR,
VERONIKA GREENFIELD, and ALAN D. ADLER

NEW ENGLAND INSTITUTE
RIDGEFIELD, CONN. 06877
DEPARTMENT OF INTERNAL MEDICINE
UNIVERSITY OF MIAMI SCHOOL OF MEDICINE
MIAMI, FLORIDA 33152
AND
LABORATORIES FOR HEMATOLOGICAL RESEARCH
HOWARD HUGHES MEDICAL INSTITUTE
MIAMI, FLORIDA 33152

Summary

An equilibrium method for determining macromolecular association by gel permeation is described. A thermodynamic description of the simplest case, dimerization, is presented in terms of the equilibrium constant for association, and equilibrium partition coefficients for species present. The theoretical analysis yields an equation relating the equivalents of monomer in the external phase to total equivalents, from which the equilibrium constant for association may be obtained.

Those parameters critical in experimental design, gel swelling time, and centrifugal collapse point are determined for Sephadex gels of porosity G-25 to G-200.

Application of the method to human hemoglobin gives a value of $K'_\alpha = 5 \times 10^{-6}$ for dissociation of oxyhemoglobin at pH 7.0 in 0.1 M buffer.

INTRODUCTION

A common feature of many biopolymers, such as enzymes or other protein molecules, is the presence of quaternary structure, i.e.,

---

* Presented at the ACS Symposium on Gel Permeation Chromatography sponsored by the Division of Petroleum Chemistry at the 159th National Meeting of the American Chemical Society, Houston, Texas, February, 1970.

the native molecule is an assembly of subunits. The hemoglobin molecule is a classic example, consisting of a tetramer formed from two pairs of similar polypeptide chains (1). Numerous other enzymes have been shown to be n-mers of identical or closely similar units (2).

It is also well established in many cases that the molecule is in rapid association-dissociation equilibrium (3), and that this dynamic subunit structure may be of utmost importance in an analysis of reactivity and structure (4). Hence, measurement of this equilibrium is of basic importance.

Similar equilibria are also encountered in nonbiological systems; thus considerations presented here are of general application.

The simplest case to consider is that of a monomer-dimer equilibrium in which rates of association and dissociation are rapid with respect to the technique used for measurement.

$$2M \overset{K}{\rightleftharpoons} D$$

where $K = [D]/[M]^2$.

Let χ be the fraction monomer, $\chi = [M]/([M] + [D])$ where $[M] + [D] = C$ (total concentration); then

$$KC = \frac{1 - \chi}{\chi^2}$$

and a plot of $f(\chi)$ vs. C may be analyzed to obtain the value of K. It is possible to obtain χ experimentally, since it is a direct function of apparent molecular weight. Such considerations have been applied by several investigators to the analysis of ultracentrifugation of associating molecular systems (5–7).

It is well established that gel permeation, or molecular sieve, chromatography can also determine an apparent molecular weight (8). Application of this experimental method to associating systems has also been analyzed, especially in a series of papers by Ackers and co-workers (9). Analysis of the chromatographic system is difficult however, since it is a transport phenomenon, and a suitable theoretical model must be chosen. Also, there is a concentration profile within the column, and the relationship of elution volume to initial concentration is complex.

An alternate method to employ gel permeation materials is in an equilibrium system (10). All relevant concentrations are now well defined and experimentally accessible. Dry gel is added to the solution containing the macromolecule and, as the gel takes up solvent, the macromolecular species are distributed between internal and external solvent spaces. The gel volume penetrated is a function of the molecular weight of the solute species. Measurement of concentration in the supernatant (external phase) and knowledge of the initial total amount allows the internal available volume to be calculated.

THEORY

Consider again the simplest case as described above, a monomer-dimer equilibrium under conditions of gel permeation equilibrium partition.

The system is essentially:

Volume V_α Volume V_β
(External phase)   (Internal phase)

$$M_\alpha \xrightleftharpoons{\kappa_M} M_\beta$$

$$K_\alpha \qquad\qquad\qquad\qquad\qquad\qquad\qquad\qquad\qquad K_\beta$$
$$2M \rightleftharpoons D \qquad\qquad\qquad\qquad\qquad\qquad\qquad 2M_\beta \rightleftharpoons D_\beta$$

$$D_\alpha \xrightleftharpoons{\kappa_D} D_\beta$$

where the external phase (α) has volume V_α and the internal phase (β) volume V_β.

The relevant equilibrium constants are defined in terms of amounts of the various species as

$$K_\alpha = \frac{[D_\alpha]}{[M_\alpha]^2} = \frac{D_\alpha V_\alpha}{M_\alpha^2} \tag{1a}$$

$$\kappa_M = \frac{[M_\beta]}{[M_\alpha]} = \frac{M_\beta V_\alpha}{M_\alpha V_\beta} \tag{1b}$$

$$\kappa_D = \frac{[D_\beta]}{[D_\alpha]} = \frac{D_\beta V_\alpha}{D_\alpha V_\beta} \tag{1c}$$

Defining N as equivalents of monomer, N_0 = total equivalents; then:

$$N_0 = N_\alpha + N_\beta = M_\alpha + 2D_\alpha + M_\beta + 2D_\beta \tag{2}$$

Substituting for M_β and D_β from Eqs. (1b) and (1c), and defining

$$\mu = 1 + \frac{\kappa_M V_\beta}{V_\alpha} \tag{3a}$$

$$\delta = 1 + \frac{\kappa_D V_\beta}{V_\alpha} \tag{3b}$$

then

$$N_0 = \mu M_\alpha + 2\delta D_\alpha \tag{4}$$

Combining Eqs. (1a) and (4) and rearranging

$$\frac{2\delta K_\alpha}{V_\alpha} M_\alpha^2 + \mu M_\alpha - N_0 = 0 \tag{5}$$

The solution of the quadratic yields, after rearrangement

$$M_\alpha = -\frac{\mu V_\alpha}{4\delta K_\alpha} + \sqrt{\frac{\mu^2 V_\alpha^2}{16\delta^2 K_\alpha^2} + \frac{V_\alpha N_0}{2\delta K_\alpha}} \tag{6}$$

The positive sign of the radical is chosen since M_α must be >0. Now defining

$$\mathbf{D} = \mu/\delta \tag{7a}$$
$$\mathbf{K} = V_\alpha/2K_\alpha \tag{7b}$$

and substituting into Eq. (6)

$$M_\alpha = -\frac{\mathbf{KD}}{2} + \frac{1}{2}\sqrt{\mathbf{K}^2\mathbf{D}^2 + \frac{4\mathbf{K}N_0}{\delta}} \tag{8}$$

Consider now the external phase α, on which experimental measurements are made:

$$N_\alpha = M_\alpha + 2D_\alpha \tag{9}$$

From Eq. (1a) and the definition of $\mathbf{K}$ in Eq. (7b)

$$N_\alpha = M_\alpha + \frac{1}{\mathbf{K}} M_\alpha^2 \tag{10}$$

Combining Eqs. (8) and (10) and simplifying

$$\frac{N_\alpha}{N_0} = \frac{1}{\delta} + \frac{1}{2}\left(\frac{\mathbf{D}-1}{N_0}\right)\left(\mathbf{KD} - \sqrt{(\mathbf{KD})^2 + \frac{4\mathbf{K}N_0}{\delta}}\right) \tag{11}$$

This final equation relates N_α, equivalents in the external phase at equilibrium, to the total equivalents added, N_0. That this expression correctly describes the physical situation can be shown by examining the behavior of this function under specific limiting conditions.

For example, from Eqs. (7b) and (11)

$$\lim_{K_\alpha \to \infty} \left[\frac{N_\alpha}{N_0}\right] = \lim_{\mathbf{K} \to 0} \left[\frac{N_\alpha}{N_0}\right] = \frac{1}{\delta} = \frac{V_\alpha}{V_\alpha + \kappa_\mathbf{D} V_\beta}$$

This is as expected, since for $K_\alpha = \infty$ all material is present as the dimer, and hence N_α is related to N_0 only by the distribution parameter δ, referring to this species.

Similarly

$$\lim_{K_\alpha \to 0} \left[\frac{N_\alpha}{N_0}\right] = \lim_{K \to \infty} \left[\frac{N_\alpha}{N_0}\right]$$

This limit is indeterminate and must be evaluated by L'Hospital's rule. Taking part of the right-hand side of Eq. (11) and substituting for $\mathbf{K}$ from Eq. (7b)

$$\mathbf{KD} - \sqrt{\mathbf{K^2D^2} + \frac{4\mathbf{K}N_0}{\delta}} = \frac{\mu - \sqrt{\mu^2 + \dfrac{8K_\alpha\delta N_0}{V_\alpha}}}{\dfrac{2K_\alpha\delta}{V_\alpha}} \qquad (12)$$

$$\lim_{K_\alpha \to 0} (\text{LHS})^* = \lim_{K_\alpha \to 0} \left[\frac{-\dfrac{1}{2}\left(\mu^2 + \dfrac{8K_\alpha\delta N_0}{V_\alpha}\right)^{-\frac{1}{2}}\left(\dfrac{8\delta N_0}{V_\alpha}\right)}{\dfrac{2\delta}{V_\alpha}}\right] \qquad (13)$$

$$\lim_{K_\alpha \to 0} (\text{LHS}) = \frac{-2N_0}{\mu} \qquad (14)$$

Combining Eqs. (7a), (11), and (14)

$$\lim_{K_\alpha \to 0} \left[\frac{N_\alpha}{N_0}\right] = \frac{1}{\delta} + \left(\frac{\mu}{\delta} - 1\right)\left(-\frac{1}{\mu}\right) \qquad (15a)$$

$$\lim_{K_\alpha \to 0} \left[\frac{N_\alpha}{N_0}\right] = \frac{1}{\mu} = \frac{V_\alpha}{V_\alpha + \kappa_M V_\beta} \qquad (15b)$$

This expression relates N_α to N_0 only by the distribution parameter, μ, for the monomer and agrees with the physical situation since for $K_\alpha = 0$ all material is present as the monomer.

It is apparent that the equilibrium constant K_α may be calculated from the variation of N_α/N_0 with N_0. All other quantities in Eq. (11) may be obtained independently; for example, V_α may be measured by using as a marker a high MW molecule which cannot penetrate the gel matrix, $(V_\alpha + V_\beta)$ is the total initial volume, and V_β may be obtained by difference. Some investigators have equilibrated the macromolecular solution against preswollen gel (11, 12). In this case $(V_\alpha + V_\beta)$ may be measured using a low MW completely penetrant small molecule. The values of κ_M and κ_D may be obtained from the distribution at low and high total concentration, respectively, provided that those regions may be reached at which essentially pure monomeric

* Left-hand side.

or dimeric species exist; alternatively they may be obtained by analysis of the experimental plot.

Note that, from Eq. (11), the limit approached by N_α/N_0 as N_0 approaches either 0 or ∞ is indeterminate. Again, application of L'Hospital's rule is essential.

$$\lim \left[\frac{(\mathbf{D} - 1)\left(\mathbf{KD} - \sqrt{(\mathbf{KD})^2 + \dfrac{4\mathbf{K}N_0}{\delta}}\right)}{2N_0} \right]$$
$$= \lim \left[\frac{-(\mathbf{D} - 1)\mathbf{K}}{\delta\sqrt{(\mathbf{KD})^2 + \dfrac{4\mathbf{K}N_0}{\delta}}} \right] \quad (16)$$

and therefore

$$\lim_{N_0 \to \infty} \left[\frac{N_\alpha}{N_0} \right] = \frac{1}{\delta} \quad (17a)$$

and

$$\lim_{N_0 \to 0} \left[\frac{N_\alpha}{N_0} \right] = \frac{1}{\delta} - \frac{\mathbf{D} - 1}{\delta\mathbf{D}} \quad (17b)$$

Combining Eqs. (7a) and (17b) and simplifying

$$\lim_{N_0 \to 0} \left[\frac{N_\alpha}{N_0} \right] = \frac{1}{\delta} - \frac{\mu - \delta}{\delta\mu} = \frac{1}{\mu} \quad (18)$$

An example is given in Fig. 1 of a calculated curve for a system with an association constant K_α of 10^5 for both monomer and dimer penetrant.

This thermodynamic analysis of the equilibrium gel permeation system does not require a postulate as to mechanism of the interaction of the gel with the associating macromolecular system (restricted penetration, adsorption, partition, etc.). The form of the equations is similar to those derived for analysis of potentiometric or colligative property data (13, 14) with the gel acting as a ligand. The only requirement for variation of N_α with N_0 is that κ_M and κ_D differ; these may represent distribution coefficients, binding constants, or some multiplicative combination of individual constants.

EXPERIMENTAL

Examination of the basic equation, Eq. (11), relating N_α/N_0 to N_0 indicates that for experimental discrimination in evaluating K_α several conditions must be optimized. First, κ_M and κ_D must differ significantly [if $\kappa_M = \kappa_D$, then, from Eqs. (3a), (3b), and (7a), $\mathbf{D} = 1$, and N_α/N_0 is not a function of K_α]. This implies choosing a

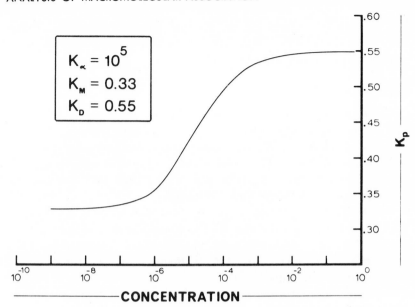

FIG. 1. Variation of equilibrium distribution with total concentration of a monomer-dimer system, theoretical curve. In analogy to gel permeation chromatography, the quantity K_p is plotted rather than N_α/N_0 where $K_p = [1 - (N_\beta V_\alpha/N_\alpha V_\beta)]$. Thus K_p varies from 1 for a completely excluded molecule to 0 for a completely penetrant molecule. In the symbolism usual for Sephadex chromatography (15), $K_p = (1 - "K_D")$, where $"K_D" = V_p/V_\beta$; V_p is defined as that fraction of the gel volume V_β which is penetrated by the molecular species considered. Thus $V_P[N_\beta] = N_\beta$ or $V_P N_\alpha/V_\alpha = N_\beta$ and $V_P = N_\beta V_\alpha/N_\alpha$. Also, the variables K_M and K_D are related to those in Sephadex chromatography; $\mathbf{K}_M = M_\alpha/(M_\alpha + M_\beta) = 1/\mu$ and similarly
$$\mathbf{K}_D = 1/\delta.$$

type of gel permeation material with suitable exclusion limits based on molecular weights of monomer and dimer.

The ratio of V_α to V_β should preferably be small (as V_α becomes large, V_α/V_{total} approaches 1 and N_α approaches N_0). If the amount of material is limiting, total volumes must be small.

For aqueous systems, the gel permeation materials available swell in the solvent. It is critical that complete swelling occur before measurements are made. An appreciable proportion of the volume V_α is within the swollen gel but external to the gel grains, and it may become necessary to use suction or pressure to recover sufficient supernatant for measurement. If this is the case, the mechanical properties of the gel must be considered in experimental design, since those grades with higher exclusion limits are soft, and may collapse under pressure, giving false low values for the internal volume. The collapse point may be as low as $30g$; this feature has been ignored in some published studies of equilibrium swelling (16).

Materials and Methods

Gel filtration materials used were the Sephadex series of dextran gels of porosity grades from G-25 to G-200 (MW exclusion limit for proteins of 5,000 and 800,000 respectively). Both powder and bead polymerized Sephadex were used without significant differences.

Blue Dextran 2000, an artificially colored high MW dextran, was procured from Pharmacia Fine Chemicals.

Gel swelling was carried out in 50 ml capped flasks in a cold room with continual agitation on an Eberbach reciprocating shaker.

Centrifugal filtration was done at 5°C in an International PR-2 refrigerated centrifuge, using the Millipore Filterfuge tube and 5μ pore diameter Millipore filters.

Spectrophotometric analysis of the supernatant was on a Zeiss PMQ-II spectrophotometer at 620 nm, the absorption maximum of Blue Dextran 2000. Measurement for hemoglobin was at 540 nm, the maximum of cyanoferrihemoglobin.

Hemoglobin was prepared from human blood by standard methods (17).

Swelling Time Experiments

A weighed amount of Sephadex, 500 mg, was added to a 10–20-ml aliquot of 0.05% Blue Dextran in 0.1 M NaCl (volume depending on gel porosity), and allowed to swell for varying times at 5° with shaking. The supernatant was collected by gravity filtration. Recovery was low but enough for analysis; centrifugal or pressure filtration was not used for initial experiments. After gel collapse points were determined (see below), time-of-swelling experiments were repeated using centrifugal filtration at appropriate speeds, with experimentally indistinguishable results.

Results are summarized on Fig. 2. As swelling proceeds, the completely nonpenetrant Blue Dextran is more concentrated in the external volume until a plateau is reached at complete swelling. The time required ranges from 1 hr for G-25 to 48 hr for G-200 (at 5° with shaking).

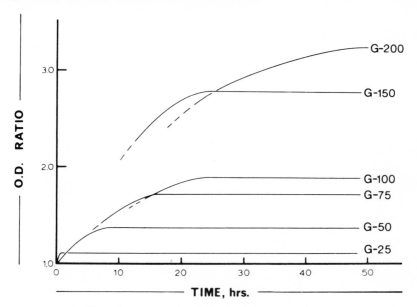

FIG. 2. Swelling time curves for Sephadex gels of varying porosity. The ordinate plotted is the ratio of optical density in V_a to optical density of the Blue Dextran stock solution.

Gel Collapse Experiments

A weighed amount of Sephadex was added to an aliquot of Blue Dextran and allowed to swell to equilibrium (see above). The supernatant was collected by centrifugal filtration at various speeds. In this case, the gel collapse point is determined by that force at which relative dilution of the supernatant occurs due to expression of solvent from the swollen gel matrix.

Results are shown in Fig. 3. Gel collapse points range from greater than 1000g for G-25 to approximately 30g for G-200.

Macromolecular Dissociation

A series of experiments was carried out on a hemoglobin model system to evaluate the utility of the gel filtration equilibrium technique in analysis of protein dissociation.

Human oxyhemoglobin, at concentrations between 5×10^{-7} and

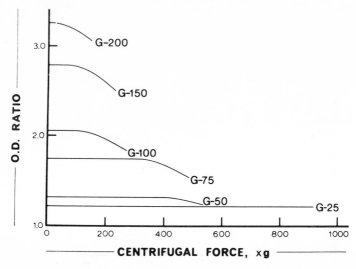

FIG. 3. Centrifugal collapse point curves for Sephadex gels of varying porosity. The ordinate variable is the same as in Fig. 2.

$5 \times 10^{-3}\,M$ in heme iron, was equilibrated in $0.1\,M$ Na-phosphate buffer pH 7.0 with dry Sephadex powder. The gel was allowed to swell for 48 hr at 5° with shaking, and an aliquot of the supernatant, V_a, was collected by centrifugal filtration at a force just below the gel collapse point.

The results are shown in Fig. 4. For a Sephadex grade which excludes the protein, there is no variation of distribution with total concentration. However, on G-200 which allows hemoglobin (MW 67,000) to penetrate, the distribution coefficient K_p (defined in the legend to Fig. 1) falls with dilution. Theoretically, at complete protein dissociation, K_p should reach another plateau characteristic of the distribution of the half-molecule; in the experiments shown this level of dilution was not reached.

The dissociation constant for human hemoglobin, conventionally expressed as

$$\alpha_2\beta_2 \overset{K_\alpha'}{\rightleftharpoons} 2\alpha\beta; \qquad K_\alpha' = 5 \times 10^{-6}\,M$$

may be obtained from Fig. 4.

Lacking data for low concentrations, this value is subject to an uncertainty of about $\pm 50\%$. It agrees well with estimates of this equilibrium constant obtained by restricted diffusion (*9*), molecular sieve

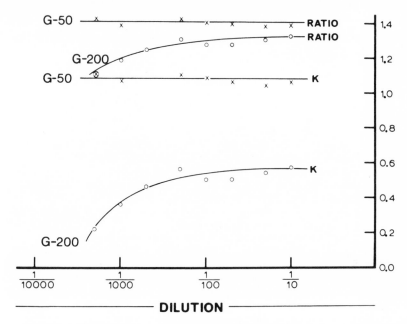

FIG. 4. Equilibrium gel partition on Sephadex G-50 and G-200 of human oxyhemoglobin, stock solution $5 \times 10^{-3} M$ in heme iron. The quantity K is identical to K_p as defined in the legend to Fig. 1, RATIO is the O. D. ratio defined in the legend to Fig. 2 under the specific experimental conditions employed.

chromatography (18), osmotic pressure (3), or ultracentrifugation (19).

Acknowledgments

This research was supported in part by Grant #661 from the Council for Tobacco Research—U. S. A. and by Project E-364 of the Connecticut Heart Association. One of us (BFC) held an Advanced Research Fellowship from the American Heart Association and is presently an Investigator of the Howard Hughes Medical Institute; another (LS) was a National Science Foundation Summer Student Fellow at the New England Institute.

We wish to thank the Blood Bank, Division of Laboratories, Danbury Hospital, Danbury, Conn., for the human blood used in the dissociation experiments.

Special thanks are due to Miss Joy Judell, who aided in computer programming, and Mrs. Yolanda Capo, who struggled with unfamiliar mathematical notation in preparation of this manuscript for publication.

REFERENCES

1. W. A. Schroeder, *Progr. Chem. Org. Nat. Prod.*, **17**, 321 (1959).
2. H. Gutfreund and J. R. Knowles, *Essays Biochem.*, **3**, 25 (1967).
3. G. Guidotti, *J. Biol. Chem.*, **252**, 3673 (1967).
4. D. E. Koshland, G. Nemethy, and D. Filmer, *Biochemistry*, **5**, 365 (1966).
5. G. A. Gilbert, *Proc. Roy. Soc.* (London), **250A**, 377 (1959).
6. E. T. Adams, Jr., *Biochemistry*, **6**, 1864 (1967).
7. D. E. Roark and D. A. Yphantis, *Ann. N. Y. Acad. Sci.*, **164**, 245 (1969).
8. J. R. Whitaker, *Anal. Chem.*, **35**, 1950 (1963).
9. G. K. Ackers and T. E. Thompson, *Proc. Nat. Acad. Sci., U. S.*, **53**, 342 (1965).
10. G. K. Ackers, *Biochemistry*, **3**, 723 (1964).
11. M. J. Stone and H. Metzger, *J. Biol. Chem.*, **243**, 5049 (1968).
12. K. H. Altgelt and J. C. Moore, in *Polymer Fractionation* (M. J. R. Cantow, ed.), Academic, New York, 1967.
13. A. D. Adler, Dissertation, University of Pennsylvania, 1960.
14. A. D. Adler, J. A. O'Malley, and A. J. Herr, Jr., *J. Phys. Chem.*, **71**, 2896 (1967).
15. P. Flodin, Dissertation, AB Pharmacia, Uppsala, 1962.
16. I. Vavruch, *Kolloid-Z. Z. Polym.* **227**, 47 (1968).
17. B. F. Cameron and P. George, *Biochim. Biophys. Acta*, **194**, 16 (1969).
18. E. Chiancone, L. M. Gilbert, G. A. Gilbert, and G. L. Kellett, *J. Biol. Chem.*, **243**, 1212 (1968).
19. M. Perutz, *J. Mol. Biol.*, **13**, 646 (1965).

Received by editor May 18, 1970

Gel Filtration Chromatography*

BRUCE F. CAMERON†

DEPARTMENT OF INTERNAL MEDICINE
UNIVERSITY OF MIAMI SCHOOL OF MEDICINE
MIAMI, FLORIDA 33152
AND
LABORATORIES FOR HEMATOLOGICAL RESEARCH
HOWARD HUGHES MEDICAL INSTITUTE
MIAMI, FLORIDA 33152

Summary

The technique of gel filtration chromatography is described, with a guide to relevant literature on methods, theory, and applications, especially the use of GFC in analysis of molecular interactions, both macromolecular and small molecule–macromolecule aggregation.

INTRODUCTION

Pores of molecular dimensions to separate molecules by size have been used for over a century, originating with the introduction of dialysis in 1861 (1). This technique, together with ultrafiltration under pressure or vacuum (2), is still of major practical importance. It is, however, only within the last few decades that chromatographic materials which separate essentially by molecular size have been developed and widely used. Such materials have been known for 40–50 years in the form of zeolites and similar molecular sieves (3), and a sieving effect in ion exchange resins (4), starch gels (5), agar gels (6), and polyacrylamide gels (7) has occasionally been utilized in macromolecular separation. This molecular sieve effect in polyacrylamide gel is an important component of the high separation power of disc gel electrophoresis (8).

* Presented at the ACS Symposium on Gel Permeation Chromatography sponsored by the Division of Petroleum Chemistry at the 159th National Meeting of the American Chemical Society, Houston, Texas, February, 1970.
† Investigator of the Howard Hughes Medical Institute.

351

In the above cases, the molecular sieve properties are secondary or incidental. In the last decade two major classes of materials for chromatography have been developed which rely primarily or exclusively on molecular size for separation. These are represented by the Sephadex series of cross-linked dextran gels (9) and highly cross-linked polystyrene gels (10).

These two classes of materials tend to be used by mutually exclusive groups and are generally called by different names. Thus the Sephadex-type materials are gels which swell in aqueous media and, under the term "gel filtration chromatography," have been extensively used in biochemistry; while the polystyrene gels are primarily used in organic solvents and, as "gel permeation chromatography," are employed in industrial and polymer chemistry.

The present conference is concerned almost entirely with gel permeation chromatography (GPC). It is hoped that in this context the present paper will provide an introduction for those familiar with GPC to the literature on "GFC," with particular reference to theories and applications that differ from those commonly used in GPC.

In essence this paper is an extension of a previous review by Altgelt and Moore (10a).

GFC MATERIALS

The materials used in GFC are characterized by their property of swelling in aqueous systems. There are two major groups of materials, the cross-linked polydextran gels (Sephadex), and cross-linked polyacrylamide (Bio-Gel P) gels (11, 12).

Both materials come in a variety of types, varying in the molecular weight range over which fractionation occurs. For example, Sephadex is obtainable in types from G-10 (fractionation range 0–700) to G-200 (fractionation range 5,000–800,000). These MW fractionation ranges are specified for globular proteins; for dextrans the equivalent ranges are 0–700 and 1,000–200,000, respectively. Particle diameters of Sephadex (a bead polymer) vary from 20–300 μ, with a superfine grade (10–40 μ) available for thin layer chromatography (13). Bio-Gel is available in types from P-2 to P-300 (fractionation ranges 200–1,800 and 60,000–400,000, respectively).

For very high MW materials, such as nucleic acids or viruses, an agarose material, Sepharose, is supplied by Pharmacia (14).

As stated above, the Sephadex-type materials are designed for use in aqueous systems, although alcohol–water mixtures, dimethylsulf-oxide, formamide, and glycol may be used. Recently Pharmacia has developed the LH series of gels which swell in organic solvents. These gels, of which LH-20 is currently available, have been used for separations of such materials as steroids (15), petroleum hydrocar-bons (16), and other organic molecules.

Although the primary mechanism of separation in GFC is by molecular sieve properties, it is important to recognize that under certain conditions (i.e., low ionic strength), adsorption, ion-exchange, or ion exclusion may occur. This is particularly noted in the retarda-tion of aromatic compounds on Sephadex (17); this phenomenon has been used to fractionate amino acids (18). Other examples of applica-tions of adsorption or ion-exchange in GFC include concanavalin purification by specific protein-carbohydrate interaction (19), frac-tionation of humic acid (20), and separation of nucleic acid com-ponents (21). It has even been possible to carry out a partial reso-lution of stereoisomers on Sephadex (22).

GFC TECHNIQUES

The usual techniques for gel preparation, column packing, and elution are summarized in relevant product literature (11, 12), and in the work of Flodin (23). The swelling kinetics of Sephadex have been the subject of several reports (24, 25), since this parameter is critical in gel preparation.

In general, simple open columns have been used, with typical diameters of 1–5 cm and lengths of 20–200 cm; closed columns with flow adaptors have been introduced by Pharmacia and several ap-paratus manufacturers. Automatic equipment has not been greatly utilized, partly because common applications of GFC, such as desalting, have not required such sophistication. However, closed systems with sample injection and provisions for recycling have been described (26), and are being used increasingly.

Sample application is one of the most critical steps in GFC, since a sharp boundary between sample and solvent is required. Several mechanical devices are available for this procedure (11), and apply-ing the sample at a density above that of the solvent is a common technique; this is achieved by addition of neutral salt or sucrose (27). In this method, care must be exercised to keep sample viscosity low, or undesirable band spreading may result (23).

In general, for protein separation, desalting, or qualitative analysis of molecular binding, extreme accuracy is not required. It is possible, however, with meticulous attention to gel preparation, packing under reverse flow, slow elution rates, and small sample size to obtain elution volumes accurate to ~0.1% (B. F. Cameron, unpublished results, see Ref. *28*). For such accuracy, volume estimation by weight may be needed (*29*).

THEORETICAL ASPECTS

This topic will be only briefly covered, since most of the theories are common to GFC and GPC, and several reviews on theoretical aspects of GPC are available (*30*).

The process of GFC is usually looked upon as a process of steric exclusion with the gel pore geometry defined in various ways (*31*), or as a process of equilibrium partition in a three-dimensional gel fiber network (*32*). Careful study of column chromatographic and equilibrium partition of macromolecules by Ackers (*33*) suggested that at least for Sephadex gels G-150 and G-200 the chromatographic separation could not be explained by partition equilibrium. An expression was derived to characterize GFC as a process of restricted diffusion in cylindrical pores, a model essential similar to a capillary exclusion model proposed earlier (*34*). It is probably a reflection on adequacy of experiment rather than correctness of the theory that all these models fit experimental data equally well (*35*). A succinct summary of thermodynamic and hydrodynamic aspects of GFC has been published (*36*).

The thermodynamic parameters of swelling of Sephadex gels are a critical factor in a theoretical analysis of GFC, and have been the subject of a series of reports by Ogston and co-workers (*37, 38*); the results on G-200 gel conflict with the conclusions of Ackers (*33*), and lend support to the equilibrium partition mechanism (*32*).

An alternate theory for molecular sieve partition has been suggested (*39*), based on a stochastic model of movement of a solute through the column (*40*). This model has not been applied to GFC, except to incorporate the formalism into an alternate expression (*41*) for general equations derived by Ackers for generalized pore geometry (*42*). However, elution profiles of both large and small molecules on Sephadex G-50 agree in shape with that proposed for the stochastic model by McQuarrie (*40*), and base widths are inversely proportional to MW as predicted by Carmichael (*39*). (Unpublished experiments, B. F. Cameron.)

APPLICATIONS

The first applications of GFC were to desalting as a substitute for dialysis, i.e., separation of a macromolecule from low MW salts, or buffer exchange; and macromolecular fractionation, particularly on Sephadex types G-75 through G-200 (*23*, *43*). Such applications remain the most important uses.

Perhaps the application of GFC to biopolymers in aqueous systems, i.e., monodisperse materials, fixed the development of analytical methods of GFC. Conversely, in GPC, the problems involve nonaqueous systems and polydisperse polymer fractions, leading to an emphasis on problems of band spreading and data correction.

Thus, the problem of molecular weight determination in GFC, while an important use of the technique, has been relatively unimportant theoretically. In general, the elution volume at the peak of a monodisperse protein is correlated with an empirical function of molecular size, and the MW of the unknown is obtained from a calibration curve obtained on the same column with proteins of similar shape (i.e., globular or random coil) and known MW.

Several parameters of molecular size have been used in correlation to MW, those most often employed being a linear relation of V_e to logarithm of MW (*27*), or of the cube root of partition coefficient to the square root of MW (*44*). An alternate calibration is in terms of the relationship of elution volume to Stokes radius given by Ackers (*33*), although this is not as generally useful. It has been pointed out that any of these empirical approaches is useful as a method of estimating MW for one of a homologous series of macromolecules (*24*, *36*).

MOLECULAR ASSOCIATION

Very early in the development of GFC the technique was applied to detection and measurement of molecular association.

The simplest system is that of binding of a small molecule to a macromolecule, where both the macromolecule and the complex are excluded from the gel. For example, GFC has been extensively used to measure free and bound small molecules, such as calcium in serum (*45*). A sample is applied to the GFC column; concentration of the bound small molecule is determined by analysis of the rapidly eluted macromolecular peak containing the complex, and concentration of the free species by difference. The method is equivalent to equilibrium dialysis of tightly bound species, and the two techniques have been shown to be quantitatively comparable (*46*).

This technique fails when the bound small molecule is freely dissociable from the macromolecule, since the complex will dissociate as it elutes and false low values of binding will be obtained. To assay such systems a variety of techniques have been used. Thus, binding may be determined as above in columns of varying length, and extrapolated to zero length (*47*), or the bound fraction may be determined by an equilibrium partition experiment rather than by chromatography (*48*, see Ref. *25*).

For the detection of such interaction, a GFC method was devised in which the macromolecule is equilibrated with a small molecule and chromatographed as a small sample on a column equilibrated with an equal concentration of the small molecule (*49*). If binding occurs, even weakly, the elution profile of the small molecule shows a positive peak at the elution volume of the macromolecule, and by conservation of mass, a corresponding trough at a greater volume. This method has been extended to quantitative estimation of equilibrium constants (*50*), although this is easier if analysis is carried out on the trailing boundary of a frontal analysis system (*51*), i.e., one in which sample volume is of the same order of magnitude as column volume such that a plateau region occurs.

The analysis of macromolecular association is the last area of major application for analytical GFC. Theoretical analyses of elution boundaries in rapidly interacting macromolecular species are based on the analysis of boundary forms in freely migrating systems (*52*). These concepts have been applied to the chromotographic case in several ways (*53–55*), all in essence relating the variation of elution volume or elution profile with concentration to the extent of interaction.

It is important to distinguish concentration dependence due to molecular interaction from that due to "physical" interaction; the latter is usually linear and positive, while the former exhibits curvature (*56*). Zonal (small sample) and frontal (large sample) techniques of measuring concentration dependence due to physical interaction have given discordant results, which have recently been reconciled by a consistent definition of concentration in zonal analysis (*57*).

In some cases it is necessary to use the GFC experiment to determine the nature of the system as well as the interaction constants. For a simple one-step polymerization reaction a method for analysis of stoichiometry has been described for zonal (58) or frontal (29) GFC. One useful feature of frontal analysis is that for reactions $nA \rightleftharpoons P$ with $n > 2$ the trailing boundary may be bimodal, and that the stoichiometric coefficient may be directly obtained from the profile (59). Such behavior has been described for chymotrypsin (60) and for soap micelles (61).

An alternative technique for analysis is to use elution data to define an apparent weight-average MW (62) and analyze the concentration dependence of this quantity by a general multinomial theory (63).

Finally, an equilibrium partition method may be used in which the partition coefficient is directly determined as a function of concentration (25, 64).

For further details, especially in respect to analytical aspects of GFC, two recent reviews should be consulted (65, 66).

A detailed study of enzyme association reactions by GFC has recently been carried out on rabbit phosphorylase (67). Phosphatase (68) has been studied at concentration levels equivalent to those used in kinetic studies; the ability of GFC to examine macromolecular association at these levels (pg/ml) is a unique advantage of the method. Many studies have been made of hemoglobin dissociation (25, 69), illustrating the utility of GFC for macromolecular systems with a very low dissociation constant ($\sim 10^{-6} M$).

REFERENCES

1. T. Graham, *Phil. Trans. Roy. Soc. London*, **151**, 183 (1861).
2. W. F. Blatt, S. M. Robinson, and H. J. Bixler, *Anal. Biochem.*, **26**, 151 (1968).
3. P. Flodin and J. Porath, in *Chromatography* (E. Heftmann, ed.), Reinhold, New York, 1961, p. 328.
4. S. M. Partridge, *Nature*, **169**, 496 (1952).
5. G. H. Lathe and C. R. J. Ruthven, *Biochem. J.*, **62**, 665 (1956).
6. A. Polson, *Biochim. Biophys. Acta*, **50**, 565 (1961).
7. S. Hjertén and R. Mosbach, *Anal. Biochem.*, **3**, 109 (1962).
8. L. Ornstein, *Ann. N. Y. Acad. Sci.*, **121**, 321 (1964).
9. P. Andrews, *Brit. Med. Bull.*, **22**, 109 (1966).
10. J. C. Moore, *J. Polym. Sci., Part A*, **2**, 835 (1964).
10a. K. H. Altgelt and J. C. Moore, in *Polymer Fractionation* (M. J. R. Cantow, ed.), Academic, New York, 1967.

11. Pharmacia Fine Chemicals, *Sephadex—Gel Filtration in Theory and Practice*, Pharmacia, Piscataway, N. J., no date.
12. Bio-Rad Laboratories, *Price List U—Ion Exchange, Gel Filtration, Adsorption*, Bio-Rad, Richmond, Calif., 1969.
13. B. G. Johansson and L. Rymo, *Acta Chem. Scand.*, **18**, 217 (1964).
14. B. Öberg and L. Philipson, *Arch. Biochem. Biophys.*, **119**, 504 (1967).
15. P. Eneroth and E. Nyström, *Biochim. Biophys. Acta*, **144**, 149 (1967).
16. B. J. Mair, P. T. R. Hwang, and R. G. Ruberto, *Anal. Chem.*, **39**, 838 (1967).
17. B. Gelotte, *J. Chromatogr.*, **3**, 330 (1960).
18. D. Eaker and J. Porath, *Separ. Sci.*, **2**, 507 (1967).
19. B. B. L. Agrawal and I. J. Goldstein, *Biochim. Biophys. Acta*, **147**, 262 (1967).
20. I. Lindquist, *Acta Chem. Scand.*, **21**, 2564 (1967).
21. S. Zadrazil, Z. Sormova, and F. Sorm, *Coll. Czech. Chem. Commun.*, **26**, 2643 (1961).
22. R. E. Leitch, H. L. Rothbart, and W. Rieman, *J. Chromatogr.*, **28**, 132 (1967).
23. P. Flodin, *Dextran Gels and Their Applications in Gel Filtration*, Pharmacia, Uppsala, 1962.
24. P. Andrews, *Biochem. J.*, **96**, 595 (1965).
25. B. F. Cameron, L. Sklar, V. Greenfield, and A. D. Adler, *Separ. Sci.*, **6**, 217 (1971).
26. W. Welling, *Sci. Tools*, **15**, 24 (1968).
27. P. Andrews, *Biochem. J.*, **91**, 222 (1964).
28. E. S. Awad and N. K. Abed, *Biochem. Z.*, **346**, 403 (1966).
29. D. J. Winzor and H. A. Scheraga, *Biochemistry*, **2**, 1263 (1963).
30. K. H. Altgelt, *Advan. Chromatogr.*, **7**, 3 (1968).
31. P. G. Squire, *Arch. Biochem. Biophys.*, **107**, 471 (1964).
32. T. C. Laurent and J. Killander, *J. Chromatogr.*, **14**, 317 (1964).
33. G. K. Ackers, *Biochemistry*, **3**, 723 (1964).
34. K. O. Pederson, *Arch. Biochem. Biophys.*, *Suppl.*, **1**, 157 (1962).
35. L. M. Siegel and K. J. Monty, *Biochim. Biophys. Acta*, **112**, 346 (1966).
36. D. M. W. Anderson and J. F. Stodhart, *Lab. Pract.*, **16**, 841 (1967).
37. E. Edmond, S. Farquhar, J. R. Dunstone, and A. G. Ogston, *Biochem. J.*, **108**, 755 (1968).
38. A. G. Ogston and P. Silpananta, *Biochem. J.*, **116**, 171 (1970).
39. J. B. Carmichael, *J. Polym. Sci., Part A-2*, **6**, 517 (1968).
40. D. A. McQuarrie, *J. Chem. Phys.*, **38**, 437 (1963).
41. J. B. Carmichael, *Biopolymers*, **6**, 1497 (1968).
42. G. K. Ackers, *J. Biol. Chem.*, **242**, 3237 (1967).
43. B. Gelotte and A. Emnéus, *Chem.-Ing.-Tech.*, **38**, 445 (1966).
44. J. Porath, *J. Pure Appl. Chem.*, **6**, 233 (1963).
45. H. M. von Hattingberg and W. Klaus, *Klin. Wochenschr.*, **44**, 499 (1966).
46. W. Scholtan, *Arzneim.-Forsch.*, **14**, 146 (1964).
47. W. Hoffmann and U. Westphal, *Anal. Biochem.*, **32**, 48 (1969).
48. P. Fasella, G. G. Hammes, and P. R. Schimmel, *Biochim. Biophys. Acta*, **103**, 708 (1965).
49. J. P. Hummel and W. J. Dreyer, *Biochim. Biophys. Acta*, **63**, 530 (1962).
50. G. F. Fairclough and J. S. Fruton, *Biochemistry*, **5**, 673 (1966).
51. L. W. Nichol and D. J. Winzor, *J. Phys. Chem.*, **68**, 2455 (1964).

52. G. A. Gilbert, *Discussions Faraday Soc.,* **20**, 68 (1955).
53. G. A. Gilbert, *Anal. Chim. Acta,* **38**, 275 (1967).
54. G. K. Ackers and T. E. Thompson, *Proc. Nat. Acad. Sci. U. S.,* **53**, 342 (1965).
55. L. W. Nichol, A. G. Ogston, and D. J. Winzor, *J. Phys. Chem.,* **71**, 726 (1967).
56. D. J. Winzor and L. W. Nichol, *Biochim. Biophys. Acta,* **104**, 1 (1965).
57. D. J. Winzor, *Biochem. J.,* **101**, 30C (1966).
58. G. K. Ackers, *J. Biol. Chem.* **243**, 2056 (1968).
59. L. W. Nichol and J. L. Bethune, *Nature,* **198**, 880 (1963).
60. M. V. Tracey, *Aust. J. Biol. Sci.,* **17**, 792 (1964).
61. H. Coll, **6**, 229 (1971).
62. G. A. Gilbert, *Proc. Roy. Soc. London,* **A250**, 377 (1959).
63. M. Derechin, *Biochemistry,* **8**, 921 (1969).
64. M. J. Stone and H. Metzger, *J. Biol. Chem.,* **243**, 5049 (1968).
65. D. J. Winzor, in *Physical Principles and Techniques of Protein Chemistry, Part A* (S. J. Leach, ed.), Academic, New York, 1969, p. 451.
66. G. K. Ackers, *Advan. Protein Chem.,* **24**, 343 (1970).
67. D. L. DeVincenzi and J. L. Hedrick, *Biochemistry,* **9**, 2048 (1970).
68. D. J. Winzor, *Biochim. Biophys. Acta,* **200**, 423 (1970).
69. E. Chiancone, L. M. Gilbert, G. A. Gilbert, and G. L. Kellett, *J. Biol. Chem.,* **243**, 1212 (1968).

IV

**APPLICATIONS OF GPC TO PROBLEMS
IN POLYMER AND PETROLEUM CHEMISTRY**

Determination of Polymer Branching with Gel Permeation Chromatography. Abstract of a Review*

E. E. DROTT and R. A. MENDELSON

MONSANTO COMPANY
TEXAS CITY, TEXAS 77590

The effect of long- and short-chain branching in polymer molecules on GPC separation is reviewed (1–4). The calculation of branched GPC curves is developed from the universal calibration techniques, which is based on the concept of hydrodynamic volume ($M[\eta]$) and previously established relationships for the effect of branching on molecular dimensions. Typical calibration curves are shown for different branching models and degrees of branching. As the branching level increases, the curves are shown to approach a limiting value. Methods of characterizing branching levels and molecular-weight distributions of fractions and whole polymers from GPC and intrinsic viscosity data are presented. An iterative computer program is described which was written to calculate the degree of branching in whole polymers. Long-chain branching in several low-density polyethylene samples was determined, using both the fraction and the whole polymer methods. Effects of various experimental errors and branching models were investigated. For polyethylene, the data show that the effect of branching in intrinsic viscosity is best described by the relationship $\langle g_3 \rangle_w^{1/2} = [\eta]_{br}/[\eta]$ where $\langle g_3 \rangle$ is the Zimm-Stockmeyer expression for trifunctional branch points in a polydisperse sample.

*Abstract of a paper presented at the Symposium on Gel Permeation Chromatography sponsored by the Division of Petroleum Chemistry at the 159th National Meeting of the American Chemical Society, Houston, Texas, February, 1970.

363

REFERENCES

1. E. E. Drott and R. A. Mendelson, *J. Polym. Sci., Part A-2*, **8**, 1361, 1375 (1970).
2. E. E. Drott and R. A. Mendelson, *Preprints*, 4th International GPC Seminar, Miami Beach, Fla., 1967.
3. E. E. Drott and R. A. Mendelson, *Preprints*, 5th International GPC Seminar, London, 1968.
4. E. E. Drott and R. A. Mendelson, *Preprints*, 6th International GPC Seminar, Miami Beach, Fla., 1968.

Received by editor May 11, 1970

Fractionation of Linear Polyethylene with Gel Permeation Chromatography. Part III*

NOBUYUKI NAKAJIMA

PLASTICS DIVISION
ALLIED CHEMICAL CORPORATION
MORRIS TOWNSHIP, NEW JERSEY 07960

Summary

Many commercial linear polyethylenes have very broad distributions of molecular weight. The high molecular weight fractions often extend beyond the highest molecular weight calibration standard of GPC. For this reason the reliability of information obtainable from GPC has been examined with attention to the average molecular weights. Calibration range is a serious limitation for the accurate determination of the weight-average and the higher averages of molecular weight. Uncertainty in the baseline at the high molecular weight region, however, does not produce a significant error. With a four-column GPC having 10^3 to 10^7 Å nominal capacity, improved resolution is needed in the high molecular weight range. In order to examine the resolution and to improve the calibration, a polyethylene standard of ca. 3–4 million molecular weight is required. With the present limitation of GPC the greatest amount of information can be obtained by examining and intercomparing the cumulative distribution curves. With this representation ca. 95% or more of the cumulative weight range is free from uncertainty in calibration and resolution. A question is raised as to whether melt index is precisely a function of the weight-average molecular weight. This question is pertinent when significantly different molecular weight distributions are involved. GPC offers an opportunity to resolve the question.

* Presented at the ACS Symposium on Gel Permeation Chromatography sponsored by the Division of Petroleum Chemistry at the 159th National Meeting of the American Chemical Society, Houston, Texas, February, 1970.

365

INTRODUCTION

In the first paper of this series (1), the usefulness of gel permeation chromatography (GPC) was examined as a tool for fractionating linear polyethylenes. The conclusions of practical importance were (a) two column GPC (10^4 and 10^6 Å) exhibits resolution comparable to extractive-column fractionation, (b) unmatched rapidity of GPC operation, (c) need for improving resolution at high molecular weights, and (d) necessity of using polyethylene fractions as the calibration standards. The questions of resolution and calibration are quite general ones with many fractionation techniques. Progress has been made along these lines with GPC; however, improvements continue almost indefinitely. For the practical utilization of GPC, therefore, one has to accept the best means available at a given time.

In the second paper (2), a four-column GPC having 10^3 to 10^7 Å nominal capacity was accepted as the practical instrument. The calibration was based on polystyrene rather than polyethylene standards. Although the latter are ideal for obtaining absolute molecular weights, the relative molecular weights based on polystyrene are quite adequate for comparing one distribution to another. The usefulness of GPC was proven by demonstrating reliable reproductibility.

One selected sample was run (2) with four different GPC units having a similar capacity. The results showed excellent reproducibility, provided sufficient care was taken in calibration and other operations. At the same time, reproducibility on a single instrument was checked and found satisfactory by making consecutive runs of a number of samples. The work not only demonstrated reproducibility but provided a control sample to be used in the future for ensuring reproducible performance of GPC.

In these earlier papers (1), (2), examples of fractionation curves were presented to show that GPC could reveal interesting details of molecular weight distribution. In this paper, an examination on the confidence limits of the information obtained from GPC fractionation is continued. Further examples of fractionation results are discussed.

RESULTS

Calculation of Moments and Average Molecular Weights

In this section, calculations of moments and average molecular weights are reviewed with respect to the confidence limits of the

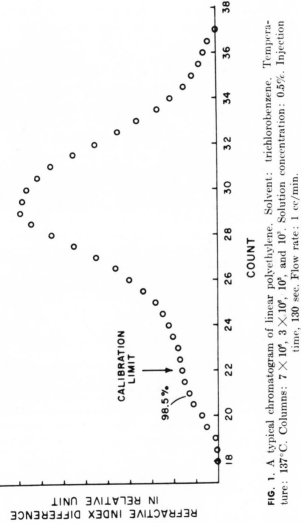

FIG. 1. A typical chromatogram of linear polyethylene. Solvent: trichlorobenzene. Temperature: 137°C. Columns: 7×10^6, 3×10^6, 10^5, and 10^3. Solution concentration: 0.5%. Injection time, 130 sec. Flow rate: 1 cc/min.

results. Specific attention is paid to the higher moments, where errors might arise from uncertainties in baseline and calibration.

A typical GPC trace of 0.4 melt index resin is reproduced in Fig. 1. The points in the curve are the input data for computer calculation (3). The computer outputs are shown in Fig. 2, as Curves a, b, c, d, and e. These five curves represent the following quantities:

$$Q_0 = \int \frac{1}{A_i} \left(\frac{dW_i}{dA_i}\right) dA_i = \int \left(\frac{dW_i}{dA_i}\right) d \ln A_i \tag{1}$$

(Curve a)

$$Q_1 = \int \left(\frac{dW_i}{dA_i}\right) dA_i = \int A_i \left(\frac{dW_i}{dA_i}\right) d \ln A_i \tag{2}$$

(Curve b)

$$Q_2 = \int A_i \left(\frac{dW_i}{dA_i}\right) dA_i = \int A_i^2 \left(\frac{dW_i}{dA_i}\right) d \ln A_i \tag{3}$$

(Curve c)

$$Q_3 = \int A_i^2 \left(\frac{dW_i}{dA_i}\right) dA_i = \int A_i^3 \left(\frac{dW_i}{dA_i}\right) d \ln A_i \tag{4}$$

(Curve d)

$$Q_4 = \int A_i^3 \left(\frac{dW_i}{dA_i}\right) dA_i = \int A_i^4 \left(\frac{dW_i}{dA_i}\right) d \ln A_i \tag{5}$$

(Curve e)

where A_i is the relative chain length based on polystyrene standards and (dW_i/dA_i) is obtained from the cumulative distribution curve as the derivative of cumulative fraction with respect to A_i. On these

TABLE 1

Moments and Average Chain Length

Quantity	Computer integration	Possible range of value resulting from baseline error (hand calculation)
Q_0	1.59×10^{-3}	Negligible
Q_1	1.01	Negligible
Q_2	8.95×10^3	Negligible
Q_3	4.77×10^8	4.52–4.85×10^8
Q_4	4.82×10^{13}	4.32–5.30×10^{13}
$\bar{A}_n$	6.3×10^2	Negligible
$\bar{A}_w$	8.9×10^3	Negligible
$\bar{A}_z$	5.3×10^4	5.05–5.4×10^4
$\bar{A}_{z+1}$	1.01×10^5	0.96–1.09×10^5

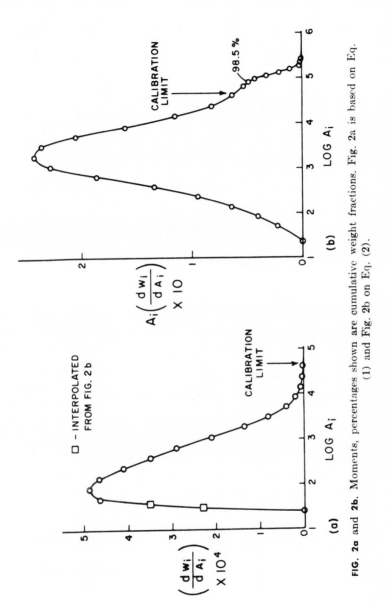

FIG. 2a and **2b**. Moments, percentages shown are cumulative weight fractions. Fig. 2a is based on Eq. (1) and Fig. 2b on Eq. (2).

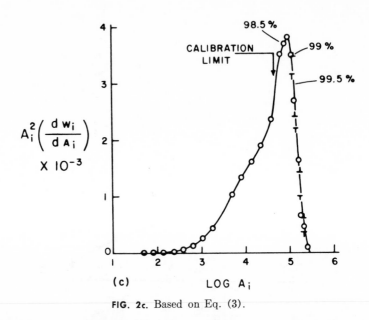

FIG. 2c. Based on Eq. (3).

curves, computer outputs are shown with circles. Except for the nine points at the highest molecular weight, every third point from the computer output is shown.

Hand calculation is also performed in the high molecular weight range, independent of the computer calculation. The input data for this calculation are taken by drawing as low and as high a baseline as possible to assess the errors involved in drawing the baseline. A graphical differentiation is used to obtain (dW_i/dA_i). The maximum ($\bot$) and minimum (T) values are shown in the graph with the symbols indicated. The results of the computer integration according to Eqs. (1)–(5) are given in Table 1, together with the limits of error in baseline drawing. Average values of chain lengths, Eqs. (6)–(9), are also listed.

$$\bar{A}_n = \frac{Q_1}{Q_0} \qquad \text{or} \qquad \frac{1}{Q_0} \qquad \text{(number average)} \qquad (6)$$

$$\bar{A}_w = \frac{Q_2}{Q_1} \qquad \text{or} \qquad Q_2 \qquad \text{(weight average)} \qquad (7)$$

$$\bar{A}_z = \frac{Q_3}{Q_2} \qquad \text{(z average)} \qquad (8)$$

$$\bar{A}_{z+1} = \frac{Q_4}{Q_3} \qquad \text{(z + 1 average)} \qquad (9)$$

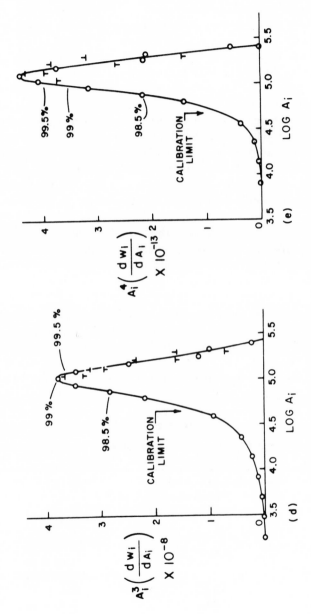

FIG. 2d and 2e. Fig. 2d is based on Eq. (4) and Fig. 2e on Eq. (5).

Conclusions from these results are:

(a) The computer outputs and the hand-calculated results are in good agreement.

(b) The errors arising from uncertainty in the base line are insignificant, even for the $(z + 1)$ average chain length.

(c) The highest molecular weight calibration standard, polystyrene of ca. 50,000 chain length, is not sufficient to provide reliable values for the weight-average molecular weight and the higher average chain length; a calibration standard having 250,000 chain length is required. This is approximately 3–4 million molecular weight polyethylene.

(d) Because the present calibration does not encompass the significant molecular weight range, $\bar{A}_w$, $\bar{A}_z$, and $\bar{A}_{z+1}$ are arbitrary values resulting from a particular data treatment. In this case, the treatment is based on the linear extrapolation of the calibration curve, i.e., a plot of logarithm of chain length versus GPC count.

(e) It appears that improved resolution is called for at the high molecular weight range. The shoulder at Count 21 in Fig. 1 may be one indication. Another is the change in the peak molecular weight from the Q_0 plot to the Q_4 plot (Fig. 2a–e and Table 2). This change becomes less and less from the lower moment to the higher moment, and behaves as if there is a cutoff at some point of the high molecular weight tail. The cut off may be real, but a lack of resolution is suspected.

Ratios of the succeeding average chain length also show the cutoff

TABLE 2

Peak Molecular Weight Corresponding to Each Moment and Ratios of Average Molecular Weight

Moment	Ratio of molecular weights	Peak chain length, A_i
Q_0		8.34×10^1
Q_1		1.8×10^3
Q_2		9.09×10^4
Q_3		1.08×10^5
Q_4		1.28×10^5
	$\bar{A}_w/\bar{A}_n$	14.2
	$\bar{A}_z/\bar{A}_w$	6.0
	$\bar{A}_{z+1}/\bar{A}_z$	1.9

TABLE 3

Melt Index and Average Chain Length

Group	Resins	Melt index	$\bar{A}_n \times 10^{-2}$	$\bar{A}_w \times 10^{-3}$	$\bar{A}_z \times 10^{-4}$	$\bar{A}_w/\bar{A}_n$
1	A	0.4	8.0	10.5	6.7	13.1
1	B	2.5	6.6	6.4	5.0	9.7
1	C	5	6.4	4.8	3.9	8.1
1	D	9	5.4	4.2	3.5	7.8
2	E	5	9.0	4.7	2.7	5.2
2	F	9	8.2	5.3	4.2	6.4
2	G	15	7.2	2.8	0.91	4.0

at the high molecular weight (Table 3). It appears that even the relative significance of the $(z + 1)$ average is very much in question.

(f) With the present state of the art, the moments and average molecular weights are merely arbitrary values, although they may be quite reproducibly determined. Any attempt to attach more than relative significance to these values requires extreme caution. This may not necessarily be true with polymers whose distribution is not extended to such high molecular weights as shown in this example. With linear high density polyethylenes, however, this example is not atypical.

Average Molecular Weights and Melt Index

In the preceding section, it has been shown that the average molecular weights determined by this technique are of relative significance only. They also provide a relative measure of polydispersity. In this section, the fractionation results on several resins are intercompared to see how much useful observation can be obtained. In Table 3 are values of average chain length for two groups of polyethylene having different melt indices. The basic difference between the two groups is in the polydispersity indicated by $\bar{A}_w/\bar{A}_n$.

With Group 1, there is a definite relation between the melt index trend and the average chain length trend. On the other hand, among the Group 2 samples the number-average chain length is the only one that shows a relationship with the melt index. By inspecting Group 1 and 2 together, however, it is evident that the melt index is not a function of number-average chain length. This is also a generally accepted observation. In Fig. 3, values of $\bar{A}_w$ are plotted against melt index. Evidently an approximate correlation is possible in this plot, as is

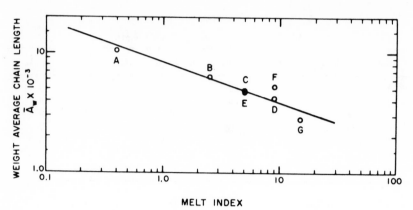

FIG. 3. Weight-average chain length and melt index.

illustrated by the straight line. From this straight line $\bar{A}_w$ value can be estimated for a given melt index. Such estimates of $\bar{A}_w$ are given in Table 4.

Whether melt index is *precisely* a function of $\bar{A}_w$ is not known. However, if we assume the correlation to be valid, errors in $\bar{A}_w$ of Resins F and G are far too excessive. Whether this is truly experimental error or not needs to be resolved.

TABLE 4

Melt Index And Weight Average Chain Length

Resins	Melt index	$\bar{A}_w \times 10^{-3}$ (observed)	$\bar{A}_w \times 10^{-3}$ (estimated)
A	0.4	10.5	11.5
B	2.5	6.4	6.2
C	5	4.8	4.9
D	9	4.2	4.0
E	5	4.7	4.9
F	9	5.3	4.0
G	15	2.8	3.4

FRACTIONATION CURVES

In the preceeding section, average molecular weights and polydispersities of seven resins have been presented. Further, an approximate correlation has been observed between melt index and weight-

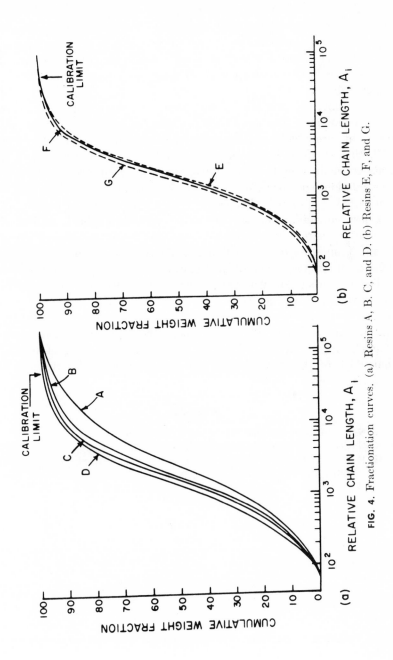

FIG. 4. Fractionation curves. (a) Resins A, B, C, and D. (b) Resins E, F, and G.

average molecular weight. The remaining question is whether GPC results have been exhaustively analyzed. In this section, the fraction curve itself is examined to see if any advantage is found in representing the result as a curve rather than as average molecular weights. In Fig. 4a, the fractionation curves of resins in Group 1 are shown in the cumulative form.

It is obvious that the chain length corresponding to a constant cumulative fraction follows a reverse order to that of melt index at practically all levels of cumulative fraction. Therefore, with this group, it is not surprising that melt index and any of the average chain lengths show the consistent trend previously noted. The fractionation curves of the Group 2 resins are presented in Fig. 4b. Compared to Group 1, this group has not only narrower distribution, but less spread among the curves at comparable melt indices. In particular, there is very little difference between the curves of Resins E and F in spite of the melt index difference of 5 and 9. This emphasizes an important need for performing the fractionation precisely.

The advantages in using the cumulative distribution curve rather than average molecular weights are (a) more detail in the molecular weight distribution may be seen, and (b) more emphasis is placed on the midportion of the curve where the data are relatively free from the uncertainty in calibration and resolution. Consequently, more than 95% of the curve in terms of the cumulative fraction can be used effectively.

REFERENCES

1. N. Nakajima, *J. Polym. Sci., Part A-2,* **5,** 101 (1966).
2. N. Nakajima, *J. Polym. Sci., Part C,* **21,** 153 (1968).
3. H. E. Pickett, M. J. R. Cantow, and J. F. Johnson, *J. Appl. Polym. Sci.,* **10,** 917 (1966).

Application of GPC in the Study of Stereospecific Block Copolymers*

R. D. MATE and M. R. AMBLER

CHEMICAL MATERIALS DEVELOPMENT
THE GOODYEAR TIRE AND RUBBER COMPANY
AKRON, OHIO 44316

Summary

Three different techniques involved in the preparation of stereospecific block copolymers were studied using gel permeation chromatography (GPC). These techniques involved the use of a monofunctional organolithium catalyst, a difunctional organolithium catalyst, and a combination of a monofunctional organolithium catalyst and a coupling technique. GPC curves were obtained on the block copolymers using two different column sequences and solvents. The ABA block copolymers studied contained styrene, vinyl toluene, or α-methyl styrene as thermoplastic monomers and butadiene as the elastomeric monomer. The results obtained showed that block copolymers prepared using monomers and catalyst systems free of impurities generally exhibited single peaked GPC curves. In systems where impurities were found to be present, small amounts of A block homopolymer and AB block copolymer were formed. In such cases, the GPC curves were observed to have two or three peaks.

INTRODUCTION

A new class of block copolymers, called the "thermoplastic elastomers," has recently achieved commercial importance. These block copolymers have been receiving considerable interest recently (1–3). The "thermoplastic elastomers" consist of ordered, block copolymers of the general structure ABA, where A is a thermoplastic block

*Presented at the ACS Symposium on Gel Permeation Chromatography sponsored by the Division of Petroleum Chemistry at the 159th National Meeting of the American Chemical Society, Houston, Texas, February, 1970.

377

polymer and B is an elastomeric block polymer. A typical example of an ABA block copolymer is polystyrene–polybutadiene–polystyrene. However, other monomers such as vinyl toluene and α-methyl styrene have been used in the A block segment and isoprene has been used in the B block segment. These copolymers exhibit rubberlike properties similar to vulcanized elastomers, but without vulcanization (4– 6). They may be formed into useful mechanical goods by modern, rapid, thermoplastic processing techniques. Furthermore, if the polymerization is carried out properly, the resultant block copolymer will have a very narrow molecular-weight distribution with a predictable and controllable molecular weight.

Techniques for the anionic polymerization of block copolymers have been adequately described (1). Hence, mechanisms of polymerization will not be discussed here. However, reference will be made to three different techniques involved in preparing block copolymers to help illustrate the usefulness of gel permeation chromatography. The several techniques involved the use of (a) a monofunctional organolithium catalyst, (b) a difunctional organolithium catalyst, and (c) a combination of a monofunctional organolithium catalyst and a coupling technique.

EXPERIMENTAL

The block copolymers were analyzed using Waters Associates' Model 100 and Model 200 Gel Permeation Chromatographs. The Model 100 instrument contained a combination of four columns of cross-linked polystyrene gels with pore sizes of 10^6, 10^5, 10^4, and 10^2 Å. Operating conditions for this instrument were: solvent, tetrahydrofuran at room temperature; pumping rate 1 ml/min. The Model 200 instrument contained five columns of crosslinked polystyrene gels with pore size of 10^7, 10^6, 10^5, 10^4, and 10^2 Å. Operating conditions for this instrument were: solvent, toluene at 70°C; pumping rate, 1 ml/min.

Reference will be made in the discussion of this paper to certain multiple peaks in the gel permeation chromatography (GPC) curves as being related to homopolymer and block copolymer peaks. Identification of these peaks was made by first fractionating and isolating the individual peaks using GPC, followed by IR spectrophotometric analyses of each fraction. Some samples known to have two or three peaks were injected into the GPC, and fractions representing each peak were collected for analysis using an IR spectrophotometer. The

IR interpretations provided the necessary information for labeling the various peaks in the GPC scans.

DISCUSSION

It is known that the strength of the thermoplastic elastomer is markedly dependent upon the perfection of the ABA block structure comprising the network and the purity of the final system (*2, 3, 7*). A primary prerequisite to the synthesis of pure block copolymers is rigorous purification of monomers and initiator systems. However, it is often difficult to obtain systems that are entirely free of reactive im-

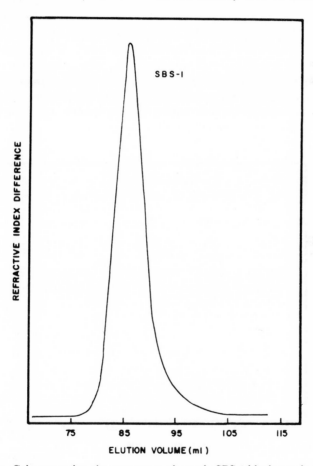

FIG. 1. Gel permeation chromatogram of sample SBS-1 block copolymer.

purities. Since, ideally, a pure ABA block copolymer is expected to have a very narrow molecular-weight distribution, its distribution curve should also be single peaked. In systems where impurities are present or side reactions occur, such as chain transfer with the solvent, small amounts of A block homopolymer and AB block copolymer may be formed. In such cases, the molecular-weight distribution curves could have two or three peaks. GPC has been used successfully in characterization studies with ABA block copolymers (7–10). The GPC chromatograms alone can provide valuable information on the synthesis of ABA block copolymers.

Figure 1 shows a chromatogram (Model 100 instrument) of a poly(styrene–butadiene–styrene) block copolymer prepared from a monofunctional organolithium catalyst. As is evident, the distribution curve is very narrow and has a single peak. The chromatogram (Model 100 instrument) of another poly(styrene–butadiene–styrene) block copolymer prepared from a monofunctional organolithium catalyst is shown in Fig. 2. In contrast to the single peaked curve for the other block copolymer, this SBS-2 sample has two peaks in its distribution curve. The first peak eluting at about 110 ml is the ABA

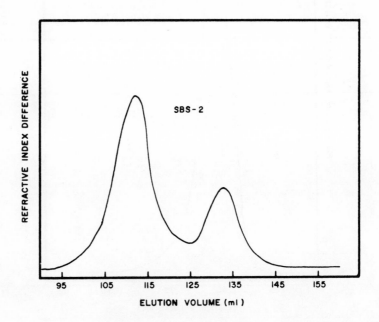

FIG. 2. Gel permeation chromatogram of sample SBS-2 block copolymer.

block copolymer, while the other peak is the A block homopolymer, which in this case is polystyrene. The polymerization of the ŠBS-2 sample was believed to have occurred normally, until the chromatogram was viewed. It was found later that a small amount of impurity in the butadiene had been sufficient to terminate some of the polystyryllithium anions formed initially. Hence, the final ABA block copolymer contained some polystyrene.

The following examples will serve to illustrate other situations which can arise in preparing SBS block copolymers. These GPC analyses were made using the Model 200 instrument. The chromatogram of Sample SBS-3, again prepared using a monofunctional organolithium catalyst, is shown in Fig. 3. Notice the two somewhat obscured and smaller peaks eluting at about 167 and 180 ml. The first peak is the poly(styrene–butadiene–styrene) block copolymer, while the last two peaks right after it are the poly(styrene–butadiene) block copolymer and polystyrene homopolymer, respectively. Figure 4 shows that the distribution curve for SBS-4 appears to be very broad and skewed in comparison to the SBS-3 block copolymer sample. These samples were run consecutively so this broadness for SBS-4

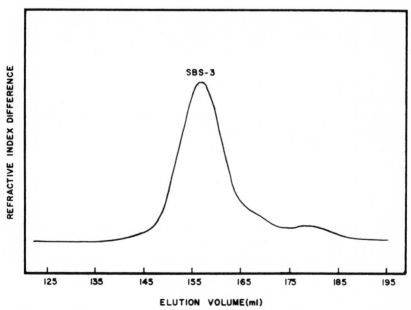

FIG. 3. Gel permeation chromatogram of sample SBS-3 block copolymer.

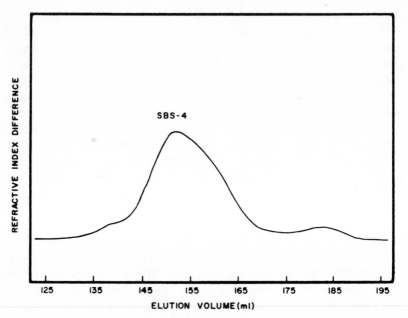

FIG. 4. Gel permeation chromatogram of sample SBS-4 block copolymer.

is believed to be due to a higher level of AB block copolymer content hidden under the trailing end of the distribution curve. The poly(styrene–butadiene–styrene) block copolymer peak in Sample SBS-4 eluted at a lower elution volume than did that for SBS-3. A similar effect was also observed for the AB block copolymer peak. However, the original intention of the synthesis work was that the monomer compositions be the same. About the only observable similarity between these two samples is the small polystyrene homopolymer peak appearing at about 180 ml.

Considering a different monomer combination, the chromatogram for a poly(vinyl toluene–butadiene–vinyl toluene) block copolymer is shown in Fig. 5. This block copolymer contained about 34% vinyl toluene and was also prepared using a monofunctional organolithium catalyst. As seen from the curve, the VBV-5 sample had three observable peaks. The main ABA peak eluted at about 160 ml, and there was a tailing polyvinyltoluene peak at about 180–185 ml. There also was an unknown component eluting at about 145–150 ml. This component is believed to be a higher molecular-weight fraction of the same composition as the main VBV material. Although the chromato-

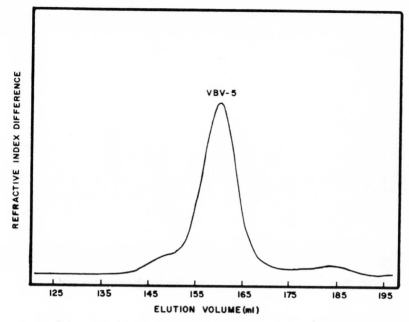

FIG. 5. Gel permeation chromatogram of sample VBV-5 block copolymer.

gram shows three distinct peaks, the polyvinyltoluene peak eluting last was relatively small. This suggested that there was only a small amount of homopolymer present in this sample. The main peak was relatively narrow and generally similar in shape to single peaked SBS curves shown earlier. In contrast to the sharp, well-defined curve for Sample VBV-5, the chromatogram for Sample VBV-6 is shown in Fig. 6. This sample contained 26% vinyltoluene, which is slightly lower than in the previous sample, and was also prepared using a monofunctional organolithium catalyst. These two VBV samples were run consecutively on the same GPC instrument. Although it was intended that the same polymerization technique be used in the preparation of both samples, a comparison of the GPC curves alone suggested that some important differences existed in the preparation of VBV-5 and VBV-6. A review of polymerization data showed that impurities were present in the monomer system for Sample VBV-6 and were slowly terminating the polymerization reaction. Also, the final polymer lacked certain physical properties associated with ABA block copolymers. Hence, in this case, it was known that sample VBV-6 was not a typical ABA block copolymer. The very broad, rolling

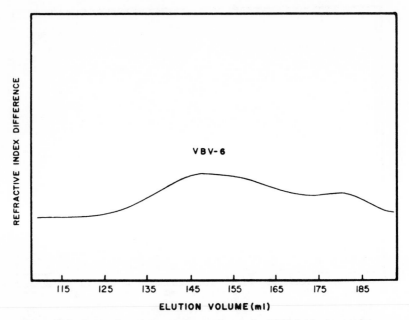

FIG. 6. Gel permeation chromatogram of sample VBV-6 block copolymer.

chromatogram showed a random distribution of high and low molecular-weight species, some of which were probably of the ABA, AB, and A block types. This type of chromatogram is generally observed with random copolymers.

As stated earlier, the preparation of ABA block copolymers can be achieved by several methods. The methods generally use lithium-based initiators in order to attain a high 1,4 chain structure in the diene block, which is important in achieving elastic behavior (7). The use of monofunctional organolithium catalyst, which involves a three-stage sequential polymerization of, for example, styrene, butadiene, and styrene, has already been discussed. Another approach involves the use of dilithium initiators. Here the sequence of monomer addition is reversed. The difunctional catalyst polymerizes the diene monomer in two directions first. The thermoplastic monomer adds last. Figure 7 shows the chromatogram (Model 200 instrument) for Sample SBS-7 which was polymerized using a dilithium catalyst. Sample SBS-7 contained approximately 35% styrene. As is evident from the chromatogram, this sample had only one observable peak and a very sharp distribution curve.

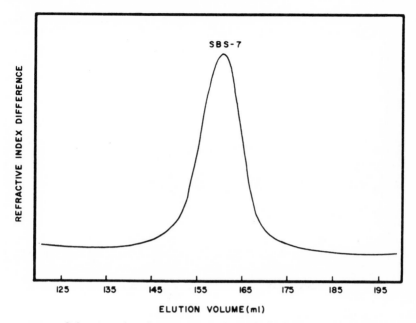

FIG. 7. Gel permeation chromatogram of sample SBS-7 block copolymer.

Figure 8 shows the chromatogram (Model 200 instrument) of sample SBS-8 which was also prepared using a difunctional lithium catalyst. Sample SBS-8 contained 46% styrene, which is slightly higher than Sample SBS-7. The main SBS peak for Sample SBS-8 eluted at about the same elution volume as that for SBS-7, near 160 ml. However, this sample had a pronounced high molecular-weight tail, which is easily seen. The physical properties, that is, the stress-strain data and the physical appearance of this sample, were similar to that expected for an ABA block copolymer. The presence of the high molecular-weight material in the GPC curve was somewhat surprising, and its origin is not completely understood. However, it is believed that an accidental side reaction had occurred late in the polymerization, resulting in incomplete termination of the system. This permitted some of the polymer molecules to continue growing to higher molecular weight.

A third approach in the preparation of ABA block copolymers involves the use of a monofunctional catalyst to prepare the AB anion, which is then linked together using a reactive halogen coupling agent (10). The chromatogram for poly(α-methyl styrene–butadiene–

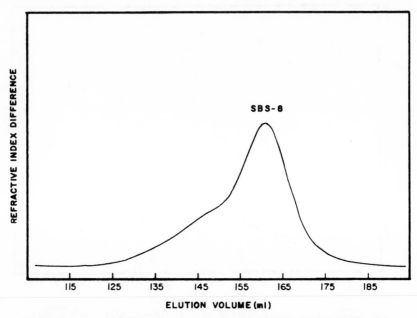

FIG. 8. Gel permeation chromatogram of sample SBS-8 block copolymer.

α-methyl styrene) block copolymer prepared using a coupling technique is shown in Fig. 9. This sample, which contained 40% α-methyl styrene, was also analyzed using the Model 200 instrument. As seen from the chromatogram, this sample had three peaks; the main ABA peak appearing at about 155 ml, the AB peak which appeared at about 165 ml as a shoulder on the tail-end of the main ABA peak, and the A block homopolymer peak appearing at about 185 ml. The presence of the homopolymer peak, as discussed earlier, probably can be attributed to system impurities which terminated the polystyryllithium anion. The presence of the AB copolymer was believed to be due to incomplete coupling of the AB anion with the coupling agent.

Heller and co-workers used GPC to study a poly(vinylbiphenyl–isoprene–vinylbiphenyl) block copolymer prepared by a coupling technique using phosgene as a coupling agent (10). Their chromatograms also contained three peaks. A compositional analysis of the block copolymer was made quantitatively by GPC from a knowledge of the refractive index increments of the homopolymer portions A and B and the over-all composition of the final product. A light-scattering study

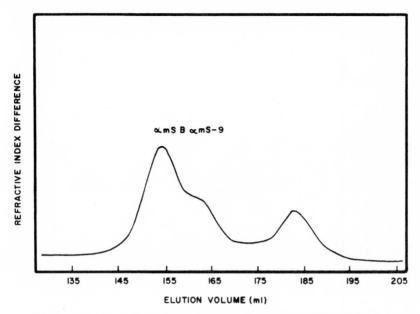

FIG. 9. Gel permeation chromatogram of sample αmSB-αmS-9 block copolymer.

provided data on the differential refractive indices, and osmometry was used to obtain molecular-weight data.

A series of six SBS block copolymers, which exhibited two or three peaks in the chromatograms, were analyzed for the polystyrene content and its molecular weight by GPC. In order to accomplish this, polystyrene standards of known molecular weight and very narrow molecular-weight distributions were injected into the GPC to obtain the peak elution volumes. A plot of the peak elution volumes versus number-average molecular weights was constructed as shown in Fig. 10. Also, various concentrations of a polystyrene standard were prepared and injected into the GPC to determine the peak area versus concentration relationship (Fig. 11). The molecular weight and polystyrene calibration data were then used to determine the number-average molecular weight and concentration of the polystyrene homopolymer in the six SBS block copolymer samples. Table 1 shows the results obtained.

The last column on the right in Table 1 lists the expected SBS composition. The results for per cent polystyrene indicate how much of the styrene content was lost to homopolymer formation. The data

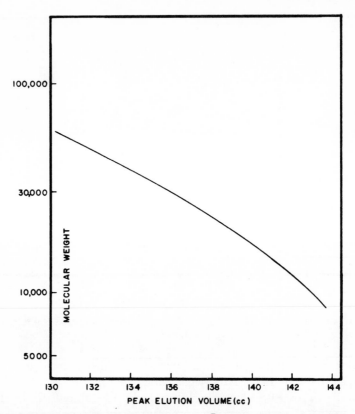

FIG. 10. Plot of peak elution volume vs. $\bar{M}_n$ for polystyrene standards.

suggest that, in general, as the styrene content was increased, the polystyrene level also increased. The high number-average molecular weights for the polystyrene portion suggested that the polystyryl-lithium anion probably was terminated at about the time the second sequential addition was made.

The same type of information can also be obtained for AB and ABA components providing standards of known molecular weight and composition are available. Although an additional study of this type is not yet complete, work along these lines is progressing.

CONCLUSION

It has been shown that ABA block copolymers prepared using monomers and catalyst systems free of impurities generally yield

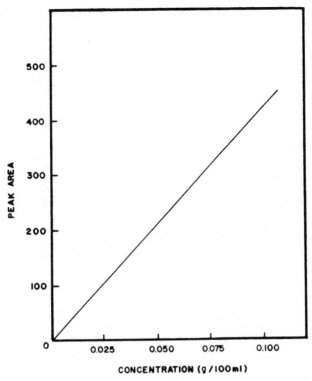

FIG. 11. Plot of peak area vs. polystyrene concentration.

TABLE 1

Analysis of Homopolymer in SBS Block Copolymer

Sample	Peak elution volume (ml)	Peak area	Concn. (g/100 ml)	$\bar{M}_n{}^a$	% Polystyrene[b]	SBS composition
A	135.4	91.00	0.0220	33,000	7.33	23-54-23
B	139.1	80.75	0.0195	21,000	6.49	15-70-15
C	132.3	266.28	0.0650	47,000	21.54	26-48-26
D	135.4	115.02	0.0275	33,000	9.17	17-66-17
E	134.7	180.88	0.0440	36,000	14.68	27-46-27
F	136.8	137.97	0.0335	28,000	11.07	21-58-21

[a] From polystyrene calibration curve.
[b] Based on total weight of polymer (SBS + P-S).

single-peaked chromatograms. Hence, GPC could be considered a candidate for a quality control technique to monitor block copolymer polymerizations, as well as for studying coupling efficiency. It is believed that this work leads to the conclusion that a total characterization of thermoplastic elastomers is very important, since performance characteristics of the copolymers are closely related to their structural features.

Acknowledgments

The authors wish to thank Dr. L. J. Fetters, Institute of Polymer Science, University of Akron, and Dr. R. E. Cunningham, The Goodyear Tire and Rubber Co., for providing most of the samples used in this work and for their helpful comments. The authors are also indebted to The Goodyear Tire and Rubber Co. for permission to publish these results.

REFERENCES

1. L. J. Fetters, *J. Polym. Sci., Part C,* **26,** 1 (1969).
2. G. Holden, E. T. Bishop, and N. R. Legge, *Ibid.,* **26,** 37 (1969).
3. E. T. Bishop and S. Davison, *Ibid.,* **26,** 59 (1969).
4. E. Fischer and J. F. Henderson, *Ibid.,* **26,** 149 (1969).
5. T. L. Smith and R. A. Dickie, *Ibid.,* **26,** 163 (1969).
6. R. F. Fedors, *Ibid.,* **26,** 189 (1969).
7. L. J. Fetters and M. Morton, *Macromolecules,* **2,** 453 (1969).
8. M. Morton, J. E. McGrath, and P. C. Juliano, *J. Polym. Sci., Part C,* **26,** 99 (1969).
9. J. F. Beecher, L. Marker, R. D. Bradford, and S. L. Aggarwal, *Ibid.,* **26,** 117 (1969).
10. J. Heller, J. F. Schimscheimer, R. A. Pasternak, C. B. Kingsley, and J. Moacanin, *Ibid., Part A-1,* **7,** 73 (1969).

Composition of Butadiene–Styrene Copolymers by Gel Permeation Chromatography*

H. E. ADAMS

CENTRAL RESEARCH LABORATORIES
THE FIRESTONE TIRE & RUBBER COMPANY
AKRON, OHIO 44317

Summary

An ultraviolet detector was added to the GPC making it possible to determine both molecular weight distribution and composition of butadiene–styrene copolymers as a function of molecular weight. After calibration, several types of commercial and experimental copolymers were analyzed. SBR was found to have a very uniform composition over most of the range of molecular weight. However, the styrene content decreased at the high end of the distribution. Representative samples of anionically polymerized copolymers were also examined. In general, the composition of these polymers varied more than SBR. Usually, the styrene content was high at low molecular weight and decreased as molecular weight increased. Several experimental copolymers that were first metalated and then grafted with styrene were studied. Incomplete grafting as well as the presence of low molecular weight homopolystyrene was readily detected. This appears to be a particularly sensitive method for studying composition of copolymers.

It is quite evident that physical properties of copolymers are dependent on their composition. However, it is not satisfactory to merely control average composition. Minor fluctuations of composition on a molecular level are also important. Thus, actual distribution of the two monomers within polymer chains must be controlled if certain properties are desired. The two extremes are pure block copolymer

* Presented at the ACS Symposium on Gel Permeation Chromatography sponsored by the Division of Petroleum Chemistry at the 159th National Meeting of the American Chemical Society, Houston, Texas, February, 1970.

on the one hand and uniform copolymer on the other. Many variations exist between these two.

Considerable information of importance can be obtained by determining the variation in composition as a function of molecular weight. Usually, solution methods (1) have been used to obtain this type of information. Fractions of copolymer were separated by addition of nonsolvent and individually analyzed for composition and molecular weight. However, fractionation depends upon solubility parameters which are a function of both composition and size so that clean separations are not obtained. These methods are also time consuming.

Gel permeation chromatography should be useful in this area (2–4). Its primary function is to separate polymer molecules of different size and thus determine molecular weight distributions of polymers. Modification of the instrumentation to determine composition of different molecular weight material as it is separated would yield the desired information. This was accomplished by adding an ultraviolet (UV) detector to the gel permeation chromatograph (GPC) in series with the standard differential refractometer normally used. Thus, the refractive index detector (RI) measures total amount of copolymer of a certain molecular weight, while the UV detector establishes its composition.

Since the GPC method of separation depends on the size of the polymer coil in solution, this method is also dependent on composition to the extent that composition influences coil size. Adequate calibration of the GPC in terms of the different compositional variations encountered will permit the results to be expressed as a function of molecular weight (4). In this study, no attempt has been made to convert to molecular weight, so the results are shown as a function of GPC counts.

EXPERIMENTAL

The GPC was modified by adding a UV detector, Uviscan III, manufactured by Buchler Instruments. It was placed in line preceding the standard RI detector, and a separate recorder was used for its output. Absorbance at 2600 Å was measured directly.

To minimize solvent effects and obtain maximum transmission by the solvent at 2600 Å, the solvent, tetrahydrofuran (THF), was purified before use in the GPC. It was refluxed over sodium for 4 hr and, after discarding the forerun, was distilled directly into the GPC solvent reservoir. A blanket of nitrogen was maintained during the distillation and use in the GPC. The purity and transmission of the THF was checked periodically with a Beckman DKU Spectrophotometer before use.

Both detectors were calibrated in terms of amount of polybutadiene and polystyrene. It was verified that neither 1,2- or 1,4-polybutadiene absorbed at 2600 Å. Thus, the UV detector is sensitive to only the amount of styrene while the RI detector measures both styrene and butadiene. Consequently,

$$r_{\mathrm{UV}} = K_s' G_s$$
$$r_{\mathrm{RI}} = K_c G_c$$

where r_{UV} and r_{RI} are the outputs of the UV and RI detectors, respectively. K_s and K_c are the proportionality constants, and G_s and G_c are the grams of styrene (s) and copolymer (c). K_c is assumed to be linearly dependent upon composition as follows:

$$K_c = (1 - W_s)K_b + W_s K_s$$

where W_s is weight fraction of styrene and K_b the proportionality constant between the RI detector readings and grams of butadiene. Thus, at each GPC count interval, the fraction of styrene $(W_s)_i$ can be calculated in the following manner:

$$(W_s)_i = R_i \frac{K_b}{K_s' - \Delta K R_i}$$

with

$$R_i = r_{\mathrm{UV}}/r_{\mathrm{RI}}$$

and

$$\Delta K = K_s - K_b$$

The actual calibration was achieved by injecting various amounts of polystyrene, PCC 51,000, obtained from Pressure Chemical Co., and also an anionically polymerized polybutadiene. The results are listed in Table 1. Only one concentration of 1,4-polybutadiene was used to calibrate the RI detector for this polymer. The areas under the curves were determined by graphical integration and used to calculate the proportionality constants listed in Table 1. The output of the UV detector was found to be linear with polymer concentration over the range of interest. Values for the constants inherently depend upon the unit of count interval (0.5 count) used to sample

TABLE 1

Calibration of UV and RI Detectors

Amount (g)	Area[a]	Peak maximum count
Polystyrene (PCC-51,000)		
RI Detector		
0.000834	208.2	32.11
0.000834	208.3	32.22
0.000417	108.2	32.20
0.000209	52.7	32.16
UV Detector		
0.000834	154.6	31.80
0.000417	79.2	31.78
0.000209	39.1	31.79
Polybutadiene		
RI Detector		
0.00125	236.9	28.08

$$K_s' = 1.88 \times 10^5 \ (UV)^a$$
$$K_s = 2.54 \times 10^5 \ (RI)^a$$
$$K_b = 1.90 \times 10^5 \ (RI)^a$$
$$\Delta K = 6.45 \times 10^4 \ (RI)^a$$

[a] These values are dependent upon the count interval used to sample the curves. The interval is 0.5 count.

the curves. The experimental equation used to calculate the styrene fraction

$$(W_s)_i = \frac{R_i}{0.989 - 0.341 R_i}$$

is independent of this count interval because of the use of ratios.

Because the two detectors are in series, their outputs do not coincide on the count axis. The UV curve appears 0.38 counts previous to the RI curve. This was taken into consideration when sampling the two curves. The introduction of the UV detector has added a volume to the system, amounting to 0.265 counts (1.33 ml). The molecular weight calibration has to be modified to account for this if conversion to molecular weight is desired.

DISCUSSION

A number of copolymers of butadiene and styrene were run on the GPC equipped with the two detectors. Several examples are shown graphically in the figures. The normalized GPC curves, showing the molecular weight distribution as a function of counts, are indicated by circles. The ordinate to the left relates to this curve; the units are arbitrary, depending upon the count interval used. The data points represented by squares give the polymer composition (fraction of styrene) as a function of counts (log molecular weight); the scale on the right ordinate is to be used with the square points.

Figure 1 shows the data for a typical SBR. A free radical initiator was used to prepare this polymer, and because the reactivity ratios of the two monomers are different, composition of the copolymer changes with conversion. However, because of relative rates of initiation, propagation, and termination, the molecular weight after very early conversion is practically independent of conversion. Theoretically, this would produce a polymer which would have very little differences in its composition of different molecular-weight species. However, at any one molecular weight, a difference in composition exists depending on whether it was produced early or late in the reaction. In this case, though, only the average value for molecular weight would be measured. Because of the rapidity of the propagation reaction, the composition of any one molecule is very uniform and random. Thus, from the accepted mechanism of free radical polymerization, SBR should have a very uniform composition as a function of molecular weight. Previous work has indicated this (5); this study also shows it to be true over the major amount of copolymer. The observed decrease in the amount of styrene at high molecular weight, however, is unexpected and will have to be verified. Although it is true that these data are less precise because the fraction of styrene is calculated from the ratio of outputs of the two detectors and both outputs are small at the extremes of the curves, this decrease appears to be greater than experimental error. It could possibly be related to branching, which would mean that branching would involve material formed at an earlier time in the polymerization and thus contains a lower amount of styrene. More data are needed to clarify this point.

The situation is quite different for an anionic initiator such as butyllithium. In this case, both composition and molecular weight vary with conversion, the former because of the reactivity ratio of the two monomers, and the latter because of the long lifetime of the growing chain. This system produces so-called "live polymers" in the absence of a termination reaction. In the extreme case of a very fast anionic initiation reaction, monodispersed polymer of a

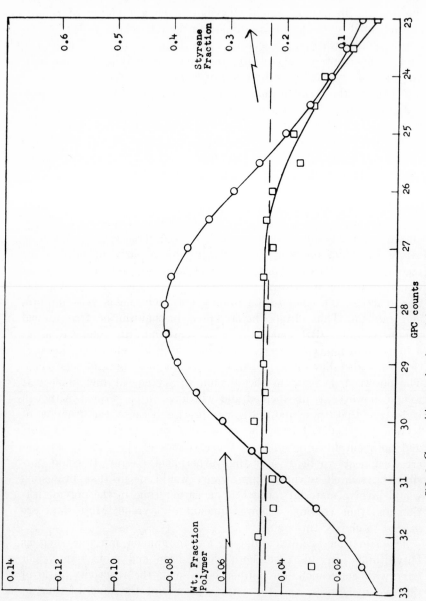

FIG. 1. Composition–molecular weight distribution of SBR, 23.5% styrene.

uniform composition would be produced. In reality, some breadth to the distribution is obtained because initiation and termination reactions do occur. These effects cause a variation in composition as a function of molecular weight. Thus, anionic copolymers of narrow molecular weight would be expected to have a more uniform distribution of styrene as a function of molecular weight than do broad polymers. However, the composition of individual chains varies along their length in contrast to the case of SBR produced by radical initiators.

Figures 2 and 3 are examples of the variation in composition that occurs in uniform copolymers made with anionic initiators. In general, these polymers have a more heterogeneous composition than does SBR which increases with the breadth of the distribution. These polymers were prepared under conditions designed to prevent the formation of blocks of polystyrene. Polymers Nos. 1 and 2 are representative of polymers made under reasonably good control and under "out of control" conditions, respectively. It is obvious that the polymer made under poor control conditions is the more heterogeneous and that the low molecular weight species have higher amounts of styrene. Duplicate runs are shown for Polymer No. 1. The near coincidence of the data indicates the reproducibility of the method.

The effect of adding a randomizing agent, such as an ether or amine, upon compositional heterogeneity is shown in Fig. 4. A uniform composition is obtained even though the molecular weight distribution is rather broad. Again, the amount of styrene decreases rapidly at the high molecular weight end of the distribution, possibly indicating the presence of branching.

Figures 5 and 6 refer to examples of SB (styrene/butadiene) and SBS (styrene/butadiene/styrene) type block copolymers, respectively. The SB polymer (Polymer No. 4) is relatively narrow in distribution and also of very uniform average composition. This indicates that practically all of the initial blocks were active during polymerization of the second monomer to produce essentially SB type polymer. In the case of SBS type (Polymer No. 5), both molecular weight distribution and composition cover a wider range. It appears that the lowest molecular weight material contains a higher percentage of styrene than the bulk of the copolymer. This could be explained by the possible presence of a small amount of low molecular weight homopolystyrene.

Thus, it would appear, experimentally, that the compositional

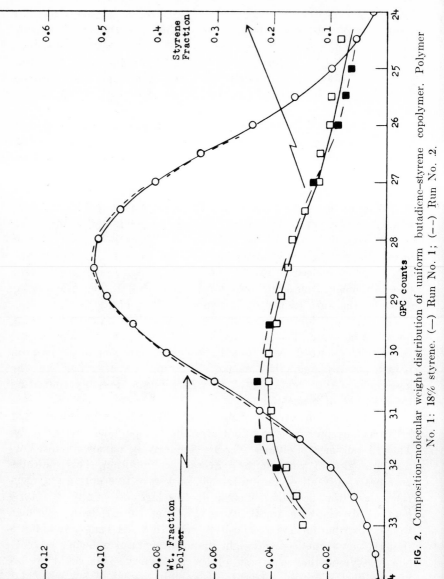

FIG. 2. Composition–molecular weight distribution of uniform butadiene–styrene copolymer. Polymer No. 1: 18% styrene. (—) Run No. 1; (– –) Run No. .2.

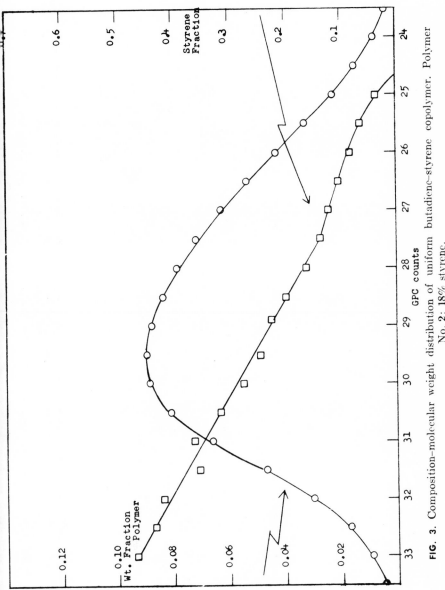

FIG. 3. Composition–molecular weight distribution of uniform butadiene–styrene copolymer. Polymer No. 2: 18% styrene.

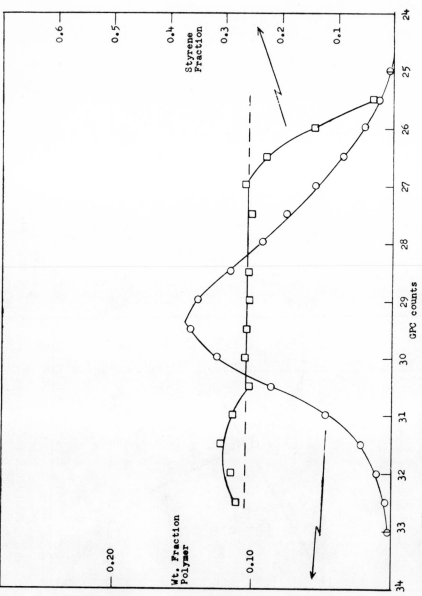

FIG. 4. Effect of randomizing agent upon the composition–molecular weight distribution of butadiene–styrene copolymer. Polymer No. 3: 25% styrene.

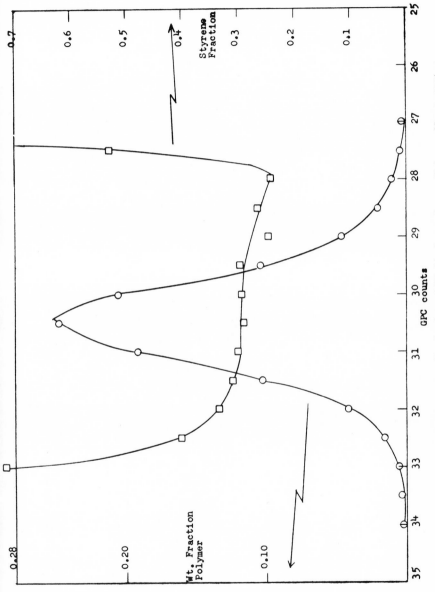

FIG. 5. Composition–molecular weight distribution of SB block copolymer. Polymer No. 4: 25% styrene.

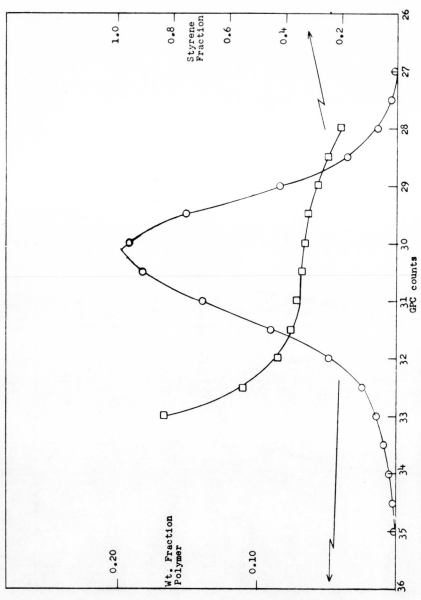

FIG. 6. Composition–molecular weight distribution of SBS block copolymer. Polymer No. 5: 30% styrene.

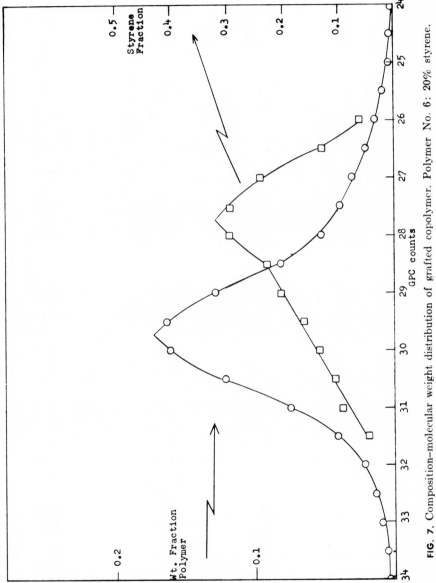

FIG. 7. Composition–molecular weight distribution of grafted copolymer. Polymer No. 6: 20% styrene.

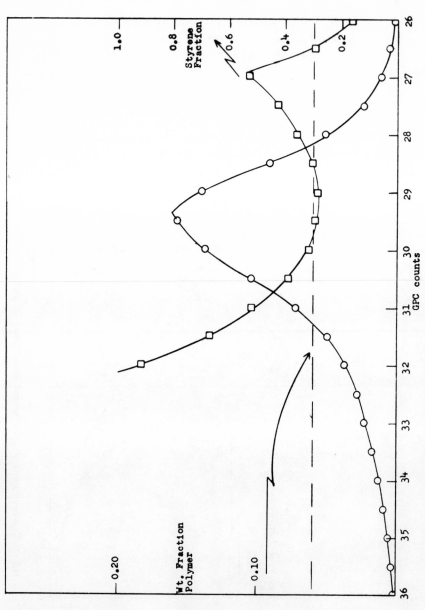

FIG. 8. Composition–molecular weight distribution of grafted copolymer. Polymer No. 7: 30% styrene.

heterogeneity of the anionic polymers roughly parallels the breadth of the molecular weight distribution.

Figures 7 and 8 are indicative of another type of material studied, namely, graft copolymers. These are experimental polymers which have been metalated and then have had styrene grafted on the polybutadiene backbone. The shape of the curves in the two figures depicting the distribution of styrene suggests that metalation and subsequent grafting were incomplete. There must have been considerable homopolybutadiene present that escaped metalation. The grafted polymer would, of course, have a higher average molecular weight than the homopolybutadiene. Polymer No. 7 differs from No. 6 in that, in addition to being a mixture of homopolybutadienes and grafted copolymer, it has some low molecular weight homopolystyrene present.

These examples serve to illustrate the versatility of the method. With a relatively inexpensive modification, the GPC can be used to obtain information concerning the composition of copolymers in addition to the usual molecular weight distribution. Useful information can be gained concerning the application of various polymerization and grafting techniques. The method should also be useful in studying relationships between physical properties and compositional heterogeneity. As usual, the simplicity and rapidity of the GPC method make it attractive.

Acknowledgments

The author wishes to thank The Firestone Tire & Rubber Company for permission to publish this paper and Dr. B. L. Johnson for many helpful suggestions.

REFERENCES

1. J. A. Yanko, *J. Polym. Sci.*, **3**, 576 (1948).
2. E. G. Owens II and J. G. Cobler, *Proceedings of Fourth International Seminar on Gel Permeation Chromatography*, Miami Beach (1967).
3. H. J. Cantow, J. Probst, and C. Stojanov, *Rev. Gen. Caoutchouc Plast.*, **45**, 1253 (1968).
4. J. R. Ruyon, D. E. Barnes, J. F. Rudd, and L. H. Tung, *J. Appl. Polym. Sci.*, **13**, 2359 (1969).
5. A. R. Kemp and W. G. Straitiff, *Ind. Eng. Chem.*, **36**, 707 (1944).

A Direct GPC Calibration for
Low Molecular Weight Polybutadiene, Employing Dual Detectors*

JAMES R. RUNYON

THE DOW CHEMICAL CO.
MIDLAND, MICHIGAN 48640

Summary

The calibration of gel permeation chromatograph (GPC) for a given polymer type is usually done by running well-characterized polymer samples of the same type. The available polystyrene and polyglycol standards with molecular weights below 5000 do not give parallel calibration curves, and the points for polystyrene are not on an extension of the higher-molecular-weight portion of the curve. In order to check the calibration in this low molecular-weight region and to establish an independent curve for polybutadiene, a polybutadiene sample prepared by α-methylstyrene tetramer-Na$_2$ initiation was chromatographed. In doing this, dual detection was used, consisting of a differential refractometer, and an ultraviolet absorption spectrophotometer.

The UV signal from the spectrophotometer was assumed to represent the tetramer portion of the polymer, and its percentage was calculated at each volume increment, using externally determined response factors. This percentage and the tetramer molecular weight gives the molecular weight of the polymer eluting at each volume increment. After correction for the aromatic portion of the polymer, these data and the points obtained with the polybutadiene standards were found to give a reasonable calibration curve for the entire range. The present paper describes the application of this technique to the examination of low molecular-weight polybutadiene.

* Presented at the ACS Symposium on Gel Permeation Chromatography sponsored by the Division of Petroleum Chemistry at the 159th National Meeting of the American Chemical Society, Houston, Texas, February, 1970.

407

DISCUSSION

The most commonly used standards for calibrating GPC columns are the anionic polystyrenes distributed by Pressure Chemical Company which have been described repeatedly in the literature. While prior work in this laboratory has been largely with polystyrene in the molecular weight range of 100,000 to 2 million, polybutadiene and its copolymers became of interest, particularly material in the lower molecular-weight range from monomer to 20,000.

Figure 1 shows the usual log molecular weight versus elution volume calibration curves. In the higher range of molecular weight the polystyrene (PS) standards and the few polybutadiene (PBD) standards available from Phillips Petroleum Company give smooth, parallel calibration curves separated by about $1\frac{1}{2}$ counts. This relationship would be predicted from the fact that tetrahydrofuran is a good solvent for both polymers.

As there were no polybutadiene standards available below molecular weight 17,000, consideration was given to extending the polybutadiene curve parallel to polystyrene. Low molecular-weight standards are available for the latter polymer. However, when GPC data for these were plotted using their nominal weights, the points made a sharp and unexplainable deviation from the rest of the calibration curve. The question arose as to the advisability of using this wandering curve for polystyrene much less for polybutadiene. That polyglycol standards which are available gave a smooth curve through this region suggested that something was wrong.

Dual detectors for styrene–butadiene copolymers had been used for some time when a sample labeled polybutadiene was received which gave a response at 260 mμ in the UV (Fig. 2). On investigation, it was found that this sample was one of the α-methylstyrene tetramer-Na$_2$ initiated polymers described by Dennis (1). This meant that each molecule contained one tetramer unit besides the butadiene chains of varying length. At low molecular weight, the aromatic portion becomes greater in spite of the decreasing total signal.

It was evident that with the proper calibration, this sample would provide a continuous polybutadiene molecular-weight calibration for the GPC columns. Such a calibration is seen in Fig. 3. The sample solution consisted of equal weights of a polybutadiene, a polystyrene, and the α-methylstyrene tetramer (AMST).

Two relative response factors are needed in this application: (a)

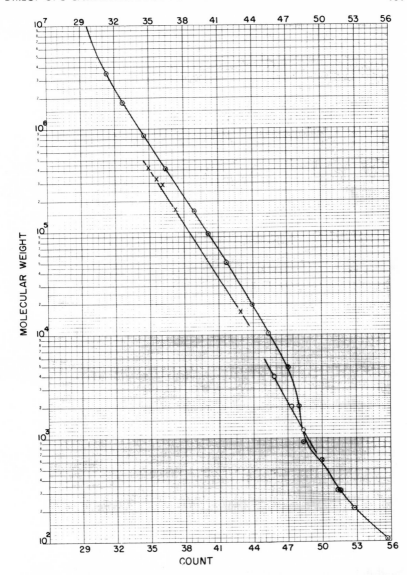

FIG. 1. GPC calibration curves from commercially available standards. ⊙, Pressure Chemical polystyrenes; ✕, Phillips Petroleum polybutadienes; ○, polyglycols. Columns used were 5×10^6 Å, 10^6 Å, mixed gel, 5×10^4 Å, 5×10^4 Å, and 300 Å. The points at 104, 208, and 312, are styrene monomer, dimer, and trimer, respectively.

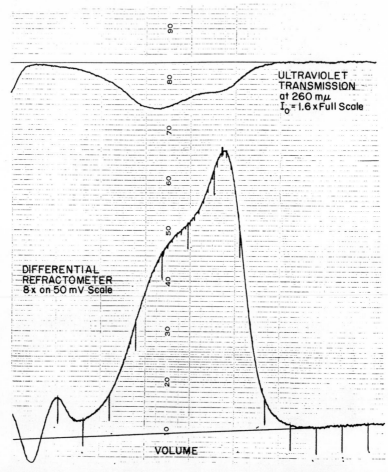

FIG. 2. Dual detector GPC trace of α-methylstyrene tetramer-Na₂ initiated polybutadiene. In the upper trace the pen is offset to correct for volume between detectors. The scale of the abscissa is 2½ min per division. Marks on the chromatogram are 5 ml counts.

the differential refractometer response to the tetramer as compared to polybutadiene; and (b) for the tetramer alone, the amount of signal to be expected in the refractometer for a given optical density (OD). Comparison of the integrals of the GPC curves has been shown to be a reliable way to obtain these response factors (2). Table 1 summarizes the average values used in the present calculations. The

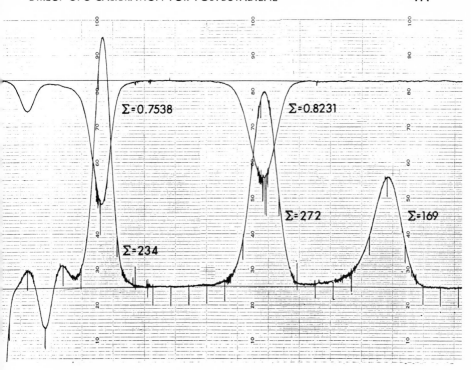

FIG. 3. Dual detector GPC calibration. The axes, etc. are the same as in Fig. 2. The sample was 0.8 ml of a solution containing equal weights of a polybutadiene (MW 423,000), a polystyrene (MW 51,000), and the α-methylstyrene tetramer (hydrogen terminated, MW 474); 0.64 mg of each polymer was injected. The areas of the refractometer peaks at $8\times$ were 169, 272, and 234 "mV ml"; the areas of the UV transmission curves for the polystyrene and the tetramer after conversion to optical density were 0.8231 and 0.7538 "O.D. ml," respectively.

TABLE 1

Relative Responses on Differential Refractometer

PS/PBD	1.40
PS/PAMS	1.0
AMST/PBD	1.47

Relative "Areas:" $\Sigma\Delta$ Refractive Index/Σ Optical Densities

PS	320 mV/OD
PAMS	440 mV/OD
AMST	310 mV/OD

polystyrene data is included for general interest. The single sample of poly-α-methylstyrene (PAMS) available at the time gave about the same refractometer response as polystyrene, but considerably less UV absorption. Because the tetramer presumably retains its identity in the center of the polybutadiene molecule, the tetramer response factors were used rather than the poly-α-methylstyrene factors.

In order to calculate the true molecular weight, one must first determine the weight fraction of tetramer. The OD at a given volume increment, multiplied by the appropriate response factor, gives the portion of the refractometer signal that is due to the tetramer:

$$\text{tetramer} = \text{OD} \times 310$$

The rest of the refractometer signal (RI) is due to polybutadiene, but must be multiplied by the relative response of tetramer to obtain a comparable weight number:

$$\text{PBD} = (\text{RI} - \text{tetramer}) \times 1.47$$

The "molecular weight" of the tetramer itself divided by its weight fraction gives the total molecular weight:

$$\text{MW} = 472 \div \frac{\text{tetramer}}{\text{PBD} + \text{tetramer}}$$

At these low weights, the size of the tetramer portion becomes significant, so some correction must be made for it. The correction used here is the same as was used for the styrene–butadiene calculation (2), which assumes that a polystyrene molecule has twice the weight for a given size that a polybutadiene molecule has. This

TABLE 2

Low Molecular-Weight Polybutadiene Calibration via Composition Analysis of AMST-Na$_2$-Initiated Polybutadiene. Data Taken from Fig. 2

Volume	ΔRI[a]	OD	Wt fraction tetramer	Calc butadiene MW	Cor butadiene MW
45.16	4.26	.0029	0.154	3065	2793
46.02	20.16	.01035	0.1141	4136	3864
47.06	24.98	.01702	0.1542	3061	2789
48.10	20.10	.02115	0.2479	1904	1632
49.13	16.92	.02556	0.3744	1261	989
50.00	9.64	.01919	0.5233	902	630

[a] Units = mV at 8 $\times$ sensitivity.

may be a somewhat doubtful assumption, but the error will not change the conclusions. Table 2 shows the actual readings and calculations for the part of the distribution curve where the optical density was great enough to be reliable.

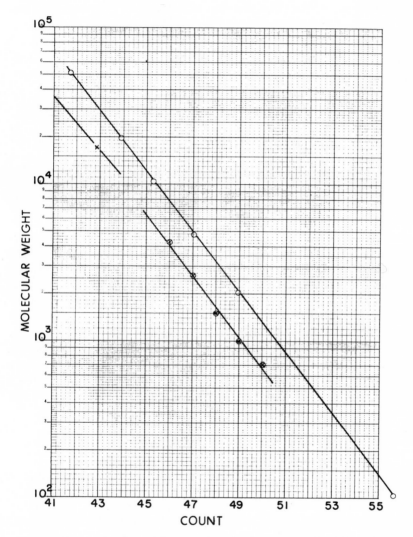

FIG. 4. Low molecular weight extension of polystyrene and polybutadiene calibration curves; ○, PS standards; ✕, PBD standard (slope from Fig. 1); ⊗, PBD calculated.

TABLE 3

Low Molecular-Weight Polybutadiene Calibration. Averages from Seven Samples

Volume	Calc MW	Cor MW
44	5300	5000
45	(3800)	—
46	4600	4300
47	2900	2600
48	1800	1500
49	1300	1000
50	1000	700

·The original sample was a blend of three materials; each of the three, and several broad distribution polymers, were chromatographed alone. The largest molecular-size portion of each sample was found to give low calculated molecular weights, that is, too much tetramer for the RI: these points were not included in the averages given in Table 3 which shows the results from seven samples. Figure 4 is a plot of these points along with the low molecular-weight end of the polystyrene and polybutadiene calibration curves of Fig. 1. The upper portion of the polystyrene curve is the same as it is in Fig. 1, but the present evidence indicates that if the polystyrene calibration is to be used as a size calibration, it must be extended (in this case) as a straight line from about molecular weight 4800 to monomer, rather than as indicated by the low molecular-weight samples.

Estimates of the molecular weight at which the peak of the various low molecular-weight polystyrene standards ought to be plotted are listed in Table 4.

TABLE 4

Recommended "Styrene Sizes" for Low Molecular-Weight Polystyrene Standards

Nominal MW	Source of standard[a]	Size scale
4800	2	4600
3600	1	4600
2030	2	3200
900	2	2500
600	2	1300

[a] 1: Arro Chemical Co. 2: Pressure Chemical Co.

In his work with polyethylene in o-dichlorobenzene at 130°C, Williams (*3*) found that normal paraffins fell on a straight line extrapolation of his higher calibration curve. He concluded that the nominal 4800 molecular-weight polystyrene ought to be plotted at 4350 ± 200, and the 900 molecular-weight polystyrene at 2100 ± 100. This is in good agreement with the present data.

REFERENCES

1. K. Dennis, U. S. Patent 3,346,666 (February 10, 1967).
2. J. R. Runyon, D. A. Barnes, J. F. Rudd, and L. H. Tung, *J. Appl. Polym. Sci.*, 13, 2359 (1969).
3. T. Williams, University of Bristol, U. K., Private Communication to L. H. Tung of The Dow Chemical Company.

Quantitative Determination of Plasticizers in Polymeric Mixtures by GPC*

DAVID F. ALLIET and JEANNE M. PACCO

MATERIALS ANALYSES AREA
XEROX CORPORATION
ROCHESTER, NEW YORK 14603

Summary

Organic plasticizers were added to plastics to improve flow and reduce brittleness by lowering the glass transition temperature. The amount of plasticizer added to the base resin determined its efficiency in bringing about these desired changes in properties. Analytical gel permeation chromatography (GPC) was utilized to quantitatively determine the amount of organic plasticizers in poly(styrene) mixtures. The internal standard method was applied to the determination of triethyleneglycoldibenzoate and tricyclohexylcitrate over the concentration range of 5.0 to 30.0 wt-% in poly(styrene). Linear calibration curves and excellent precision between measurements was demonstrated over the concentration range investigated. GPC analysis has the advantage over spectrophotometric techniques in its ability to separate low molecular weight plasticizers from higher molecular weight resins. In addition to the potential of making quantitative measurements from the detected peak, the associated material can be separated from the polymer, collected, and separately analyzed by UV or IR techniques.

INTRODUCTION

Organic plasticizers were added to polymers, primarily amorphous, to improve flow and reduce brittleness by lowering the glass transition temperature. The amount of plasticizer added to the base resin determined its efficiency in bringing about these desired changes in

---

*Presented at the ACS Symposium on Gel Permeation Chromatography sponsored by the Division of Petroleum Chemistry at the 159th National Meeting of the American Chemical Society, Houston, Texas, February, 1970.

properties. Therefore, good quantitative analytical methods are needed to control the amount of plasticizer necessary for optimum materials performance.

Several authors have successfully applied gel permeation chromatography (GPC) to the quantitative measurement of various components in polymers and other mixtures. Mate (1) applied GPC to the analysis of oil content in oil-extended elastomers. Larsen (2, 3) used the technique to determine butyl cellosolve in copolymer reactions, unreacted trimethylol-propane (TMP) in polyesters, and diphenyl-methylsilanol in silicones. Recently, Limpert (4) quantitatively measured the amount of asphaltines, solvent distillates, and oils extracted from coal.

Our recent efforts have produced a further quantitative application in GPC analysis. The technique was applied to the quantitative determination of triethyleneglycol dibenzoate and tricyclohexyl citrate plasticizers in poly(styrene) over a 5–30% concentration range.

EXPERIMENTAL

Equipment and Supplies

GPC analyses of the plasticizer–polymer mixtures were carried out using a Waters Associates Model 200 Gel Permeation Chromatograph. The equipment was operated at 40°C, with Eastman tetrahydrofuran as the carrier solvent (flow rate, 1 ml/min). An automatic injection system was used in conjunction with four columns in series; a listing of the column characteristics is given in Table 1.

TABLE 1

Column Characteristics

Designated porosity (Å)	Plate count (plates/ft)
10,000	1230
1,000	1285
250	575
60	900

Materials

Two low molecular-weight plasticizers, tricyclohexyl citrate (TCHC) and triethyleneglycol dibenzoate (TEGDB), were chosen for this study because of their purity and elution volumes. A high molecular-weight ($\simeq 50K$) poly(styrene) was used to prepare the physical mixtures to insure complete separation of the polymer and plasticizer peaks. Ultra-pure benzil was used as the internal standard because of its long elution time and high purity. Chromatograms of the TCHC, TEGDB, and benzil standard are shown in Figs. 1–3, re-

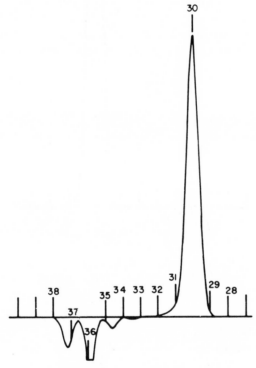

FIG. 1. Chromatogram of TCHC plasticizer.

spectively. The monodispersity of these compounds makes them ideal for quantitative application. Negative peaks at counts 35–37 are associated with dissolved water and air.

Standard Solutions

All standard analytical solutions were prepared by dissolving varying amounts of plasticizer mixed with poly(styrene). The plasticizer concentrations were varied over a range of 5–30% of the total sample

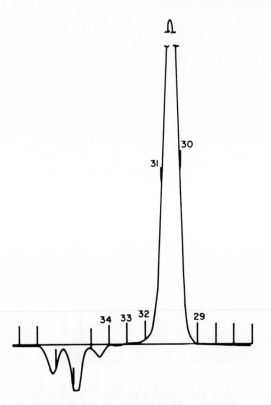

FIG. 2. Chromatogram of TEGDB plasticizer.

weight. Stock solutions were made to minimize weighing errors. The final solution concentrations were 0.5% (500 mg in 100 ml THF).

Analytical Procedures

Each of the standard solutions were analyzed three times to establish the precision of the GPC data. One milliliter of a 1% benzil solution was added to a 10-ml aliquot of each stock solution and prefiltered before injection into the automatic injection system.

Peak area measurements of the plasticizer and internal standard peaks were made using the triangulation method. This method involves the use of the product of peak height and the width at half-height. It is fast and gives reasonable accuracy when the peaks are symmetrical. The heights and widths at half-height were measured as accurately as possible to the nearest hundredth of an inch.

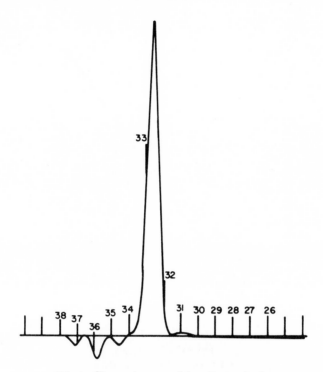

FIG. 3. Chromatogram of benzil standard.

RESULTS AND DISCUSSION

Accurate quantitative chromatographic procedures require a thorough knowledge of the operating characteristics of the detector. Therefore, the refractometer detector system of the GPC was evaluated to determine the linearity of its response within the expected concentration ranges. This was accomplished by analyzing each plasticizer over the range of 5–30% by weight in the polymeric mixture. The results, plotted as weight of plasticizer vs. area of plasticizer peak (Figs. 4 and 5), show that the detector response was linear within the working range, and therefore it was not necessary to apply empirical correction factors.

The internal standard procedure is used extensively in gas chromatography (5) to compensate for minor variations in column operation over an extended period of time. This procedure allows for instrumental variations by compensating through the use of an internal stand-

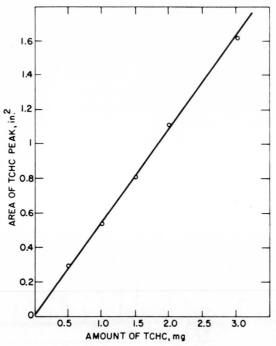

FIG. 4. Effect of TCHC concentration on refractometer response.

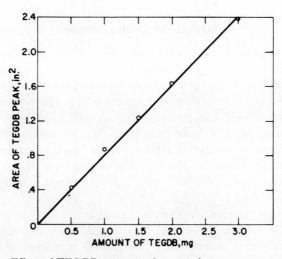

FIG. 5. Effect of TEGDB concentration on refractometer response.

ard. Harvey and Chalkley (6) outlined the requirements of an internal standard as follows: (a) yields a completely resolved peak, (b) should be eluted close to component to be measured, and (c) the ratio of peak area to that of the component should be close to unity.

The first two requirements set by Harvey and Chalkley were met by benzil. This organic compound yields a sharp, narrow chromatogram (Fig. 3) and elutes close to both plasticizers studied.

Because of the amount of internal standard added and the refractive index of the plasticizers, the third requirement set by Harvey and Chalkley was deviated from at lower concentrations of plasticizer. However, excellent precision was maintained over the entire range of plasticizer concentration studied. The results of the internal standard measurements for the TCHC and TEGDB plasticizers are summarized in Tables 2 and 3, respectively.

Linear calibration curves (Figs. 6 and 7) were obtained by plotting the ratio, internal standard peak area:plasticizer peak area vs. actual

TABLE 2

Precision of Internal Standard Method for Tricyclohexyl Citrate Plasticizer at Various Concentrations

Solution code	Wt-% TCHC plasticizer added to mixture	Ratio $\dfrac{\text{TCHC peak}}{\text{Standard peak}}$
A	5.0	0.113
		0.105
		0.114
		Average = 0.1106 ± .0017[a]
B	10.0	0.204
		0.210
		Average = 0.2070
C	15.0	0.297
		0.297
		0.305
		Average = 0.2996 ± .0015
D	20.0	0.414
		0.404
		0.428
		Average = 0.4153 ± .0053
E	30.0	0.572
		0.599
		0.603
		Average = 0.5913 ± .0056

[a] Standard deviation.

TABLE 3

Precision of Internal Standard Method for Triethyleneglycol Dibenzoate
Plasticizer at Various Concentrations

Solution code	Wt-% TEGDB plasticizer added to mixture	Ratio $\dfrac{\text{TEGDB peak}}{\text{Standard peak}}$
F	5.0	0.160
		0.159
		0.160
		Average = 0.1596 ± .0003[a]
G	10.0	0.310
		0.318
		0.313
		Average = 0.3137 ± .0019
H	15.0	0.445
		0.445
		0.443
		Average = 0.4443 ± .0003
I	20.0	0.600
		0.601
		0.598
		Average = 0.5996 ± .0007
J	30.0	0.910
		0.922
		0.895
		Average = 0.9090 ± .0072

[a] Standard deviation.

concentration of plasticizer in the physical mixture. Chromatograms of the physical mixtures containing 20% TEGDB and 30% TCHC in poly(styrene) are shown in Figs. 8 and 9, respectively. The chromatograms demonstrate the excellent separation between the high molecular-weight poly(styrene) and low molecular-weight plasticizers.

The sensitivity of the detector to the compound being studied is important in quantitative GPC. Because the refractive index of the TEGDB component is higher than that of THF, while that of TCHC is closer to THF, one can measure lower concentrations of the TEGDB plasticizer using THF as the carrier solvent.

GPC analysis has an advantage over spectrophotometric techniques in its ability to separate low molecular-weight plasticizers from higher molecular-weight resins. This eliminates spectral interference problems which have to be compensated for in the analysis of mixtures. GPC also offers advantages over other quantitative methods such as

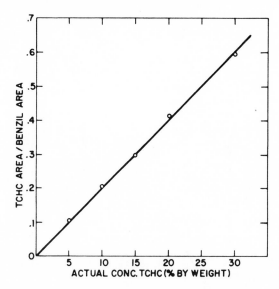

FIG. 6. Calibration curve for per cent TCHC plasticizer using internal standard method.

gas chromatography. First, it eliminates separation of the polymer component before analysis, and second, thermally unstable materials can be analyzed at ambient conditions.

In addition to being able to make quantitative measurements from the detected peak, one can separate the associated material from the polymer and, after collection, analyze it by UV or IR techniques. This is particularly advantageous when one wishes to identify the chemical structure of an unknown plasticizer in a physical mixture.

CONCLUSIONS

Analytical GPC has been successfully applied to the quantitative determination of low molecular-weight organic plasticizers in polymeric mixtures. The internal standard method proved to be the best procedure for this application, as it provides compensation for experimental variations over a long period of time. GPC offers definite advantages over other analytical techniques in its ability to separate high from low molecular-weight materials, and therefore, looks highly promising for quantitative applications to monomers, additives, and catalysts in polymer systems.

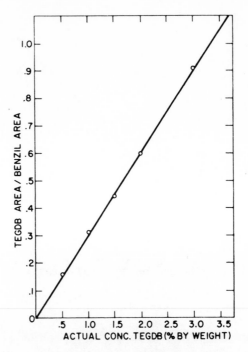

FIG. 7. Calibration curve for per cent TEGDB plasticizer using internal standard method.

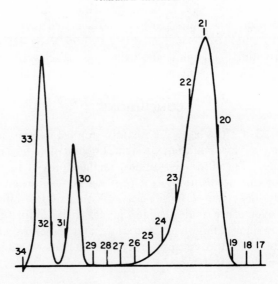

FIG. 8. Chromatogram of 20% TEGDB in physical mixture.

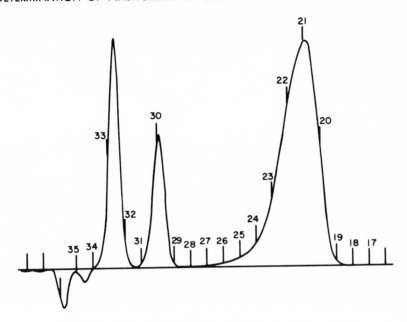

FIG. 9. Chromatogram of 30% TCHC in physical mixture.

Acknowledgments

The authors wish to gratefully acknowledge the help of Dr. J. Holman and J. Short for their assistance in the development of the procedures used in this work.

REFERENCES

1. R. D. Mate and H. S. Lundstrom, *J. Polym. Sci., Part C,* **21,** 317 (1968).

2. F. N. Larsen, *J. Appl. Polym. Sci. Symposia,* **8,** 111 (1969).

3. F. N. Larsen, *Amer. Lab.,* p. 10 (October, 1969).

4. R. J. Limpert and E. L. Obermiller, Paper presented at Sixth International GPC Symposium, Miami Beach, 1968

5. S. Dal Nogare and R. S. Juvet, Jr., *Gas-Liquid Chromatography,* Wiley (Interscience), New York, pp. 256–264, 1962.

6. D. Harvey and D. E. Chalkley, *Fuel,* **34,** 191 (1955).

Evaluation of Pulps, Rayon Fibers, and Cellulose Acetate by GPC and Other Fractionation Methods*

W. J. ALEXANDER and T. E. MULLER

EASTERN RESEARCH DIVISION
ITT RAYONIER, INC.
WHIPPANY, NEW JERSEY 07981

Summary

Chain length distribution of a broad spectrum of wood celluloses and cellulose derivatives was determined by gel permeation chromatography. Relative amounts of short and long chain-length species were characterized, and uniformity indices were calculated. Prefractionation was found to be a desirable approach to amplify low- and high-DP regions. This was accomplished using a 55/45 ethyl acetate/ethyl alcohol mixture to yield the low-DP fraction and with a varying composition acetone/water system to obtain high-DP material. Fractions of regenerated cellulose from rayon obtained by treatment with 6.5 and 10% sodium hydroxide and by acid hydrolysis were characterized. Wood celluloses and rayons were analyzed in their nitrate form, whereas cellulose acetates were studied directly. This work was aimed primarily at elucidating the gel fraction that appears in the form of a peak of apparently high-DP material, resulting in a bimodal distribution.

INTRODUCTION

The application of gel permeation chromatography (GPC) to the molecular weight fractionation of wood cellulose, using the trinitrate derivative, was first reported in 1966 (1). There have since been a number of papers dealing with cellulose derivatives demonstrating growing interest in the use of GPC for cellulose analysis (2–7).

* Presented at the ACS Symposium on Gel Permeation Chromatography sponsored by the Division of Petroleum Chemistry at the 159th National Meeting of the American Chemical Society, Houston, Texas, February, 1970.

† Contribution No. 115 from the Research Divisions of ITT Rayonier, Inc.

429

A prime objective of this work was to define real differences in chain length uniformity between various grades of unmodified cellulose and the resultant end-products. This paper deals with the resolution of differences in the degree of polymerization (DP) distribution of wood pulp, rayon, fiber, and cellulose acetate as measured by GPC directly or on prefractionated samples. Numerical criteria have been selected for a variety of celluloses to permit evaluation of their relative DP uniformity for comparison with process history and various analytical characterizations.

Whereas relatively large and distinct differences in uniformity exist between paper-grade and dissolving-grade wood pulps, differences among dissolving grades may be small, primarily in average DP level. Small differences in the amount of short-DP material, however, are of major significance. Accordingly, particular emphasis was given to the characterization of regions at the extremes of polymer chain-length. The approach involved prefractionation by one of the conventional techniques used to isolate short and longchain-length fractions and their examination by GPC. With cellulose, as with other polymers, low molecular weight species are regarded as detrimental to physical strength while high molecular weight moieties influence the rheological behavior of polymers in solution.

EXPERIMENTAL

The work is divided into four groups. The first group outlines the experimental technique and calibration employed. The second group describes pulp analysis, the third relates to rayon fibers, and the fourth to cellulose acetates.

The main operating conditions employed are shown in Table 1. The chromatograph was a Waters Model 200 instrument, equipped with

TABLE 1

GPC Operating Conditions

Instrument: Waters Associates Model 200
Columns: Styragel 1×10^6, 3×10^6, 1×10^5, 3×10^4
Solvent: THF
Temperature: Ambient
Flow rate: 1 ml/min
Sample injection: Automatic
Refractometer null glass: 1/64 in.

four columns (2 columns of 10^6, 1×10^5, and a 3×10^4). The solvent was tetrahydrofuran (THF); flow rate 1 ml/min. Temperature was ambient, and a 1/64 in. null glass was used in the refractometer.

Pulps and rayons were analyzed after conversion to the trinitrate derivatives as reported previously (3). Cellulose acetates were in the conventional diacetate form with an acetyl level of 38–40%.

The amount of sample injected for a test was based on the DP level

TABLE 2

Schedule for Selection of Test Solution Concentration[a]

Concentration (g/100 ml)	DP Level
0.100	<1200
0.050	1200–2000
0.025	>2000

[a] Filtration apparatus: Hypodermic syringe, Swinny adapter. Media: Millipore Mitex (Teflon), Type MF-LS.

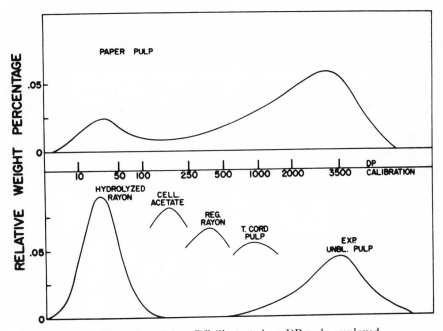

FIG. 1. Range in cellulose DP illustrated on DP scale employed.

of the material, as shown in Table 2. The volume of solution was 2 ml. An automatic sample injector was used for the most part.

RESULTS AND DISCUSSIONS

The DP range as applied to the actual GPC elution curves is shown in Fig. 1. With two exceptions which involve different columns, all figures showing elution curves have the same scale, ranging from 2 to >3500 DP. The top curve represents paper pulp, covering virtually the entire DP range. The bottom curve is intended to place in perspective the various celluloses to be discussed, high-DP wood or cotton celluloses, dissolving pulps, rayon fibers, cellulose acetates, and hydrolyzed rayon which is referred to as levelling-off DP or LoDP cellulose. The scale shown in the center was arrived at from the calibration curve shown in Fig. 2.

Most of the points in Fig. 2 represent cellulose nitrate fractions separated by fractional precipitation of predominantly cotton cellulose. There are two exceptions; the members of lowest DP are cello-

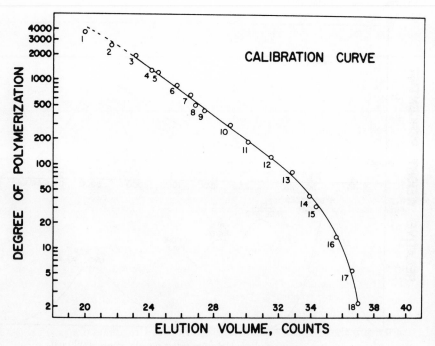

FIG. 2. GPC calibration curve.

pentose and cellobiose. The DP's given are nitrate DP values obtained from intrinsic viscosity determinations in ethyl acetate at 25°C (8).

Elution curves for fractions employed for the calibration are shown in Fig. 3. Nearly all of these tests were made using solutions with a concentration of 0.05 g/100 ml. Amplification was reduced for the narrow distribution cellobiose and cellopentose. The two highest DP fractions were analyzed at a concentration of 0.025 g/100 ml; these were obtained by fractionating selected celluloses into four or five fractions to cover the DP range.

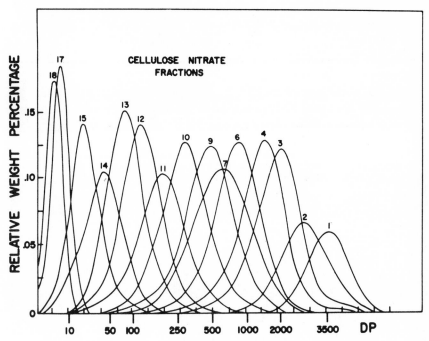

FIG. 3. GPC elution curves for cellulose nitrate fractions employed for calibration.

The prefractionation procedure is outlined in Table 3. Aliquots of water were added to the sample dissolved in a mixture of 91% acetone and 9% water. The amount of initial sample dissolved varied from 0.10 g for a 3000 DP cotton to 1.00 g for two celluloses obtained from aged alkali cellulose with average DP values below 100.

Figure 4 shows a series of celluloses obtained from aged alkali cellulose derived from wood pulp. A similar series was prepared from cot-

TABLE 3

Precipitation Fractionation Scheme for Cellulose Nitrate
Solvent: 91% acetone, 9% water
Amount: 200 ml
Nonsolvent precipitant: water

Sample weight (g)	DP Level
0.100	3000
0.150	2000
0.300	1000
0.500	250
1.000	100

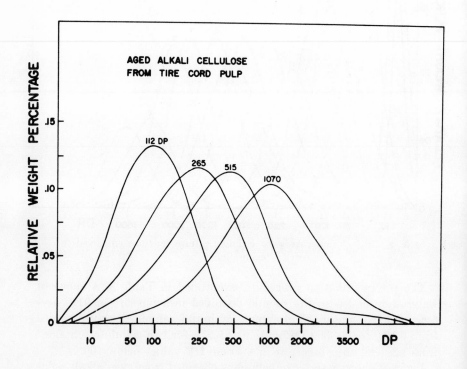

FIG. 4. GPC elution curves for aged alkali cellulose.

ton linters to provide samples for fractionation to calibrate in the DP range below 600.

Figure 5 shows curves for the fractions of one of these samples with an average DP of 77, from which calibration points corresponding to 120, 80, and 40 DP were obtained.

The next group of figures illustrates differences among the chain-length distributions of a variety of wood pulps.

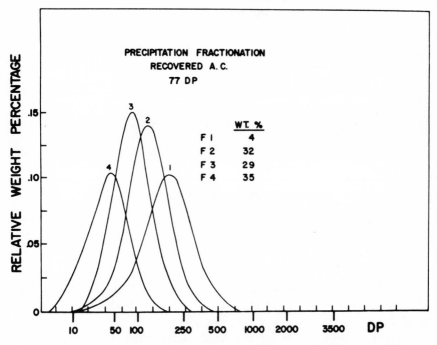

FIG. 5. GPC elution curves for cellulose nitrate fractions from low DP recovered alkali cellulose.

Figure 6 depicts three sulfite pulps. Numbers 1 and 2 are paper pulps which have virtually the same hemicellulose content. The fraction soluble in 10% sodium hydroxide at 20°C (S_{10}) is 13%. This percentage corresponds approximately to that portion classified as <100 DP by the GPC curve. A 10% difference in average DP level (the principal difference between these pulps) appears primarily in very high DP material. Pulp No. 3 is a low-average-DP pulp in which a high percentage of the hemicellulose component was retained as in the paper pulps. The location of the hemicellulose peak is shifted from

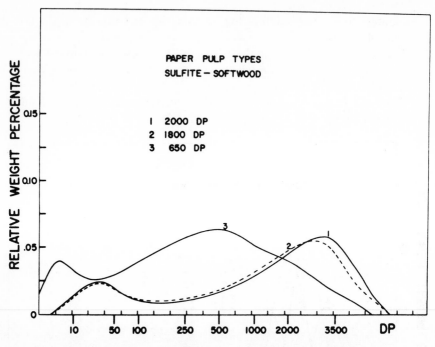

FIG. 6. GPC elution curves for three softwood paper pulp types.

DP 50 to the less than 25 DP level. The fraction classified by GPC as <100 DP has been increased to 24%, corresponding to a S_{10} value of 23.8%. A larger area under the curve for Pulp 3 may be accounted for by the solution concentration used (0.1% vs. 0.05%) for Pulps 1 and 2.

Figure 7 shows curves for two kraft paper pulps. The wood source for one was a blend of southern pine and hardwoods in which pine is the major component. The other one is 100% hardwood. These pulps have a distinct third peak at the high-DP end of the scale in addition to peaks representing the main cellulose and hemicellulose portions. This peak appears at the exclusion volume of the columns and is considered to be the result of a "gel fraction." This is not evident in the curves for the pulps shown in the previous figures. A significant difference in xylan content exists among these pulps. The hardwood pulp contains 16% xylan, the predominantly pine pulp 7%, while the pulps discussed previously had 2% or less. Some correlation seems to be implied between gel fraction and xylan residue. The residue may nitrate

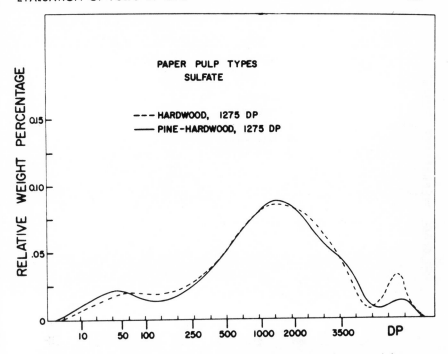

FIG. 7. GPC elution curves for two sulfate paper pulps containing hardwoods.

to the dinitrate but not to the trinitrate derivative. The extent and exact nature of this connection will be the subject of a separate study.

Figure 8 depicts four dissolving grade pulps. Listing these in the order of decreasing average DP, they are: acetate-, tire cord-, rayon staple-, and cellophane-grade pulps. The curves reflect differences in the content of short-DP material among these pulps; the tire cord and acetate grades show the least and the cellophane grade pulp the most. Three factors may be obtained from the GPC curves giving numerical values for establishing distribution uniformity. These values are relative areas under the curve corresponding to specific DP classifications: (a) the percentage of <50 DP and (b) of <100 DP material to evaluate low-DP content, and (c) the percentage >2 times (2×) the average DP to reflect differences in high-DP material.

Table 4 lists values obtained by this procedure for these four pulps. The figures for amount of <100 DP material correspond roughly to the respective S_{10} levels and reflect a low amount of short-chain-

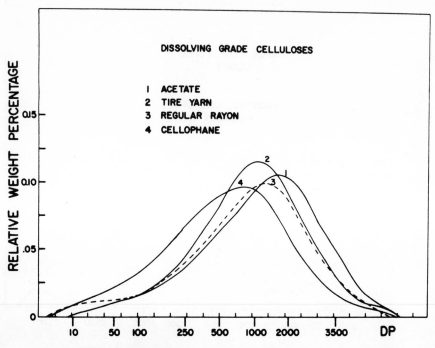

FIG. 8. GPC elution curves for several grades of dissolving grade celluloses.

length material present in top quality tire cord and acetate pulps. The percentages of ">2× average-DP" material fall within a relatively narrow range for these dissolving pulps. This is not characteristic of all pulps, as shown in Table 5. This table lists DP uniformity data for several types of celluloses at the 700 average-DP level. The last column gives the percentages classified by GPC as >2× average-DP, showing a range from 11 to 39. Although S_{10} values are not listed,

TABLE 4

Comparison of DP Uniformity Indices for Several Dissolving Grade Celluloses

Pulp type	% <50 DP	% <100 DP	% >2 × DP avg
Tire-cord grade	1.3	3.2	22
Acetate grade	1.4	3.3	23
Rayon grade	3.7	6.4	24
Cellophane grade	4.5	8.6	24

TABLE 5

Comparison of DP Uniformity Indices for Several Celluloses at 700 DP Level

Cellulose	DP	% <50 DP	% <100 DP	% >2 × DP avg
Recovered alkali cellulose viscose linters	700	1.3	4.2	11
Recovered alkali cellulose wood pulp	700	2.5	6.2	18
Wood pulp cellophane grade	760	4.5	8.6	24
Wood pulp experimental	760	5.6	10.1	39

there is a reasonable correlation between the S_{10} level and the <100 DP material. GPC curves for three of these celluloses are shown in Fig. 9.

Direct comparison of the curves is possible because of the similarity in DP level. It is quite apparent that there is a difference in the per

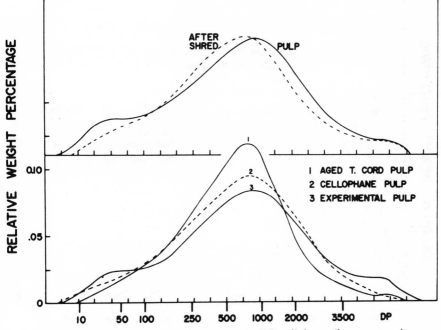

FIG. 9. GPC elution curves for three 700 DP celluloses (lower curves) and pulp before and after steeping and shredding steps in viscose process.

cent material classified as <100 DP and $>2\times$ average-DP. The upper set of curves shows the effect of steeping an experimental pulp in 18% sodium hydroxide on its chain-length distribution; there is an indication of the removal of some low-DP material. With more highly refined pulps this effect of steeping is hardly noticeable.

A single-stage fractional extraction of nitrated cellulose was carried out to permit a closer examination of the low-DP components. The conditions for the fractionation scheme employed are given in Table 6. The fractionation involves a 48-hr extraction of the cellulose nitrate with a mixture of 45% ethanol (95%) and 55% ethyl acetate at ambient temperature.

GPC fractions obtained from three prefractionated pulps are shown in Fig. 10. Curves 1 and 1R represent soluble and insoluble fractions from an experimental, high-DP, high-hemicellulose pulp. The two components represent a distinct separation of the hemicellulose fraction from the cellulose. The curves show almost no material present in the DP range of 250 to 750.

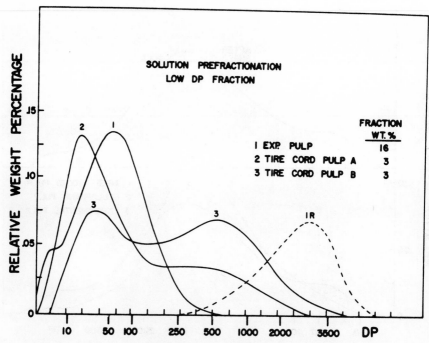

FIG. 10. GPC elution curves for low DP fractions isolated by prefractionation—broad range GPC columns.

TABLE 6

Single-Stage Solution Fractionation Scheme for Cellulose Nitrate

Solvent: 45% ethanol (95%), 55% ethyl acetate
Amount: 200 ml
Sample wt: 2.00 g
Extraction: 48 hr at ambient temperature
Filtration: Sintered glass funnel
Solute recovery: Aliquot evaporated in water bath

Curves 2 and 3 represent the low DP fractions (3% by weight), isolated from two dissolving pulps with an average DP of 1100. The principal difference between these two highly refined pulps is that No. 3 had received a cold caustic extraction to reduce low-DP material. Two points may be noted: (a) extraction with cold caustic did reduce the percentage of material classified as <100 DP, and (b) it increased accessibility, allowing the extraction of a higher percentage of longer (>100 DP) chains.

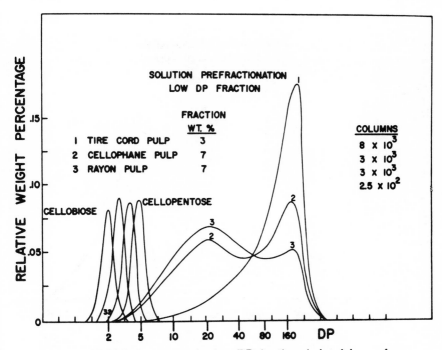

FIG. 11. GPC elution curves for low DP fractions isolated by prefractionation—low DP range GPC columns.

Curves obtained for low DP fractions isolated from tire-cord-, cel-
lophane-, and rayon-staple-grade pulps are compared in Fig. 11. These
curves were obtained using a set of columns selected for lower DP
celluloses. The calibration scale covers the 2 to 160 DP range. Resolution of cellobiose, cellotriose, cellotetrose, and cellopentose is shown
by the peaks on the left. Extraction isolated a fraction 7% by weight
from the cellophane and rayon pulps and only 3% from the tire-cord
pulp. The chain-length composition differs among the three fractions.
The rayon-pulp fraction has the highest percentage at the 25 DP level
and the highest S_{18} or gamma values. The cellophane-pulp fraction
shows a higher percentage at the 150 DP level than does the rayon
fraction. The S_{10}–S_{18} percentage (or beta fraction) is the highest for
this pulp. The fraction isolated from the tire-cord pulp represents only
3% of the pulp by weight, it is almost free of material in the <25 DP
range, and it has a substantial amount of >100 DP material.

An abridged version of the ASTM fractional precipitation method

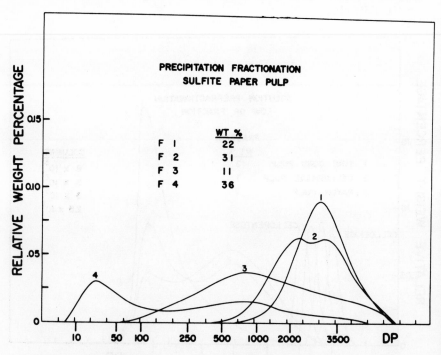

FIG. 12. GPC elution curves for precipitation fractions from sulfite
paper pulp.

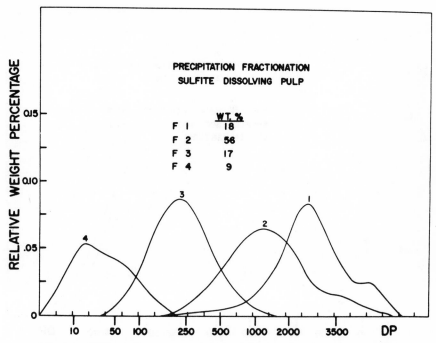

FIG. 13. GPC elution curves for precipitation fractions from sulfite dissolving pulp.

for cellulose nitrate was used to permit examination of high-DP portions in greater detail. The technique proved to be useful at the two extremes of the DP range.

Figure 12 shows separation of a 2000-DP paper pulp into four fractions. The No. 1 fraction has a symmetrical distribution peaking at about 3000 DP. The No. 4 fraction contained essentially all of the hemicellulose components.

Figure 13 shows curves representing fractionation of a 1250-DP sulfite dissolving pulp. The No. 1 fraction contains 18% of the sample and has a shoulder or the beginning of a second peak. The No. 4 fraction represents 9% of the sample and is comprised primarily of chains <100 DP.

Figure 14 compares the first precipitated fractions from a sulfate hardwood paper pulp, a sulfite paper pulp and a 2000-DP cotton linters. The linters and sulfite-paper-pulp fractions have similar distributions, neither one having a gel fraction. The sulfate hardwood

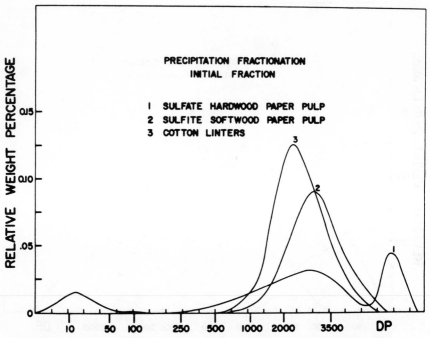

FIG. 14. GPC elution curves for initial precipitation fraction from three pulps.

paper-pulp fraction is rather unique in that it shows that this particular fraction did not separate according to DP, but according to solubility differences attributable to other factors. The exclusion volume component apparent in the unfractionated pulp is shown to be concentrated in the first two fractions. In previous studies, analyses of denitrated fractions separated by the same procedure applied to pulps with a high xylan content revealed the major portion of xylan to fractionate in the first two fractions.

Figure 15 compares the DP distribution of four regenerated cellulose fibers: regular rayon (No. 1), a polynosic rayon (No. 2), Fortisan (No. 3), and a high-wet-modulus (HWM) type (No. 4). The latter fiber has an average DP over two times that of the regular rayons. The percentage classified by GPC as <100 DP ranges from 19% for the regular rayon to 8% for the Fortisan and the Experimental High Performance-type (No. 4), Table 7. Relatively large differences between the high-DP regions of the curves are evident. For example,

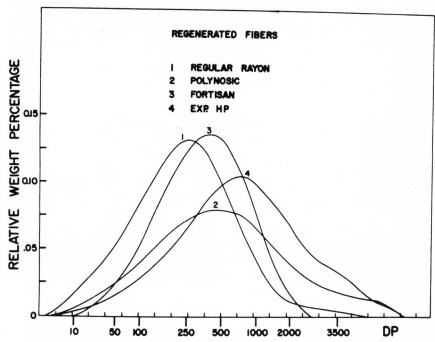

FIG. 15. GPC elution curves for several types of regenerated cellulose fibers.

a value of 28 was obtained for the >2× average-DP for the polynosic, compared to values of 15 for the Fortisan and regular rayon samples.

Data for these fibers and a tire yarn examined by GPC are shown in Table 8. The fibers range from 300 in average DP for regular rayon to 700 for an improved high performance, HWM type. Their resist-

TABLE 7

Comparison of DP Uniformity Indices for Regenerated Cellulose Fibers

Type	DP	% <50 DP	% <100 DP	% >2 × DP avg
Regular rayon	300	10	19	15
Tire yarn	525	4	9	18
Polynosic	500	5	11	28
Exp. HP	700	3	8	18
Fortisan	400	3	8	15

TABLE 8

Analytical Data for Regenerated Cellulose Fibers

	Regular rayon	Polynosic staple	Exp. HP staple	Tire yarn	Fortisan
DP	300	500	700	525	400
Caustic solubility					
$S_{6.5}$, %	23.5	8.3	1.8	19.5	0.6
S_{10}, %	50.0	20.2	13.8	50.0	2.8
Acid hydrolysis					
Limit I.V. (cuene)	0.19	0.26	0.28	0.17	0.28
Yield, %	84.6	92.0	94.0	88.7	92.8

ance to caustic varies as indicated by the respective solubilities in 6.5 and 10% sodium hydroxide, ranging from <1 to 50%. Yield and leveling-off intrinsic viscosity data for cellulose residues obtained from these fibers after a standardized acid hydrolysis treatment are in-

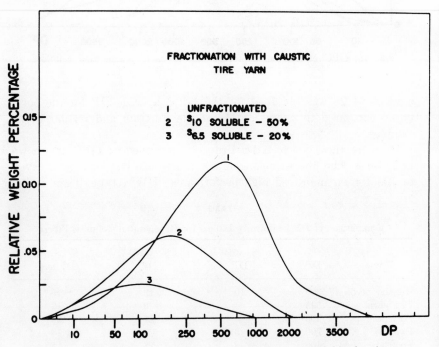

FIG. 16. GPC elution curves for tire yarn and caustic-soluble fractions.

cluded. GPC analysis has been combined with caustic solubility and acid hydrolysis to gain a better appreciation of their meaning and significance.

With pulps, the amount soluble in 10% sodium hydroxide correlates approximately to the percentage classified as <100 DP. This does not follow for regenerated fibers which are far more susceptible to swelling in caustic. For example, 50% of a regular-type rayon or a conventional tire yarn may be dissolved in 10% sodium hydroxide while the soluble fraction of HWM types may be less than 10%. The resistance to caustic solution, particularly to 6.5% sodium hydroxide at 20°C, has proved to be a useful technique for classifying fibers. The HWM types, in which crystallite regions are relatively large and more highly oriented, are characteristically more resistant to sodium hydroxide.

Figure 16 shows the DP distribution of $S_{6.5}$- and S_{10}-soluble fractions of a tire yarn fiber. The relative areas of the two fractions have been adjusted by selecting concentrations to correspond to the relative

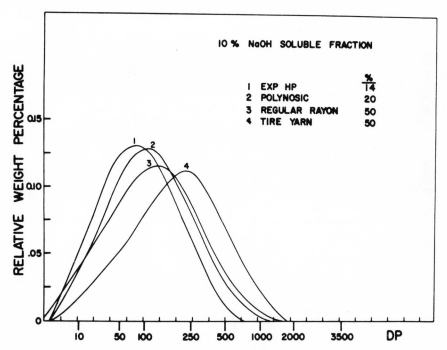

FIG. 17. GPC elution curves for 10% NaOH-soluble fraction from four types of rayon fiber.

weight percentages. It is evident that caustic fractionation is controlled by fiber accessibility as well as DP. A more concentrated caustic solution has extracted more and longer chains. A number of fibers have been examined in this manner, and they were found to follow a similar pattern with shifts on the scale corresponding to variations in average-DP level or the amount of caustic-soluble material.

Direct comparison of the S_{10} fraction from the four fibers is shown in Fig. 17, indicating a difference in average-DP level. The lower-DP sample is from the HWM fiber from which the lowest weight-percentage was dissolved. The fraction with the highest-DP represents the tire yarn fiber with the highest weight-percentage soluble in 10% sodium hydroxide.

Leveling-off DP celluloses obtained by subjecting rayon fibers to a standard acid hydrolysis treatment show differences in the DP of the acid-resistant residues. These differences may reflect, to some degree, the relative size of the crystallite or most highly oriented regions of

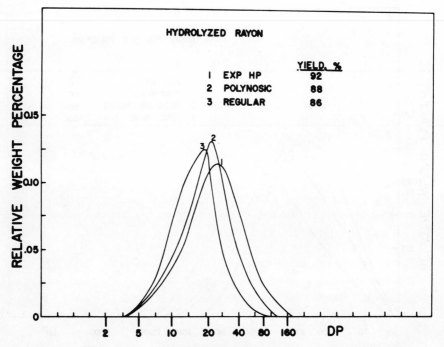

FIG. 18. GPC elution curves for hydrolyzed rayon—limit or leveling-off DP cellulose residues.

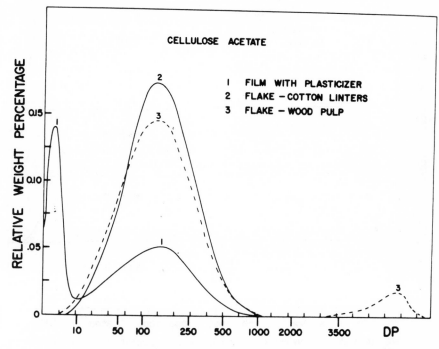

FIG. 19. GPC elution curves for cellulose acetates.

the fiber. Although the DP levels are low (in the 20 to 30 DP range), the LoDP cellulose from HWM fibers may have an average DP of 50% or more, greater than that from a regular rayon or conventional tire yarn.

Figure 18 compares the curves of three hydrolysis residues representing a high-performance, HWM fiber (No. 1), polynosic (No. 2), and a regular rayon (No. 3). These were obtained on the columns selected for low DP celluloses.

The balance of this work relates to cellulose acetates. Figure 19 compares a plasticized acetate film and acetate flakes derived from cotton linters and from wood pulp. In Curve 1 the low molecular-weight plasticizer is resolved as a distinct peak. Curves 2 and 3 show cotton and wood pulp acetates to have a similar DP distribution. However, Curve No. 3 from wood pulp shows a small peak at the exclusion volume of the columns. This peak was also designated the "gel fraction."

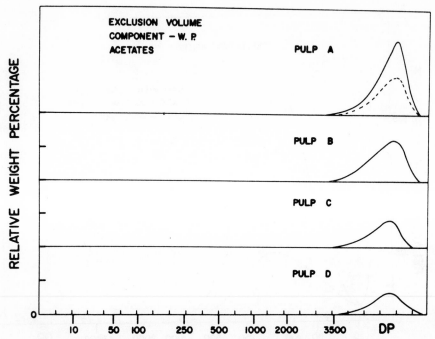

FIG. 20. GPC elution curves for gel fraction component in several wood pulp acetates.

The relative size of this fraction may vary with the pulp grade from which the acetate is derived, as shown in Fig. 20. The size of the gel fraction is smaller in acetates derived from more highly purified grades of pulp. Pulps A through D represent acetate grades containing progressively less hemicellulose. The dotted curve for Pulp A shows a reduction in the gel fraction in the GPC test solution, resulting from substituting a Krueger filter pad for the Millipore Mitex filter that was generally employed in preparing the GPC test solutions.

The "gel fraction component" is removed in the first (least soluble) fractions isolated by precipitation-fractionation of the acetate. Figure 21 shows curves obtained for fractions isolated from an acetate made from wood pulp. The fractionation scheme employed a solvent/nonsolvent system of acetone and water. The first fraction represented 3% and the second 15% of the sample. All of the gel fraction was removed in these two fractions. In previous studies where fractions were separated by the same method, the first one or two fractions precipitated were found to account for the major portion of the xylan, mannan, and carboxyl present in the sample.

Figure 22 shows that the "gel fraction" is absent in the fractions obtained by precipitation from a cotton linters acetate.

Figure 23 compares the GPC curves of a cellulose acetate and of the nitrate derivative of this same sample after deacetylation. The nitrated sample peaks at a higher DP than the original one. It might be expected that after saponification with caustic and subsequent nitration there would be a slight shift to a lower DP level. It is more likely, however, that for a given DP the cellulose nitrate chain has a larger volume in THF than does the cellulose acetate. This would be consistent with their respective DP/intrinsic viscosity relationships. It also calls attention to the fact that the DP calibration used is not applicable to cellulose acetates without some adjustment. One additional point to note is the presence of a "gel fraction peak" in the

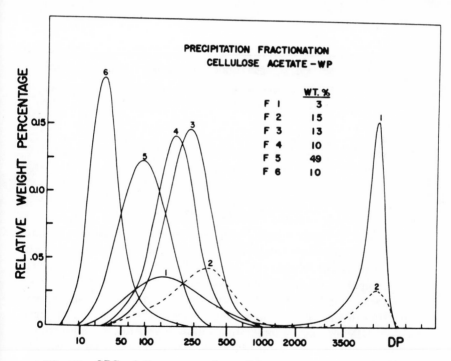

FIG. 21. GPC elution curves for cellulose acetate fractions from wood pulp.

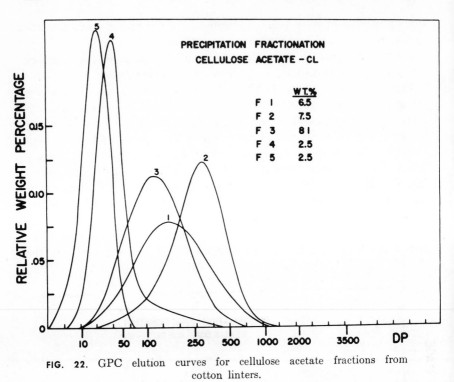

FIG. 22. GPC elution curves for cellulose acetate fractions from cotton linters.

nitrate sample not present in the acetate, suggesting that all gel fractions do not derive from a common origin.

CONCLUSIONS

Numerous cellulosic samples have been fractionated on a gel column system calibrated with cellulose nitrate fractions. These samples have been evaluated in terms of relative DP, and DP uniformity levels were compared. Conventional fractional-solution and fractional-precipitation methods have been utilized effectively in conjunction with GPC to improve the definition of the DP distribution of high- and low-DP portions isolated from pulps, rayon fibers, and cellulose acetates. GPC has made possible a clearer understanding of the DP distribution of celluloses than was previously available from other noninstrumental fractionation methods.

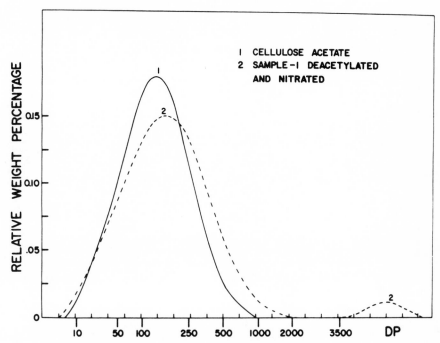

FIG. 23. GPC elution curves for cellulose acetate and the cellulose nitrate derivative of the saponified acetate.

REFERENCES

1. L. Segal, *J. Polym. Sci., Part B,* **4**, 1011 (1966).
2. G. Meyerhoff and S. Joranovie, *J. Polym. Sci., Part B,* **5**, 495 (1967).
3. W. J. Alexander and T. E. Muller, *J. Polym. Sci., Part C,* **12**, 283 (1968).
4. L. Segal, *J. Polym. Sci., Part C,* **21**, 267 (1968).
5. R. J. Brewer, L. J. Tanghe, S. Bailey, and J. T. Burr, *J. Polym. Sci., Part A-1,* **6**, 1697 (1968).
6. R. Y. M. Huang and R. G. Jenkins, *Tappi,* **52**, 1503 (1969).
7. L. Segal and J. D. Timpa, *Tappi,* **52**, 1669 (1969).
8. W. J. Alexander and R. L. Mitchell, *Anal. Chem.,* **21**, 1487 (1949).

Characterization of the Internal Pore Structures of Cotton and Chemically Modified Cottons by Gel Permeation*

L. F. MARTIN, F. A. BLOUIN, and S. P. ROWLAND

SOUTHERN REGIONAL RESEARCH LABORATORY†
NEW ORLEANS, LOUISIANA 70119

Summary

The pore structures of cotton and formaldehyde-modified cottons have been characterized by applying the principles of gel permeation to chromatographic columns formed of the finely divided fibrous cottons. Measurements were made of the relative elution volumes of suitable solutes which cover a range of known, discrete molecular weights. The solutes chosen were sugars ranging in molecular weight from 198 to 738, the elution volumes of which are in an inverse linear relationship to their molecular weights. Extrapolations of this linear relationship provide measures of the effective internal solvent volumes and the permeability limits of the cotton celluloses. Changes in these parameters are indicative of the nature and extent of the alterations of the pore structure produced by cross-linking.

Marked differences have been found in the pore structures of cotton cross-linked to progressively higher levels with the fibers in a collapsed state by a bake-cure process and with the fibers in a semidistended state by reaction in an acetic–hydrochloric acid medium. The technique will be useful for relating changes in pore structure to the modifications of physical and performance characteristics of fabrics produced by various chemical treatments.

* Presented at the ACS Symposium on Gel Permeation Chromatography sponsored by the Division of Petroleum Chemistry at the 159th National Meeting of the American Chemical Society, Houston, Texas, February, 1970.

† One of the laboratories of the Southern Utilization Research and Development Division, Agricultural Research Service, U. S. Department of Agriculture.

455

INTRODUCTION

Cotton cellulose strongly sorbs water and in doing so undergoes swelling to permit permeation by an appreciable volume of liquid; there is an extensive system of pores within the supramolecular organization of its elementary and microfibrillar units (1). The presence of such a pore structure provides access of chemical reagents to the microstructural units which exhibit the crystalline organization of the cotton cellulose. Techniques of gel permeation chromatography are useful in determining certain characteristics of this pore structure. For this purpose, finely divided samples of cellulose are employed as the media in chromatographic columns and the elution volumes of suitable solutes of a range of known, discrete molecular weights are measured. It has been shown by this method (2) that cotton cellulose decrystallized by ball-milling is permeated by (sugar) solutes up to a molecular weight of approximately 1900. Chopped fibrous cotton contains some larger pores as indicated by a permeability to approximately 2900 molecular weight. However, the specific internal pore volume of water-swollen, fibrous cotton is only one-third of that of the decrystallized cotton. This technique has proven effective for investigating changes in pore structure caused by cross-linking decrystallized cotton (2), by mercerizing fibrous cotton (3), and by cross-linking fibrous cotton (4).

The object of the present work was to study the relation of changes in gel permeation characteristics of fibrous cotton as a function of the levels of formaldehyde cross-linking achieved under two different sets of reaction conditions. In one process, the bake-cure conditions effect reaction of formaldehyde with the collapsed fibers of the cotton. The second process employs an acetic–hydrochloric acid medium in which the cellulosic structure is semidistended, but the fibers are not as swollen as in the aqueous wet-cure process.

EXPERIMENTAL

Preparation of Cross-linked Cottons

Cross-linking with formaldehyde in the acetic–hydrochloric acid medium and by the bake-cure process was applied to cotton fabric. The fabric used was 80 × 80 printcloth, 3.2 oz/sq yd; it had been desized, scoured, and bleached.

The procedure described by Chance et al. (5) was used for the cross-linking in the acetic–hydrochloric acid medium. In this case, the fabric was cut into 6 in. × 10 in. pieces, and approximately 70 g batches of fabric were treated with the reagent solution in shallow pans at room temperature for each cross-linked sample. The reagent composition was: CH_2O (paraformaldehyde) 5.5%, HCl 5.5%, CH_3COOH 72.9% and H_2O 16.1%. Proportions of 13.5 ml of solution/g of cotton (air dried) were used for all four samples, D-1 through D-4, and progressively higher levels of cross-linking were obtained by increasing the times of reaction as shown in Table 1. After thorough washing and air drying, the batches of fabric were cut into 1-in. squares and reduced to the particle size required for column packing by Wiley milling successively through the 20- and 60-mesh filter screens. Untreated fabric was milled similarly for the control sample, WM-6.

The bake-cure cross-linking reaction with formaldehyde was carried out as described by Fujimoto et al. (6) with 12 in. × 18 in. pieces of fabric which were padded with a solution of paraformaldehyde and magnesium chloride catalyst. In this case, progressively higher levels of cross-linking were obtained by increasing the concentration of the reagent solution while maintaining the ratio of $CH_2O:MgCl_2 \cdot 6H_2O$ at 3.75:1. All of the batches of fabric were padded to a wet pickup of approximately 130%. Drying and curing were effected in a single step, employing a forced draft oven preheated 175–178°C. Introduction of six fabric pieces, suspended to permit adequate circulation, reduced the temperature to 135–140°C. The temperature reached 161–163°C in 2 min and was maintained at this temperature for 3 min for curing. The treated fabrics were rinsed immediately in hot, running tap water,

TABLE 1

Gel Permeation Characteristics of Cotton Cross-Linked in the Acetic–Hydrochloric Acid Medium

Samples	Reaction time (min)	CH_2O (%)	Total internal H_2O (V_r)(ml/g)	Internal solvent volume (V_i)(ml/g)	V_i/V_r	Permeability limit (M_x) MW
WM-6	0	0	0.351	0.303	0.86	2830
D-1	5	0.49	0.344	0.291	0.85	1810
D-2	10	0.74	0.347	0.228	0.66	1400
D-3	15	0.81	0.239	0.206	0.86	1500
D-4	30	1.39	0.232	0.163	0.70	1180

followed by several rinses in distilled water, and finally by 15 min boiling to remove hemiformals and unreacted formaldehyde.

The samples of fabric were reduced to fine fiber particles by milling successively through the 20-, 60-, and 80-mesh screens in the Wiley mill. An untreated control sample, WM-13, was similarly prepared for packing the chromatographic column. A control sample, C-C, which was subjected to the bake-cure in the presence of the catalyst (no formaldehyde) was also included in this series. The paraformaldehyde concentrations of the padding solutions and the resulting formaldehyde contents of the fabrics, C-1 through C-4 are given in Table 2.

TABLE 2

Gel Permeation Characteristics of Cotton Cross-Linked in the Bake-Cure Process

Samples	Reagent concn CH_2O (%)	Sample CH_2O (%)	Total internal H_2O (V_r) (ml/g)	Internal solvent volume (V_i) (ml/g)	V_i/V_r	Permeability limit (M_x) MW
WM-13	0	0	0.334	0.289	0.87	2430
C-C	0	0	0.249	0.237	0.95	2130
C-1	2.5	0.47	0.231	0.170	0.74	1280
C-2	4.5	1.21	0.167	0.096	0.60	1120
C-3	6.0	1.47	0.152	0.069	0.45	1290
C-4	7.5	2.20	0.107	0.047	0.44	1280

Gel Permeation Chromatographic Measurements

Details of the instrumentation, of the packing of sample columns, and of the techniques for making the gel permeation measurements have been described previously (7). Adaptation of the method for the study of chopped fibrous cotton also has been reported (3). The sugars used as solutes were glucose monohydrate, maltose monohydrate, raffinose pentahydrate, and stachyose tetrahydrate; they covered the range of molecular weight from 198 to 738. Relative peak elution volumes were determined with reference to the column void volumes measured by dextran of molecular weight $\bar{M}_w = 10,000$. The dextran and each of the four sugars were eluted individually at intervals sufficiently spaced to prevent overlapping of the elution peaks. They were injected from a sample loop of 0.25 ml volume as 4% solutions (solutions containing 40 mg/ml). Elution with distilled water was carried out at a linear flow rate of approximately 3 cm/hr. The inverse linear relationship of peak elution volumes to molecular weights of the characteristic hydrates of these sugars was found to apply in chromatography of the samples of cotton investigated in these experiments.

RESULTS

Data obtained by the successive elution of dextran and each of the four sugars from columns containing cottons cross-linked by the bake-cure process are presented in Fig. 1, together with results of measurements of an uncross-linked control sample, WM-13. The spread of triplicate measurements with each of the four sugars on each sample

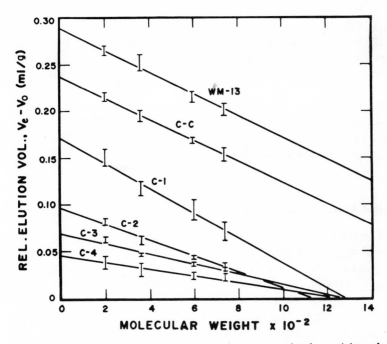

FIG. 1. Relationship of relative elution volumes to molecular weights of the hydrates of glucose (198), maltose (360), raffinose (594), and stachyose (738) for the bake-cure cross-linked cottons, C-1 to C-4, for uncross-linked sample, C-C, and for unmodified cotton sample, WM-13. Bars show spread of triplicate measurements of the specific elution volumes of the four sugars, relative to the void volumes, V_0, measured by dextran, on each of the cellulose sample columns.

column shows the reproducibility of determinations of these volumes. The lines are plotted from the best fit by least squares of the combined data for each sample to an inverse linear equation relating relative elution volumes to molecular weights. Comparable results were obtained from the samples of cotton cross-linked in the acetic–hydrochloric acid medium.

The volume occupied by the water-swollen fibrous cotton particles is readily calculated (on a specific basis) from the difference between the known total column volume occupied by sample material and the void volume measured by the peak elution volume of dextran. It is assumed that the molecular chains of cellulose forming the "gel" have the specific volume 0.629 which is calculated from the dimensions of the unit cell of crystalline cellulose (8). Substracting this value from the specific gel volume yields the total internal water content, V_r, given in the fourth columns of Tables 1 and 2.

Extrapolation of the chromatographic measurements relating relative elution volumes to molecular weights (e.g., see Fig. 1) provides values for the specific internal solvent volume of the cellulose, V_i (at zero molecular weight), and the limit of permeability in terms of molecular weight, M_x (at zero relative elution volume). Changes in these parameters with increasing modification of the cellulose structure by cross-linking reveal the nature of alterations of the supramolecular structure and internal pore volume distributions produced by cross-linking under the two different reaction conditions.

Cotton Cross-linked in the Acetic–Hydrochloric Acid Medium

The relation of the internal volumes and permeability limits to the levels of cross-linking carried out in the acetic–hydrochloric acid medium are shown graphically in Fig. 2. Under these conditions, there is a progressive elimination of larger pores, i.e., a reduction of the size of internal spaces available to larger solutes. This is shown by the continuing decrease of the permeability limit, M_x, with increasing formaldehyde content of the cotton samples. Little reduction of the internal volumes, either total water content, V_r, or effective solvent water, V_i, occurs until a formaldehyde content of 0.6% is reached. At a formaldehyde content of 0.7% both of these volumes decreased markedly and tended to level off. There is not much more reduction in V_r or V_i up to a formaldehyde content of 1.4%. Since these samples were prepared by reaction for periods of time ranging from 5 to 30

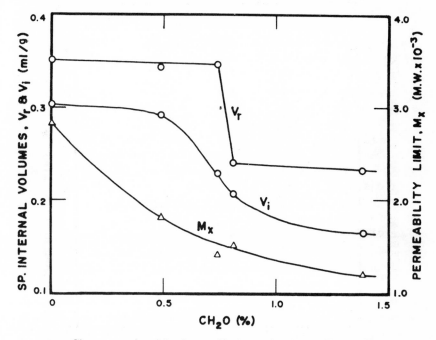

FIG. 2. Changes produced in the specific internal volumes, V_r and V_i, and in the permeability limit, M_x, of fibrous cotton by acetic-hydrochloric acid cross-linking to progressively higher levels of CH_2O content.

min, the marked decreases of the internal pore volumes above 0.6% formaldehyde content are a function of the time of exposure to the medium as well as to the progressively increasing extent of cross-linking.

Cotton Cross-linked in the Bake-Cure Process

Reduction of the values of the gel permeation parameters relative to the percentages of formaldehyde introduced by the bake-cure reaction conditions are shown in Fig. 3. These changes are not a function of reaction time, as all of the samples were cured for the same length of time under the same conditions. As higher percentages of formaldehyde are introduced, there is a progressive and uniform decrease of the total internal water content, V_r, and a roughly parallel reduction of the effective internal solvent volume, V_i. This set of reaction conditions produces a striking change in the permeability

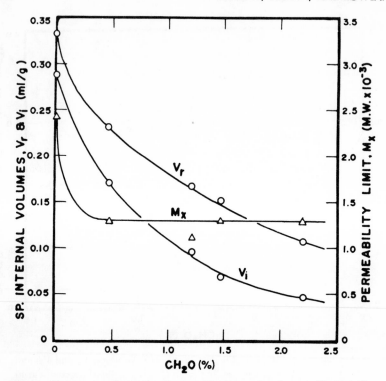

FIG. 3. Changes produced in the specific internal volumes, V_r and V_i, and in the permeability limit, M_x, of fibrous cotton by cross-linking to progressively higher levels of CH_2O content in the bake-cure acid medium.

limit, M_x; this parameter is reduced to half of the value found for untreated cotton at the lowest level of formaldehyde examined in these compositions. As cross-linking proceeds to higher levels of formaldehyde, the permeability limit remains essentially constant (to formaldehyde contents of 2.2%). These changes of the supramolecular structure of the cotton cellulose polymer as a result of cross-linking of fibers in the collapsed state are in marked contrast to those shown in Fig. 2 for comparable levels of cross-linking of the fibers in a semidistended state.

DISCUSSION

The pattern of changes in gel permeation parameters illustrated in Figs. 2 and 3 are indicative of different behavior of the cellulose and perhaps the introduction of different reagent residues during cross-linking of the cellulose polymer under the two sets of reaction conditions. In the acetic–hydrochloric acid medium, formaldehyde reacts with accessible hydroxyl groups on surfaces of the larger pores, progressively reducing the permeability limit of the cellulose. Since there is little reduction of the internal volumes, V_r and V_i, up to about 0.6% formaldehyde content, cross-links must be entering in such a manner that large pores are subdivided or that spaces between surfaces of large pores are progressively reduced. Beyond this level of reaction, the structure appears to undergo a rather rapid transition either as the result of cross-linking of the microfibrillar units or the effect of the reagent solution in collapsing the structure after reaction for 5–10 min. In the earlier stages of reaction, there is a redistribution of pore sizes with little decrease in the total internal pore volume. When sufficient cross-linking has occurred to cause a large decrease in the internal volumes, rapid reduction in the volume distributed in small pores occurs as there is only small additional diminution of the limit of permeability.

The marked difference in the progress of alteration of the pore structure of cotton with progressive cross-linking by the bake-cure process is illustrated by the data presented in Fig. 3. In this case there is a uniform decrease in the internal volumes, V_r and V_i, as the percentage of formaldehyde increases. On the other hand, the reduction of the permeability limit occurs in the very early stages of the cross-linking. All of the decrease does not occur as a result of collapse of the cellulose polymer structure by heating and drying, as shown by the relatively small change in this parameter from the control sample, C-C, which was subjected to curing with the catalyst alone. The limit of permeability is reduced to approximately half the value found for untreated cotton at some percentage of formaldehyde equal to or less than 0.47%. With progressively higher levels of cross-linking, the limit of permeability remains essentially unaltered, while both the total internal water content and the effective internal solvent volume are reduced progressively. These results indicate that half of the internal volume comprised of large pores is eliminated by the heating and drying and a minimal level of formaldehyde, which fixes the value of the permeability limit. Further reaction proceeds randomly in the pores and there is no selectivity of the reduction of volume composed of small pores by the cross-linking process. While the internal pore volumes decrease progressively, the distribution of pore sizes within this internal volume remains unaltered. This is true of the pores of sizes accessible to solutes such as the sugars, constituting the effective internal solvent volume. It is also true of the total internal water content which includes pores accessible only to water, i.e., pores too small to accommodate molecules of the nature of the solutes used.

The results of these experiments demonstrate that the gel permeation technique provides detailed information for characterizing pore structure of cotton cellulose and changes in pore structure of cotton cellulose; this information is not obtainable from other methods of measurement such as water of imbibition. Water of imbibition measures only the total water content in and on and between the particles of water-swollen cellulose. The values obtained by water of imbibition are high by comparison with chromatographic determinations of total internal water content, V_r (3). The gel permeation method is expected to provide a means of developing a better understanding of changes in the supramolecular structure of cotton. It is expected also to provide a means for investigating the dependence of physical and performance properties of modified cottons upon the nature of the supramolecular structure and its alteration by chemical treatment.

REFERENCES

1. J. O. Warwicker, R. Jeffries, R. L. Colbran, and R. N. Robinson, *A Review of the Literature on the Effect of Caustic Soda and Other Swelling Agents on the Fine Structure of Cotton*, Shirley Institute Pamphlet No. 93, Didsbury, Manchester, England, December, 1966.
2. L. F. Martin, N. R. Bertoniere, F. A. Blouin, M. A. Brannan, and S. P. Rowland, *Text. Res. J.*, **40**, 8 (1970).
3. F. A. Blouin, L. F. Martin, and S. P. Rowland, *Text. Res. J.*, **40**, 809 (1970).
4. F. A. Blouin, L. F. Martin, and S. P. Rowland, *Text. Res. J.*, **40**, 959 (1970).
5. L. H. Chance, R. M. Perkins, and W. A. Reeves, *Amer. Dyestuff Reptr.*, **51**, 583 (1962).
6. R. A. Fujimoto, R. M. Reinhardt, and J. D. Reid, *Amer. Dyestuff Reptr.*, **52**, 329 (1963).
7. L. F. Martin, F. A. Blouin, N. R. Bertoniere, and S. P. Rowland, *Tappi*, **52**, 708 (1969).
8. P. H. Hermans, *Physics and Chemistry of Cellulose Fibers with Particular Reference to Rayon*, Elsevier, New York, 1949, p. 20.

Application of GPC to Studies of the Viscose Process. I. Evaluation of the Method*

L. H. PHIFER and JOHN DYER

RESEARCH AND DEVELOPMENT
AMERICAN VISCOSE DIVISION
FMC CORPORATION
MARCUS HOOK, PENNSYLVANIA 19061

Summary

This is the first part in a series of papers describing the application of GPC to studies of the viscose process. In this part the procedures used are described. For these studies THF solutions of nitrocellulose were used. Modifications were made to the standard nitration procedures, resulting in improved efficiency. The importance of sampling homogeneity was studied and precision was obtained. In addition, a computer program to handle the data was developed.

INTRODUCTION

Although chromatography has been used extensively to separate mixtures of organic compounds, only recently has it been used to separate polymers. In conventional liquid–liquid chromatographic systems, the separation is generally based on differences in the chemical affinity of the materials being separated for each of the liquid phases involved in the system. With large polymer molecules, the differences in affinity are so small that efficient separations cannot be achieved.

The idea that separations on the basis of size alone were possible in a chromatographic-type system was first suggested by the observation

* Presented at the ACS Symposium on Gel Permeation Chromatography sponsored by the Division of Petroleum Chemistry at the 159th National Meeting of the American Chemical Society, Houston, Texas, February, 1970.

that macromolecules (i.e., cellulose xanthate, lignin sulfonic acid, etc.) were not retained on ion exchange resins (1) of the polyvinyl benzene-styrene type even though they had ionic functional groups. Similarly, certain sugars, glycols, and other organic molecules with no ionic groups were held up or completely absorbed in ion exchange columns (2). Studies of the relation between the method of preparation of the resin and these phenomena led to the conclusion that small molecules could be retained in the interstitial portion of the resin bead and, conversely, total rejection of the molecule would occur if the molecular dimensions exceeded the dimension of the interstitial portion (3).

Experimentation then turned to studies of controlling the pore or interstitial size of polystyrene gels. This can be done by controlling the concentration of diluent during the cross-linking of the resin. By varying the amounts and nature of the diluent, it is possible to produce rigid cross-linked polymers with any desired interstitial dimensions (4). Calculated average pore size dimensions on commercially available materials of this type vary from 45 to 10^7 Å. Similar materials can also be prepared from polydextrans, polyacrylamide, etc. These materials will separate macromolecules on the basis of molecular size (5).

The mechanism of separation is not entirely clear. There is some evidence of swelling when the resin bead is placed in a solvent. This leads to the title "gel," although in the normal sense of the word it is not a gel. As the liquid phase containing dissolved polymer passes through a column filled with these swollen resin beads, the polymer molecules permeate into those parts of the beads not mechanically barred by size restrictions. The higher molecular weight molecules with their larger dimensions have a greater restriction as to their path, whereas the lower molecular weight molecules can permeate into the beads and are thus "held up" longer on the column. The result is a separation with the higher molecular weight material eluting from the column first. The pattern of retention resembles a logarithmic decay mechanism. For the separation to occur, the "gel" and solvent must be similar in polarity to avoid the possibility of selective association with either phase. Specific strong association between the polymer and the resin beads also must be avoided.

The usual chromatographic column system consists of a series of columns loaded with polymer beads of decreasing interstitial dimensions. Thus in the first column, all of the molecules may permeate into the beads, with the probability that some rejection of the higher molecular weight molecules occurs. In the next column, total rejection of the highest molecular weight molecules may occur with some rejection of intermediate molecular weights, and so on through the columns. The choice of the proper range of columns depends on the ranges of molecular dimensions being separated.

As in all liquid–liquid chromatographic systems, a suitable detector for the presence and amounts of polymer in the effluent of the column is a problem. One method which has gained wide acceptance uses a differential refractometer. Available commercial units are capable of indicating reproducibly a change in the refractive index of 10^{-7}. Even this remarkable sensitivity is not adequate unless a sufficient difference in refractive index exists between the solvent and solute. The generality which can be made is that a difference of at least 0.1 must exist, a situation which generally can be met with most polymers.

FRACTIONATION OF CELLULOSE

It has long been recognized that a knowledge of the molecular weight distribution is necessary for the complete characterization of cellulose. A large number of attempts have been made to secure this information on various cellulose derivatives. The methods generally have been those of fractionation into fractions of narrower molecular weight distribution, which are then characterized by the available methods, and the construction of the distribution curve using the weight and molecular weight of these fractions. For various reasons which will be discussed later, these results have been unimpressive. Almost all of the work on fractionation methods was done before 1945 with relatively little success. It was realized that a new approach to the problem had to be developed if any progress was to be made, owing to the inadequacies of the existing techniques.

The first major problem encountered with fractionation of cellulose is that it is soluble in a very limited number of solvents, and rapid degradation occurs in most of these. Those in which it is soluble with minimal degradation are complex mixtures—cuene, cadoxene, zincene, ferric tartrate complexes, etc. It is difficult to find cellulose nonsolvents which are miscible with these systems and which will result in a precipitation on a molecular distribution basis. A limited success has been obtained using glycerin or one of the glycols as the nonsolvent, but it is difficult to remove the precipitated fraction due to the high viscosity of the system. Turbidimetric techniques have been unsuccessful because the refractive index of the solvent-nonsolvent mixture is very close to that of cellulose, and it is very difficult to see the precipitated material. Many of the difficulties cited above can be eliminated if a soluble cellulose derivative is used, and most attempts to fractionate cellulose have been made using the nitrate derivative.

The use of the nitrate derivative is justified by the claim that *no* degradation occurs during nitration. This is questionable since the procedures used to study degradation of cellulose are relatively insensitive to small changes, and it is known that almost anything done to cellulose can result in degradation. The operations in the nitration which could lead to degradation are the mechanical opening of the sample and the dilution which occurs during washing to remove the acid.

The procedure for fractionation of cellulose which has gained the widest acceptance is to dissolve cellulose nitrate in ethyl acetate, acetone, or a similar solvent and then fractionate into approximately 30 fractions by precipitation using ethanol or water as the nonsolvent (6). As discussed previously, this leads to a series of distributions which are weighed and characterized by measurement of the intrinsic viscosity. These fractions can have a distribution almost as wide as that of the starting material. For this reason, several refractionations are usually run. The validity of the results obtained using this technique (7) is questionable. The method will show differences between two extremes, like pulp and a rayon, but is of little value when the differences are small.

Fractional solution techniques have been attempted on cellulose. These techniques, which involve successively leaching with solvents of increasing dissolving power, appear very attractive. Experimentally it has been found that equilibrium is reached very slowly and, in fact, may never be reached. There is no question that better results may be obtained using the fractional precipitation method.

A number of other possible fractionation methods have been considered. These and the techniques of fractional precipitation and solution have been reviewed by Schneider (8).

Since the introduction of gel permeation chromatography (GPC) in 1964 (5) as a rapid method of fractionating polymers, several papers describing the application of GPC to cellulose derivatives have appeared. Papers by Segal (9) and by Rinaudo and Merle (10) report the results of studies of nitration methods with the conclusion that the nitrating acid composition has little or no effect on the GPC fractionation. Detailed studies of calibration methods have been made by Segal (*11*), and Meyerhoff (*12*), and Huang and Jenkins (*13*). The idea of a universal calibration for GPC has received considerable attention, although in 1965 Meyerhoff (*14*) had concluded that the GPC behavior of polymers of different structure cannot be generalized on the basis of chain length. Huang and Jenkins (*13*) prepared samples for GPC calibration by precipitation fractionation of cellulose nitrate and gamma-ray irradiation of wood cellulose. Their calibration curve differed significantly from the curve derived using polystyrene standards.

Applications of GPC to cellulose and cellulose derivatives have been made by Alexander and Muller (*15*), Segal (*16*), Brewer and co-workers (*17*) and Rinaudo and Merle (*10*). In these papers the molecular weight distribution curves for cellulose and cellulose derivatives from different sources have been compared.

The objective of the research reported in this series of papers was to study the changes in the molecular weight distribution during the viscose process and in the use of the products. Therefore the concern was for reproducibility and significance of observed changes in the GPC curves, rather than measurement of the absolute distribution. The need for a data presentation format which was numerical instead of graphical was considered to be of prime importance. In addition, it was also desirable to reduce the personnel time per analysis to as low a level as possible. This paper describes the preparation of cellulose nitrate especially for GPC, an evaluation of the variable in the procedure, and a description of a computer program to handle the data.

EQUIPMENT

A Waters Gel Permeation Chromatograph Model 200 with the automatic sample injector was used for all the studies. Four high resolution columns containing Styragel of porosity 10^6, 10^5, 10^4, and 10^3 Å were used. The eluting solvent was tetrahydrofuran (THF) containing the normal level (0.1%) of stabilizer. The THF was not recycled. Since gelatination occasionally occurs in the sample loops, an extra 7 μ sintered steel filter has been added in the line between the sample loop and the front end of the first column to protect the columns.

METHODS

Nitrating Acid Preparation

The ASTM procedure (*18*) has been modified to eliminate some of the safety hazards, to permit more rapid mixing, and to avoid the preparation of excessive amounts of the acid. The actual procedure used is as follows.

a. The contents of a 1-lb bottle (453 g) of 90% nitric acid are poured carefully into a 2-liter Erlenmeyer flask.

b. The flask is placed in a 3-liter stainless steel beaker, and ice is packed around it.

c. The beaker with the flask is placed on a magnetic stirrer, and a Teflon stirring bar is added.

d. After cooling with stirring for at least 10 min, 187 g of phosphorus pentoxide is added very slowly. This step should take at least 30 min. Be careful not to add any P_2O_5 which shows evidence of being moist.

e. After the P_2O_5 addition is completed, the flask is stirred for at least 1½ hr, adding ice around the flask as necessary.

f. Transfer the nitrating acid to a 1-liter glass-stoppered reagent bottle, and refrigerate until use.

g. This mixture should be used with 48 hr. Any solid separating or noticeable darkening indicates that the acid should be discarded.

Cellulose Sample Preparation

If a pulp sheet is being used, shred or open in a Waring Blendor several grams of sample. If yarn or film, cut into quarter-inch pieces. Powders are used as is. The samples are dried overnight at 50° in small beakers or weighing bottles in an air oven.

Nitration

a. Weigh 50 mg of sample into a 100-ml weighing bottle.

b. Add 25 ml of ice-cold nitrating acid.

c. Place in an ice bath for 60 min.

d. Swirl at 10 min intervals to insure good mixing.

e. Rapidly *but carefully* pour the contents of the weighing bottle into a 60-ml coarse sintered filter funnel, sucking the acid off as rapidly as possible into a 2-liter suction flask.

f. Fill the weighing bottle with cool distilled water, and pour rapidly over the sample on the filter. Any additional sample adhering to the weighing bottle is rinsed into the funnel with additional water.

g. Wash the sample on the filter with 1 liter of distilled water, adding the water to the funnel with the suction off and then applying a vacuum to the suction flask to remove the water.

h. Place the sample on the sintered glass filter funnel in the oven at 50° and allow it to dry overnight.

i. If yields are desired, a tared filter may be used.

Solution Preparation

Weigh 50 mg of the nitrated cellulose on weighing paper and transfer to a 60-ml glass-stoppered Erlenmeyer flask. Add 50 ml of THF from the storage reservoir of the GPC instrument. For rapid solution, stir on a magnetic stirrer—otherwise allow to dissolve overnight with occasional shaking. Examine the solution for gel particles and discard if further mixing does not result in a good solution. Filter, using the pressure filter apparatus supplied by Waters Associates, and collect the sample in a 60-ml glass-stoppered bottle. Rinse the filter with THF before and after each sample to avoid contamination of samples.

Sample Injection

The syringe is rinsed with a small amount of the sample. Approximately 4 ml of the sample is drawn up into the syringe, and all air bubbles are removed. The solution is injected slowly into the sample loop, displacing the solvent which is in the loop. The 4 ml of sample is sufficient to wash and fill the loop. The sample advance button is pressed, and when the next loop is in position, the next sample is injected. This is continued until all six loops are filled.

STANDARD CURVE

The standard curve used for this study was prepared by nitration and chromatography of a series of pulp, yarn, and Avicel microcrystalline cellulose samples. The elution volume of the peak of the GPC curves for each sample was plotted against log $\overline{DP}$ values obtained on the samples by the viscometric method (ASTM). This curve is reproduced in Fig. 1. It was recognized that the samples were of relatively wide distribution and that this method is not an ideal way to establish the standard curve. Some samples showed evidence of

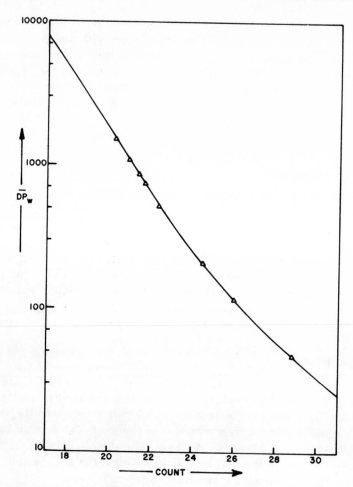

FIG. 1. GPC standard curve for cellulose.

being bimodal, and these values were not considered in plotting the curve. Weight-average DP calculated from GPC curves using this standard curve is in the same order of magnitude as that obtained by the viscometric method (Table 1).

DATA HANDLING

Like any other chromatographic curve representing the separation of a mixture of components, visual comparison of GPC curves is

TABLE 1
Comparison of Weight-Average DP Using GPC and
$\overline{\mathrm{DP}}$ as Determined Viscometrically

	$\overline{\mathrm{DP}}_w$	
Sample	GPC	Viscometric
A	1173	1040
B	1168	1045
C	1046	970
D	1011	890
E	934	755
F	903	740
G	789	660
H	654	600
I	641	535
J	498	350
K	215	245
L	178	200

qualitative. Quantitative information about the molecular weight distribution and any changes that may result during processing can be obtained from a detailed statistical analysis of the curves. This analysis was accomplished using a computer program originally devised by Chevron which has been extensively modified. The parameters calculated include:

the average molecular weights calculated as the first moment, $\mu^{(1)}$, of the particular distribution functions, $N(M)$, $Z(M)$, etc.

$$\mu^{(1)} = \bar{M}_x = \frac{\Sigma N_i M_i^{(x+1)}}{\Sigma N_i M_i^x} \qquad \text{when } x = 0,\ \bar{M} = \bar{M}_n$$
$$1,\ \bar{M} = \bar{M}_w$$
$$2,\ \bar{M} = \bar{M}_z$$
$$3,\ \bar{M} = \bar{M}_z + 1$$

the polymolecularity functions,

$$U_x = \frac{\bar{M}_{(x+1)}}{\bar{M}_x} - 1$$

and the distribution width indices

$$\sigma_x = \sqrt{M_x M_{(x+1)} - M_x^2}$$

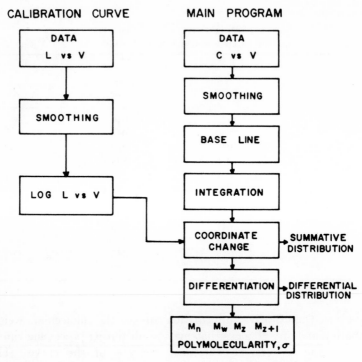

FIG. 2. Schematic of computer program for handling GPC data.

The analysis also included calculation of the integral weight curve and weight and number differential curves for the sample.

The general outline of the program is given in Fig. 2.

Calibration data are entered in a separate program which uses a standard Lagrangian method to smooth the curve, expressing it in the standard form log DP vs. peak elution volume. The compiled data is stored in an out file where it is available for use in the main program.

In the main program, data are entered as the intensity of the curve as a function of the peak elution volume. After smoothing, corrections are made by subtracting the baseline, set as a straight line joining the two extremes of the curve at which points the concentration is zero. These points are also taken as integration limits, and the area beneath the curve is calculated. Coordinates are changed, DP replacing peak elution volume and concentration replacing the intensity. A cumulative distribution curve is then calculated by integration of the curve with the total concentration set at unity. Differentiation of the integral curve yields weight distribution data from which the various

CN. NO.   892
BASELINE SET AT COUNT 38
ALL VALUES AFTER BASE LINE SETTING HAVE BEEN SET AT 0

MN	MV	MW	MZ	M(Z+1)	RED AREA
188.844	336.704	347.291	651.0	1233.4	1.3225

UN	UW	UZ	SIGMA N	SIGMA W	SIGMA Z
0.839	0.874	0.895	173.0	324.7	615.7683

MOL WT	CUM AMT	D AMT/DM	DN/DM
23.1	0.0000	-0.000030	-130.948
26.9	-0.0000	0.000071	264.746
31.4	0.0010	0.000382	1216.623
36.6	0.0041	0.000832	2273.060
42.7	0.0106	0.001259	2948.094
49.8	0.0207	0.001556	3125.874
58.1	0.0345	0.001765	3040.173
67.7	0.0524	0.001950	2880.899
78.9	0.0753	0.002116	2681.682
92.0	0.1039	0.002258	2453.464
107.3	0.1393	0.002389	2227.077
125.1	0.1827	0.002493	1992.946
145.8	0.2347	0.002509	1720.372
170.0	0.2946	0.002436	1432.382
198.3	0.3617	0.002304	1162.274
231.2	0.4351	0.002129	921.058
269.5	0.5129	0.001900	704.755
314.3	0.5921	0.001624	516.719
366.4	0.6698	0.001320	360.377
427.2	0.7406	0.001024	239.799
498.1	0.8044	0.000757	151.954
580.8	0.8565	0.000535	92.055
677.2	0.8996	0.000357	52.652
789.5	0.9311	0.000226	28.661
920.6	0.9544	0.000136	14.754
1073.3	0.9701	0.000079	7.347
1251.4	0.9807	0.000044	3.503
1459.1	0.9874	0.000024	1.624
1701.3	0.9916	0.000012	0.732
1983.6	0.9943	0.000007	0.351
2312.8	0.9961	0.000004	0.193
2696.6	0.9975	0.000003	0.113
3144.1	0.9986	0.000002	0.064
3665.8	0.9994	0.000001	0.030
4274.2	0.9999	0.000000	0.010
4983.4	1.0000	0.000000	0.001
5810.5	1.0000	-0.000000	-0.000
6774.7	1.0000	-0.000000	-0.000
7899.0	1.0000	0.000000	0.000
9209.8	1.0000	0.000000	0.000
10738.1	1.0000	0.000000	0.000
12520.2	1.0000	0.000000	0.000
14597.9	1.0000	0.000000	0.000
17020.4	1.0000	0.000000	0.000

TOTAL NO. OF MOLECULES PR UNIT WT.=   531419.

LINE   100:   END OF DATA

FIG. 3. Typical output of GPC program.

average molecular weights are calculated. Distribution width indices and polymolecularities are then calculated. Finally the total number of molecules in unit weight is obtained by integrating the number distribution curve.

Usually the standard calibration data is entered and retained within the saved program. The values of the retention volume are also entered and retained. Thus it is only necessary to enter for each sample, the sample number and the intensity values corresponding to the previously entered peak elution volumes.

An illustration of the typical output is given in Fig. 3. This is a complete statistical description of the DP distribution, cumulative weight, and number differential. Curves can be plotted from the data either manually or by using a standard computer plotting routine.

RESULTS AND DISCUSSION

Homogeneity of Samples

In considering the use of GPC in studies of the viscose process it was recognized that sample uniformity would be a problem. With cellulose, one is dealing with a polymer, the composition of which is determined by the wood source and production conditions.

Usual pulping methods involve as nearly complete dispersion as possible in large quantities of water and extensive mixing prior to laying down the pulp sheet. This would be expected to homogenize the product so that no significant variation exists across the pulp sheet. A commercial pulp sheet was cut into 2-in. squares from which random samples were nitrated and chromatographed. The molecular weight distribution curves from this experiment were laid out on a "mock-up" of the original pulp sheet in the locations from which the samples originated. The results are given in Fig. 4. It can be observed that the changes in distribution within the sheet are probably gradual rather than abrupt (the appearance of the trimodal curve in the middle portion of the sheet has distributions near it, also showing evidence of trimodal distribution. The bimodal nature of the distribution changes gradually, etc.). It should be noted that the $\overline{DP}$ as determined viscosimetrically and the calculated weight-average molecular weight from these GPC curves gave no indication of the degree of nonuniformity in this pulp sheet. These results are given in Table 2. Similar results were obtained on a second pulp sheet. The sampling technique which has been adopted is to take as large a sample as possible, even

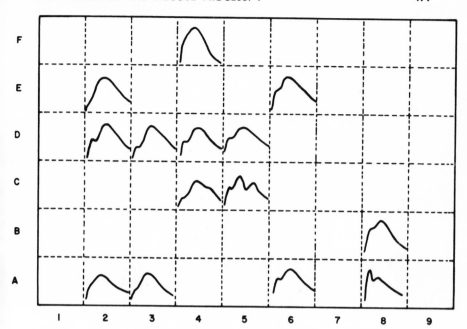

FIG. 4. Differential distribution curves at different points on the pulp sheet. Ordinate is intensity and abscissa is log DP.

TABLE 2

Weight-Average DP and Viscometric $\overline{DP}$ on Individual Segments of Pulp Sheet

$\overline{DP}$	$\overline{DP}_w$
1052	1111
1045	1056
1069	1006
1071	1002
1046	1092
1037	1044
1047	1118
1070	1134
1047	1169
1035	955
	1072
	1114
	1133
	1145

a whole pulp sheet, open it in a Waring Blendor, and then sample from the homogenized sample.

The normal method of baling pulp is to use alternate lots for alternate sheets in the bale. Although no significant differences were observed in the DP distribution between alternate sheets in five bales of different pulps (see Table 3), it is possible that widely different

TABLE 3

DP Distributions in Alternate Sheets in Pulp Bales

Pulp		$\bar{M}_n$	$\bar{M}_w$	$\bar{M}_z$	$\bar{M}_{z+1}$
I	A	306	1118	2056	3045
	B	362	1162	2056	3041
II	A	446	1309	2260	5383
	B	472	1346	2358	3578
III	A	251	1027	2187	3483
	B	261	1006	2189	3624
IV	A	348	1348	2483	3377
	B	355	1417	2527	3490
V	A	303	1243	2317	3305
	B	224	1194	2335	3319
σ between back-to-back sheets		32	34	41	84

distributions could exist in alternate sheets if basic DP were the only factor considered in the blending operation. Consequently this should be considered when sampling from a bale of pulp, i.e., the sample should be taken from pairs of adjacent sheets selected at random. There were significant variations in the molecular weight distribution of pulp lots manufactured over a period of time, as illustrated in Table 4.

The processing used in the manufacture of viscose and in the final

TABLE 4

DP Distributions of Different Pulp Lots from the Same Manufacturer

Variable	Mean	Standard deviation	Standard error	Range
$\bar{M}_n$	536	34	15	86
$\bar{M}_w$	1278	45	20	120
$\bar{M}_z$	2232	125	55	280
$\bar{M}_{z+1}$	3190	251	112	639

product results in drastic changes in the molecular weight distribution of the cellulose. Prior to actual solution of the xanthated crumb, the process is entirely heterogeneous and is particularly subject to non-uniform processing conditions. For example, exposure to oxygen is not uniform during the aging of alkali crumb, temperature variations exist within the steeping tank or aging bins, etc. Consequently the change in distribution will not be uniform throughout the batch. In mixing viscose, however, the distribution nonuniformities are averaged out. These factors must be considered in both experimental design and sampling.

PRECISION OF DATA

The precision of the GPC method when used to determine the molecular weight distribution of cellulose is related to nitration, solution preparation, injection, instrument operation, and data transfer. Each of these variables was considered when establishing the precision of the whole procedure.

Examination of the calculation methods indicated that $\bar{M}_n$ and $\bar{M}_{z+1}$ were most sensitive to small changes in the data at the extreme ends of the GPC curve; $\bar{M}_w$ should be relatively sensitive only to shifts in the maximum peak position and gross changes in the extreme ends of the curve. Precision of these parameters thus represents a reasonable indication of the precision of the whole GPC method.

For the precision study, large samples of commercial pulp and rayon were homogenized, the pulp being opened and mixed in a Waring Blendor and the rayon being cut into quarter-inch pieces with scissors and then blended in the Blendor. Samples of 50 mg were taken from these samples, nitrated, and chromatographed over several weeks time. The data was calculated using the computer program and then statistically analyzed. The data for the two types of samples are compiled in Table 5. There is no significant difference in the precision figures for the two types of samples. Comparison of these data with those obtained by repetitive injections of samples from the same solution suggested that approximately half of this error was related to the analytical sampling, nitration, and solution steps.

The precision figures given in Table 5 indicate that a single sample is adequate if major differences are expected between samples

TABLE 5

Statistical Analysis of GPC Data from Pulp and Rayon Samples

	Variable	Mean	Standard deviation	Standard error	Range
Pulp	$\bar{M}_n$	454	16	5	43
	$\bar{M}_w$	1145	42	13	107
	$\bar{M}_z$	2118	68	22	200
	$\bar{M}_{z+1}$	3050	86	27	256
Rayon	$\bar{M}_n$	293	5	2	14
	$\bar{M}_w$	595	18	7	49
	$\bar{M}_z$	1296	66	27	183
	$\bar{M}_{z+1}$	2357	192	78	541

under study. If only small differences exist, multiple samples must be run, and calculated standard deviations or errors must be used to establish the significance of the differences from the data.

REFERENCES

1. O. Samuelson, *Svensk Papperstidn.,* **46,** 583 (1943).
2. F. H. Yorston, *Pulp Paper Mag. Can.,* **50**(12), 108 (1949).
3. O. Samuelson, *Ion Exchangers in Analytical Chemistry,* Wiley, New York, 1953.
4. W. G. Lloyd and T. Alfrey, *J. Polymer. Sci.,* **62,** 159, 301 (1962).
5. J. C. Moore, *J. Polymer. Sci.,* **A2,** 835 (1964).
6. J. N. Bronsted, *Z. Phys. Chem., Bodenstein-Festband,* **1931,** 257. See also, E. Ott, H. M. Spurlin, and M. W. Grafflin, *Cellulose,* 2nd ed., Part III, Wiley (Interscience), New York, 1955, p. 1176.
7. T. E. Timell and E. C. Jahn, *Svensk Papperstidn.,* **54,** 831 (1951).
8. N. S. Schneider, *J. Polymer. Sci., Part C,* **8,** 179 (1965).
9. L. Segal, J. D. Timpa, and J. I. Wadsworth, *J. Polymer. Sci., Part A-1,* **8,** 25–35 (1970).
10. M. Rinaudo and J. P. Merle, *Eur. Polym. J.,* **6,** 41–50 (1970).
11. L. Segal, *J. Polym. Sci., Part B,* **5,** 495 (1967).
12. G. Meyerhoff, *Makromol. Chem.,* **89,** 282 (1965).
13. R. V. M. Huang and R. G. Jenkins, *Tappi,* **52,** 1503–1507 (1969).
14. G. Meyerhoff and S. Jovanovic, *J. Polym. Sci., Part B,* **5,** 495–499 (1967).
15. T. E. Muller and W. J. Alexander, *J. Polym. Sci., Part C,* **21,** 283 (1968).
16. L. Segal, *J. Polym. Sci., Part C,* **21,** 267 (1968).
17. R. J. Brewer, L. J. Tanghe, S. Bailey, and J. T. Burr, *J. Polym. Sci., Part A-1,* **6,** 1697–1704 (1968).
18. *Cellulose Chain Length Uniformity by Fractional Precipitation of Cellulose Nitrate,* ASTM designation D1716-62.

Application of GPC to
Studies of the Viscose Process.
II. The Effects of Steeping and Alkali-Crumb Aging*

JOHN DYER and L. H. PHIFER

RESEARCH AND DEVELOPMENT
AMERICAN VISCOSE DIVISION
FMC CORPORATION
MARCUS HOOK, PENNSYLVANIA 19061

Summary

Changes in the distribution of degree of polymerization (DP) of cellulose during the steeping and alkali-crumb aging steps of the viscose process have been studied. A dissolving-grade pulp was used, and two mechanisms of degradation—oxidation and hydrolysis—were considered. Alkaline hydrolysis was relatively unimportant during the steeping, the major change in the DP distribution being a result of the much faster oxidative degradation. The extent of oxidation was determined by the solubility of oxygen in the alkali solution. Both oxidation and hydrolysis cause degradation during the aging of alkali crumb. Under the conditions used, hydrolysis was a slow, pseudo-first-order reaction, and oxidation a much faster zero-order reaction. Since there was a large difference in the rates of these reactions, the kinetics of degradation approached zero order.

INTRODUCTION

The production of fibers and film from cellulose by the viscose process is a multistep operation. Cellulose is converted to an alkali-soluble derivative by steeping pulp in aqueous sodium hydroxide and after aging the sodium cellulosate (alkali crumb), reacting it with

* Presented at the ACS Symposium on Gel Permeation Chromatography sponsored by the Division of Petroleum Chemistry at the 159th National Meeting of the American Chemical Society, Houston, Texas, February, 1970.

carbon disulfide. The derivative (sodium cellulose xanthate) is dissolved, yielding a viscous solution (viscose). After aging, the viscose is extruded into an acid bath where the cellulose is recovered by coagulation and decomposition of the derivative.

The distribution of degree of polymerization (DP) is a variable which changes during the viscose process. Previously, the nature and extent of the changes had been inferred from measurements of the basic DP. Information about the DP distribution could only be obtained by combining several tedious determinations, e.g., osmometry, viscometry, and ultracentrifugation, or from complicated fractionation techniques. Gel permeation chromatography (GPC) affords a rapid method of studying these changes and evaluating the importance of this variable. The GPC method and its limitations have been described in Part I of this paper (1). In this Part II, the changes in DP distribution during steeping of pulp and aging of alkali crumb are considered.

The main objectives of the steeping process are (a) to convert the cellulose to a more reactive sodium cellulosate and (b) to remove the alkali-soluble material. In commercial practice, this is achieved either by a batch process where the pulp sheets are steeped in large steep presses or by a continual slurry steeping. An 18% aqueous sodium hydroxide solution is usually used in both systems.

Aging of alkali crumb results in a reduction of the basic DP. This is necessary because the solubility of the cellulose derivative is an exponential function of the reciprocal DP. The aging is usually carried out in large bins at controlled temperatures.

It is recognized that oxidation and hydrolysis cause degradation in both steeping and alkali crumb aging. Although there have been many studies of these process steps (2), the mechanisms of degradation are not understood.

RESULTS AND DISCUSSION

Steeping

To interpret the changes of DP distribution in cellulose during steeping, the chemical and physical reactions of cellulose in the presence of alkali solution and the design of the experiments were considered.

It has been established by x-ray diffraction that when cellulose is steeped in alkali, changes in its structure occur (3). Some of the possible transitions are outlined in Scheme 1. The changes are related to the alkalinity and temperature. In Scheme 1, there are nine forms of cellulose which may behave differently.

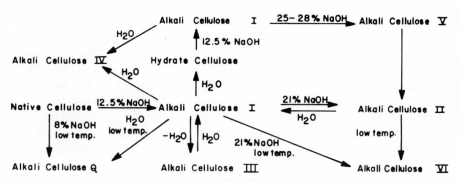

SCHEME 1. Some transitions of cellulose in alkali.

In addition, heat is evolved during steeping (4). The heat effect ΔH_T can be separated into parts according to its origin.

(a) Heat of wetting ΔH_a
(b) Heat of reaction (formation of alcoholate) ΔH_b
(c) Heat of possible lattice transition ΔH_c
(d) Heat of other reactions—degradation, ΔH_R
     oxidation, etc.

$$\Delta H_T = \Delta H_a + \Delta H_b + \Delta H_c + \Delta H_R$$

The heat of steeping, approximately 2 cal/g with 18% NaOH, may cause local temperature variation of several degrees in the system. Dissipation of this heat to the steep liquor is a variable influenced by rate of filling the steep tank and packing of the pulp sheets, etc.

During steeping, the pulp sheets swell and absorb the alkaline solution (5). The amounts of swelling and absorption increase to a maximum at 18% sodium hydroxide. Above 18%, the swelling decreases slightly and is then unaffected by further increase in the alkalinity. The amount of alkali absorbed behaves similarly.

An initial drop in the basic DP is observed when pulp is steeped in alkali. This is most probably due to an oxidative degradation. It will thus depend on the concentration of oxygen both in the steep liquor and adsorbed on the pulp sheets. Although the latter factor should be similar for each pulp batch, the former depends on the alkalinity, the solubility of oxygen in alkali decreasing with increased alkalinity and temperature (6). After the initial drop in $\overline{DP}$, the degradation proceeds more slowly, and here the mechanism probably involves alkaline hydrolysis.

The removal of alkali-soluble material, hemicelluloses, and degradation products from the pulp sheets is related to the solubility of the material in the steep liquor and diffusion of this material into the steep. The solubility of cellulose in alkali is exponentially related to the reciprocal $\overline{DP}$. It is also dependent on the alkali concentration, being greatest around 10% NaOH (7). With commercial batch steeping, the extraction of alkali-soluble material is diffusion controlled, and the amount of material removed will be influenced by the temperature, alkalinity, nature of the pulp, time, etc.

Effect of Steep Alkali Concentration. To determine if alkali hydrolysis was a significant factor during steeping, a dissolving-grade pulp was immersed in aqueous sodium hydroxide solution of different dilutions. Five concentrations (6, 12, 18, 24, and 30%) were used. Each solution was saturated with nitrogen by bubbling the gas through the solution for 1 hr at room temperature to minimize oxidative degradation during the experiment.

One-gram samples of shredded pulp were steeped for 10 min in 50 ml of the alkali. After steeping, either the alkali crumb was washed with fresh portions of the original alkali to remove alkali-soluble material, or excess acetic acid was added to the mixture, neutralizing the alkali and precipitating the alkali-soluble material. Table 1 shows the average molecular weights after steeping.

The most significant change is the decrease (approx 10%) in the

TABLE 1

Influence of Steep Alkali Concentration on DP Distribution

% NaOH	$\bar{M}_n$		$\bar{M}_w$		$\bar{M}_z$		$\bar{M}_{z+1}$	
	NaOH	HOAc	NaOH	HOAc	NaOH	HOAc	NaOH	HOAc
6	455	429	978	1063	1854	2004	3000	3178
12	513	441	1056	1042	1823	2109	2818	3836
18	488		1074	1045	2012	1892	3596	3032
24	462	415	1081	1004	1836	1955		3456
30	416	416	1051	1013	2060	1858	3765	2877
Pulp	454		1145		2118		3050	

weight-average molecular weight from the pulp to the alkali crumb. This is attributed to oxidative degradation as a result of air in the pulp.

There is no evidence of any difference in the alkaline hydrolysis during the short steeping, as the $\bar{M}_n$, $\bar{M}_w$, and $\bar{M}_z$ values in the acetic-acid precipitated samples are unaffected by the alkali concentration. Extraction of the alkali-soluble material by NaOH affected the value of $\bar{M}_n$ only to an extent related to the solubility of cellulose in the alkali. There were significant differences for the number-average molecular weights, dependent on both the alkalinity of the steep and the method of sample preparation. The highest value of $\bar{M}_n$ was obtained at 12% NaOH consistent with the solubility characteristics discussed earlier.

Effect of Steeping Time. To determine the effect of steeping time on the DP degradation, only one alkali concentration was used. No attempt was made to remove oxygen from the system. The results obtained for different steeping time are given in Table 2. After the initial drop in basic DP—to an extent probably related to the consumption of oxygen in the system—degradation proceeds very slowly. Rates of hydrolysis could not be calculated because the changes observed were within the precision limits of the method.

TABLE 2

Influence of Steeping Time on DP Distribution (18% NaOH at 25°C)

Min	$\bar{M}_n$		$\bar{M}_w$		$\bar{M}_z$		$\bar{M}_{z+1}$	
	NaOH	HOAc	NaOH	HOAc	NaOH	HOAc	NaOH	HOAc
5	457	439	954	961	1783	1738	3213	2711
10	541	379	1034	901	2001	1756	3879	2925
60	504	402	1093	908	2037	1759	3437	2986
120	395	322	913	976	1933	2175	3770	3694

Alkali Crumb Aging

The largest change of basic DP and distribution of DP occurs during the aging af alkali crumb. The extent of degradation is determined by the temperature, alkali concentration, oxygen content, and handling. Normal commercial practice is to hold these constant and vary the time to attain a specific basic DP.

Effect of Aging Time. To separate the effects of oxidation and hydrolysis, the reduction of DP and the DP distribution in alkali crumb was studied in oxygen and nitrogen atmospheres. To prepare the alkali crumb, the pulp was shredded in a Waring Blendor and then steeped in 18% aqueous sodium hydroxide for 10 min. The alkali crumb was transferred to a sintered-glass filter funnel, and excess steep liquor removed by suction. The crumb was again shredded in the Waring Blendor to ensure uniformity and divided into two equal portions. The batches were placed in 1 liter Erlenmeyer flasks sealed with rubber stoppers carrying two 7 mm i.d. glass tubes. The contents of the flasks were then aged at room temperature (25°C) for 10 days. A flow of oxygen was passed through one of the flasks, and a flow of nitrogen through the other. The flow rate was approximately 1 ml/5 min. To prevent loss of moisture from the crumb during aging, the gases were bubbled through distilled water before entering the flasks. Periodically, samples were removed for analysis. Although in the presence of oxygen the degradation was faster, the alkali crumb remained white. Under nitrogen, the alkali crumb became yellow, eventually turning brown. The DP distribution results are given in Table 3. Many more low DP molecules are formed in the oxygen

TABLE 3

DP Distribution for Alkali Crumb Aged at 25°C in Oxygen and Nitrogen

Age (days)	$\bar{M}_n$		$\bar{M}_w$		$\bar{M}_z$		$\bar{M}_{z+1}$		Viscometric DP	
	N_2	O_2	N_2	O_2	N_2	O_2	N_2	O_2	N_2	O_2
0	389	330	1015	1008	2151	2264	3849	3926	1040	1045
1	396	375	936	758	1768	1504	2882	3208	970	755
2	371	304	843	567	1545	913	2444	1387	870	535
4	479	184	800	337	1382	563	2588	844	740	350
7	348	151	656	215	1133	302	2057	404	660	245
10	330	125	602	178	981	248	1430	329	600	200

environment, and with 4 or more days' aging, the DP distribution was extremely narrow.

Integral or cumulative amount curves for the samples aged in nitrogen are shown in Fig. 1. The degradation was slow. Initially, 30% of the sample had a DP greater than 1200, and 12% was less than 200. After 10 days at 25°C, there was still 11% above DP 1200 and only 15% below DP 200. The differential curves for these samples

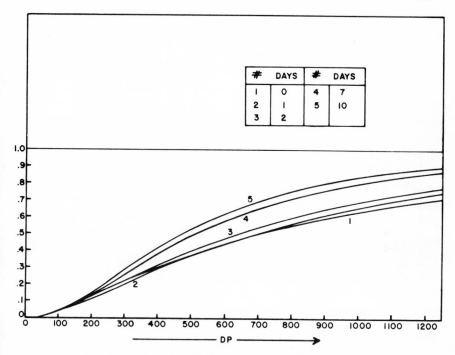

#	DAYS	#	DAYS
1	0	4	7
2	1	5	10
3	2		

FIG. 1. Integral curves, alkali crumb aged at 25°C in nitrogen.

are illustrated in Fig. 2. The distributions are all relatively broad, although the shoulder due to low-DP material present in the initial sample is gradually obscured by the formation of more small molecules by degradation.

Although there are significant differences in the distribution as a result of aging the alkali crumb under nitrogen, it is not possible to separate the effects of random scission and end clipping (end-group attack). However, as the greatest change in the differential curve was between DP 100 and DP 700, the sample probably originated from a random scission of the cellulose molecules.

In oxygen, the degradation was much faster. The integral curves (Fig. 3) show that although initially 30% of the sample had a DP above 1200 and 12% below 200, after 10 days at 25°C there were no molecules present with a DP above 700, and 68% of the sample had a DP below 200. This is reflected in the differential curves (Fig. 4); after 7 days, the distribution was narrow. The distribution shown in these curves may be the result of random scission and/or end clipping.

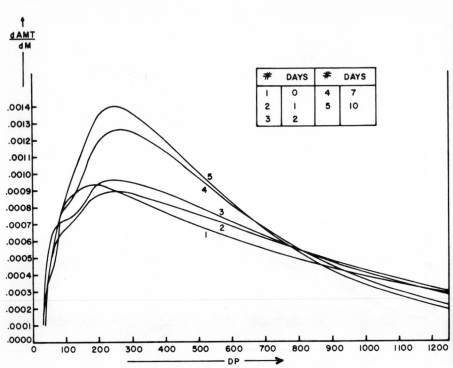

FIG. 2. Differential curves, alkali crumb aged at 25°C in nitrogen.

It must be emphasized at this point that the GPC data is only representative of the distribution in the sample as injected into the GPC equipment. It is possible that low-DP material is lost during sampling, handling, and preparation. This would be relatively unimportant if only the change in basic DP were used—it would not affect $\bar{M}_z$ or $\bar{M}_{z+1}$ and have only a slight effect on $\bar{M}_w$—there would, however, be a significant effect on $\bar{M}_n$. The effects of end clipping and random scission can only be separated if the low DP degradation products are retained in the sample.

There are at least two chemical reactions that will result in degradation of the DP; hydrolysis and oxidation. When oxygen is excluded from the system, only hydrolysis occurs. A semilogarithmic plot of the number-average DP as a function of the aging time under nitrogen was linear. Although the reaction is second order, under the conditions used it was pseudo-first order.

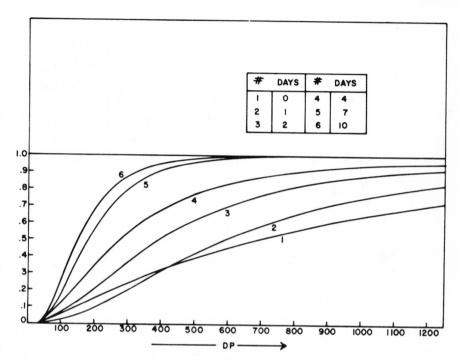

#	DAYS	#	DAYS
1	0	4	4
2	1	5	7
3	2	6	10

FIG. 3. Integral curves, alkali crumb aged at 25°C in oxygen.

In the presence of oxygen, it may be assumed that both oxidation and hydrolysis occur. The rate of oxidation is related to the solubility of oxygen in sodium hydroxide and will be almost independent of the oxygen concentration in the atmosphere around the alkali crumb. The oxidation can be represented as a reaction of zero order. Thus kinetics of degradation is the result of two competing reactions, one zero order and one second order. Hydrolysis, the second-order reaction, is slow compared to oxidation. Consequently, a zero-order plot of the data will only approach linearity as shown in Fig. 5. Results obtained for the reciprocal of the weight-average DP of the cellulose after aging the alkali crumb under nitrogen are shown for comparison. The changes are much smaller, and this curve also approaches linearity.

To obtain data for the third curve (air) in Fig. 5, alkali crumbs were aged in a sealed can and exposed to the air only when samples were removed. This represents a situation where the available oxygen was consumed and then replenished when the can was opened. The

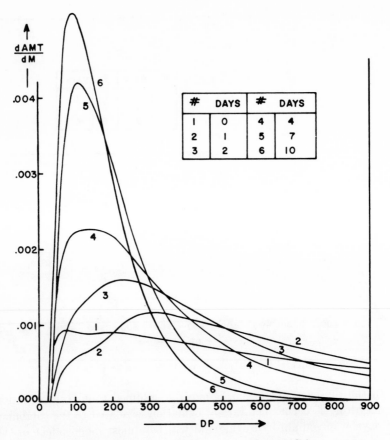

#	DAYS	#	DAYS
1	0	4	4
2	1	5	7
3	2	6	10

FIG. 4. Differential curves, alkali crumb aged at 25°C in oxygen.

reciprocal weight-average DPs were more scattered and were intermediate to those obtained in oxygen or nitrogen atmospheres.

CONCLUSIONS

Gel permeation chromatography can be used to study the distribution of DP of cellulose during the steeping and alkali-crumb aging
steps of the viscose process. Oxidation and hydrolysis are the two
mechanisms of degradation with oxidation being much faster. During
steeping, the extent of oxidation is determined by the amount of
oxygen dissolved in the system. After consumption of the available
oxygen, steeping time is relatively unimportant.

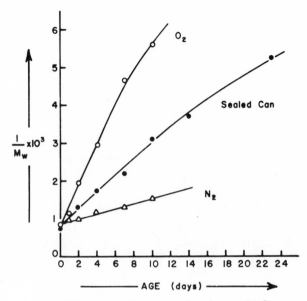

FIG. 5. Effect of alkali crumb aging time at 25°C.

In alkali-crumb aging, hydrolysis is by a pseudo-first-order reaction and oxidation by a zero-order reaction. As there is a large difference in the rates of these reactions, the kinetics of degradation under normal conditions will approach zero order.

REFERENCES

1. L. H. Phifer and J. Dyer, *Separ. Sci.,* **6**, 73–88 (1971).
2. M. M. Chochieva, N. I. Nikitin, and Yu. V. Brestkin, *Cellulose Chem. Tech.,* **1**, 313–326 (1967). A. Kantouch, and J. Meybeck, *Bull. Inst. Text. Fr.,* **23**, 189–202, 203–212, 213–222 (1969). P. Barthel and B. Philipp, *Faserforsch. Text.,* **18**, 266–273 (1967). A. M. Dowell and K. A. Kuiken, *Tappi,* **46**, 723–730 (1963). D. M. MacDonald, *Tappi,* **48**, 708–713 (1965). D. Entwistle, E. H. Cole, and N. S. Wooding, *Text. Res. J.,* **19**, 527–546, 609–624 (1949).
3. H. Sobue, *J. Soc. Chem. Ind., Japan,* **43**, B24 (1940). K. Hess and C. Trogus, *Z. Phys. Chem.,* **B4**, 321 (1929); *Ibid.,* **B11**, 381 (1930). G. Von Susich and W. W. Wolff, *Z. Phys. Chem.,* **B8**, 221 (1930).
4. B. G. Ranby, *Acta Chem. Scand.,* **6**, 101 (1952).
5. E. Heuser and R. Bartunek, *Cellulose Chem.,* **6**, 19 (1925).
6. A. Kantouch and J. Meybeck, *Bull. Inst. Text. Fr.,* **23**, 189–202 (1969).
7. G. F. Davidson, *J. Text. Inst.,* **27**, T112 (1936); *Ibid.,* **25**, T174 (1934).

Gel Permeation Chromatography Calibration. I. Use of Calibration Curves Based on Polystyrene in THF and Integral Distribution Curves of Elution Volume to Generate Calibration Curves for Polymers in 2,2,2-Trifluoroethanol*

THEODORE PROVDER,† JAMES C. WOODBREY, and JAMES H. CLARK

MONSANTO COMPANY
ST. LOUIS, MISSOURI 63166

Summary

A general method is proposed for obtaining gel permeation chromatographic (GPC) molecular weight (MW) and hydrodynamic volume (HDV) calibration curves for polymer–solvent systems where primary polymer standards are unavailable. The method is demonstrated by using a HDV calibration curve based on polystyrene in tetrahydrofuran (THF), in conjunction with integral distribution curves of elution volume for poly(methyl methacrylate) (PMMA) in THF and in 2,2,2-trifluoroethanol (TFE) for the generation of a HDV calibration curve in TFE. Transformation methods for generating secondary MW calibration curves from HDV calibration curves are discussed and applied to PMMA in THF and TFE, and to poly(trimethylene oxide), poly(vinyl acetate), and certain polyamides in TFE. The utility and reliability of the secondary calibration curves are demonstrated by comparing MW

* Presented at the ACS Symposium on Gel Permeation Chromatography sponsored by the Division of Cellulose, Wood and Fiber Chemistry at the 159th National Meeting of the American Chemical Society, Houston, Texas, February 1970.

† Present address: Dwight P. Joyce Research Center, Glidden-Durkee Division, SCM Corporation, 16551 Sprague Road, Strongsville, Ohio 44136.

averages and intrinsic viscosities obtained by GPC and by the classical methods. Molecular structural differences among the polyamide samples associated with the distribution of short- and long-chain branches are discussed in relation to their secondary calibration curves.

INTRODUCTION

The solvent most commonly used for the gel permeation chromatography (GPC) characterization of polyamides has been m-cresol. The solvent 2,2,2-trifluoroethanol (TFE) also is a good solvent for polyamides and has many more desirable properties than m-cresol. The differential refractive index of poly(methyl methacrylate) (PMMA) and polyamide polymers is greater in TFE than in m-cresol. In order to obtain the equivalent recorder response at a sensitivity of $4\times$ for a refractometer with a 0.004-in. slit, polymer concentrations >0.3 w/v-% must be used in m-cresol compared to concentrations <0.1 w/v-% in TFE. The high concentrations that must be used in m-cresol can cause column overloading and increase chromatogram peak spreading due to dispersion, skewing, and flattening effects ($1, 2$). The higher operating temperature, $>100°C$, required for the highly viscous m-cresol, compared to the $50°C$ operating temperature for TFE, has been shown to lead to polymer degradation ($3, 5$). TFE does not degrade polyamides at $50°C$. m-Cresol is subject to oxidative degradation and has additional annoying low-molecular-weight impurity peaks (4) which contribute to baseline instability and interfere with the normal chromatogram. TFE has only the normal air and water peaks. The water peak in TFE is controllable by solvent and sample drying techniques to be discussed later. Unlike m-cresol, TFE does not burn the skin and is less toxic than tetrahydrofuran (THF). However, with all these advantages, the one main disadvantage connected with using TFE as a GPC solvent is the insolubility of the readily available characterized anionic polystyrene (PS) standards. This insolubility prevents the generation of primary and secondary calibration curves in TFE.

In this paper a method will be presented that removes this difficulty. This method makes use of the hydrodynamic volume (HDV) calibration curve in THF constructed from PS standards, and integral distribution curves of elution volume in THF and TFE for uncharacterized PMMA samples, for the generation of a HDV calibration curve in TFE. Two methods will be presented for the construction of molecular weight calibration curves from a HDV calibration curve.

These methods will be applied to PMMA in THF and PMMA, poly-(vinylacetate) (PVAC), certain polyamides, and poly(trimethylene oxide) (PTMO) in TFE.

MATERIALS AND METHODS

Samples

Fourteen PMMA samples were prepared by routine free-radical bulk and solution polymerization methods (6). These samples covered a wide molecular weight range. A blend of these samples was fractionated on a Waters Associates Ana-prep GPC in THF with a Styragel column having a nominal porosity of 10^4 Å. Seven useful fractions were obtained and denoted as C, D, F, G, H, I, and J. The baseline-adjusted elution volume curves of these fractions in TFE are shown in Fig. 1. The conditions under which these and other GPC

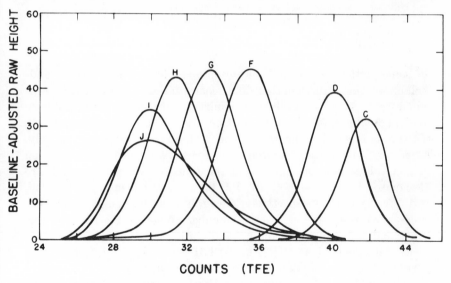

FIG. 1. Baseline-adjusted chromatograms of PMMA Fractions C, D, F, G, H. I. and J.

curves were obtained are described later in this paper. A more detailed report describing the preparation and Ana-prep fractionation of the PMMA blend, and molecular structure characterization of the resulting fractions will be published later (7).

Anionically polymerized caprolactam samples were prepared according to methods described by Gechele and Stea (8), and by Greenley, Stauffer, and Kurz (9). The synthesis conditions of these experimental polycaprolactam (PC) samples, which are designated as PC-4, PC-5, PC-6, PC-7, and PC-8, are shown in Table 1. Fractions

TABLE 1

Synthesis Conditions of Experimental Polycaprolactam Samples

Sample designation	Catalyst (concn)[a]	Initiator (concn)[b]	Polymerization temp. (°C)
PC-4	NaH (1/100)	N-AcCL[c] (1/100)	160
PC-5	EtMgBr (1/200)	N-AcCL (1/100)	130
PC-6	EtMgBr (1/796)	N-AcCL (1/398)	140
PC-7	NaH (1/600)	N-AcCL (1/600)	160
PC-8	NaH (1/67)	Is-A[d] (1/200)	160

[a] Number of moles of catalyst per mole of monomer.
[b] Number of moles of initiator per mole of monomer.
[c] N-Acetylcaprolactam.
[d] Isatoic anhydride.

of some of these PC samples were obtained by separate use of sand-column-elution fractionation and coacervation fractionation techniques with m-cresol–cyclohexane mixtures. Fractions obtained from the sand-column-elution fractionation are designated by symbols F3, F4, F5, etc., which indicate 1st, 2nd, 3rd, etc., fractions, respectively. Generally, the molecular weight increases with increasing fraction number. Fractions obtained from the coacervation fractionation are designated by the symbols P1, P2, P3, etc., which indicate 1st, 2nd, 3rd, etc., fractions, respectively. Generally, the molecular weight decreases with increasing fraction number. When a P2 fraction was fractionated further by sand-column-elution fractionation, the fractions were denoted as F1P2, F2P2, F3P2, etc. A detailed report describing the fractionation and characterization of these PC samples will be published later (10). The baseline-adjusted elution volume GPC curves of the PC samples and fractions used in this study are shown in Figs. 2–5.

A commercial PC sample made by hydrolytic polymerization methods was obtained from Allied Chemical Corporation and is designated as P-8205. A commercial sample of PVAC was obtained from Farbwerke Hoechst A.G. through Prof. H. Benoit as part of the

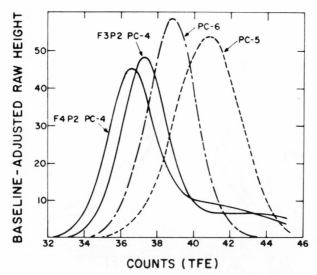

FIG. 2. Baseline-adjusted chromatograms of Samples F3P2 PC-4, F4P2 PC-4, PC-5, and PC-6.

IUPAC polymer study program. An experimental sample of PTMO made by cationic polymerization methods was provided by Dr. R. E. Wetton of the University of Technology, Loughborough, Leicestershire, England. The baseline-adjusted elution volume GPC curves of

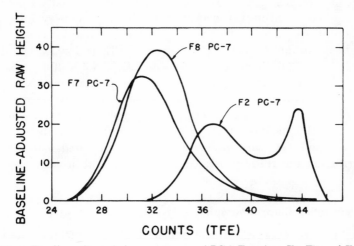

FIG. 3. Baseline-adjusted chromatograms of PC-7 Fractions F2, F7, and F8.

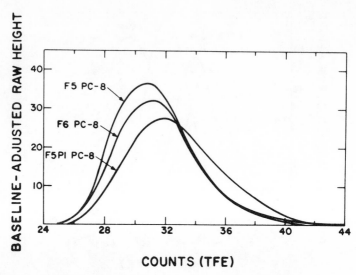

FIG. 4. Baseline-adjusted chromatograms of PC-8. Fractions F5P1, F5. and F6.

the samples P-8205, PVAC, and PTMO used in this study are shown in Fig. 5.

Solvents

Certified reagent grade THF ($n_D^{25} = 0.888$, bp $= 64$–$66°C$) obtained from Fisher Scientific Co. was used for both viscometry and GPC. The solvent contained 0.025 (w/v) % di-tertiary-butyl-p-cresol which served as an antioxidant. The solvent TFE ($n_D^{20} = 1.2907$, $d^{25} = 1.3823$, bp $= 73.6°C$, ionization constant $K_a = 4.3 \times 10^{-13}$) was obtained from Halocarbon Products Corp. in Hackensack, N. J. The GPC eluted polymer-contaminated TFE was routinely recovered by first running the solution through a 3-Å molecular sieve column to remove small amounts of water and then fractionally distilling the dried solution. Gas chromatography analysis indicated that the total impurities in the freshly distilled dry solvent were usually less than 0.1%.

Gel Permeation Chromatography

Two Waters Associates Model 200 Gel Permeation Chromatographs, each fitted with five Styragel columns, were used for the analysis of

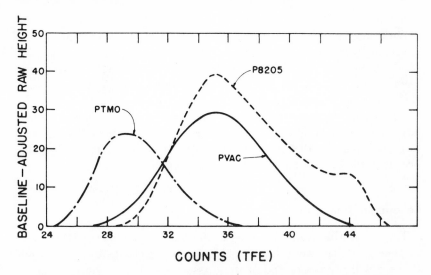

FIG. 5. Baseline-adjusted chromatograms of Samples PVAC, PS205, and PTMO.

molecular weight distributions in THF and TFE. The nominal porosity designations of the column sets used in THF and TFE were 10^6, 10^5, 10^4, 10^3, 250 Å and 10^7, 10^7, 10^6, 1.5×10^5, 1.5×10^4 Å, respectively. The column set used with THF was operated at room temperature, $24 \pm 1°C$, and had a plate count of 734 plates/foot with o-dichlorobenzene, while the column set used with TFE was operated at $50.0 \pm 0.5°C$ and had a plate count of 568 plates/foot with ϵ-caprolactam. The differential refractometer of the instrument used with THF had a 0.019-in. slit, was operated at a sensitivity of $8 \times (100)$, and monitored the effluent streams at $42°C$. The degasser was operated at $55°C$. The differential refractometer of the instrument used with TFE had a 0.004-in. slit, was operated at a sensitivity of $4 \times (100)$, and monitored the effluent streams at $54°C$. The degasser was operated at $65°C$. The solvent flow rates were controlled at better than 1.00 ± 0.05 ml/min. To eliminate errors in elution volume measurement due to variations in the rate of solvent evaporation in the siphon tube (1 count = 5.024 ml for THF, 1 count = 5.148 ml for TFE), a vapor feedback loop device similar to that of Yau, Suchan, and Malone (11) was installed. Polymer samples were dried in vacuo overnight at $60°C$, dissolved in degassed solvent taken from the GPC solvent reservoir, and then were Millipore-filtered under N_2 pressure through $0.2 \, \mu$ Metricel Alpha-8 filters. The polymer samples and PS-calibration standards were injected for 120 sec by means of the Waters Associates Automatic Sample Injection System. All the polymer sample solutions had concentrations <0.1 (w/v) %. The GPC traces were recorded digitally at 20 sec intervals by means of the Waters Associates Digital Curve Translator. Molecular weight averages, intrinsic viscosity, and integral and differential distribution curves were calculated on an IBM 360/65 computer according to the basic integral formulas given by Pickett, Cantow, and Johnson (*12*). These formulas are given later in the paper as Eqs. (7) through (10).

Calibration Standards

The calibration standards used in the construction of the HDV calibration in THF were linear polystyrene standards obtained from Pressure Chemical Co. and Waters Associates. The absolute number- and weight-average molecular weights, polydispersity ratios, and peak elution volume values, designated respectively by $\bar{M}_n(t)$, $\bar{M}_w(t)$, $P(t)$, and PEV, of the PS standards for the column set used with THF were shown in Table 1 of Ref. *1*. The Mark-Houwink intrinsic viscosity-molecular weight relation used to obtain the absolute intrinsic viscosity, $[\eta](t)$, for PS is given by (*1, 13*)

$$[\eta]_{\text{THF},25°C}^{\text{PS}} = 1.60 \times 10^{-4} \, \bar{M}_v^{0.706}, \qquad \bar{M}_v > 3000 \qquad (1)$$

The HDV calibration curve obtained from the $[\eta](t)$–$\bar{M}_w$–PEV data for polystyrene in THF is shown in Fig. 6.

Membrane Osmometry

Number-average molecular weights were determined with a Mechrolab Model 501 high-speed membrane osmometer fitted with a Hewlett-Packard variable-temperature controller and 10 mV Texas Instrument Servoriter-II recorder. The PMMA polymers were measured in toluene at 60°C and the PC, PVAC, and PTMO polymers were measured in *o*-chlorophenol (OCP) at 60°C. Schleicher and Schuell, Inc., type 08 deacetylated acetyl cellulose membranes were used for both solvents and were conditioned by the recommended method (*14*) of gradually changing the medium from water through ethanol to the desired solvent. Stable readings were usually obtained with each solvent within 5 min and $\bar{M}_n$ values as low as 5,000 and

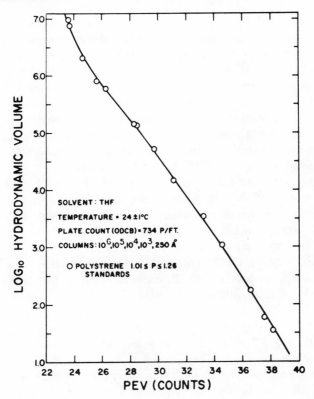

FIG. 6. Polystyrene hydrodynamic volume calibration curve in THF.

15,000 could be determined on polymer fractions in OCP and toluene, respectively, without evidence of diffusion. The membrane life in toluene was 2 weeks, while in OCP it was 4 weeks. The number-average molecular weight was obtained from the intercept (15) of a linear least-square fit of five to six values of $\{C, \sqrt{\pi}/C\}$ where π (g/cm²) is the osmotic pressure and C (g/ml) is the concentration and abscissa of the parameter set. In most cases, the experimental error in $\bar{M}_n$ was less than ±3%.

Viscometry

Viscometry measurements were made in THF at 25°C and in TFE at 50°C with uncalibrated Cannon-Ubbelohde dilution viscometers which gave solvent times greater than 100 sec. The viscometers used had centistoke ranges denoted by viscometer sizes of 50 and 75 for THF and TFE, respectively. The solvent and solution efflux times were determined by means of the Hewlett-Packard Autoviscometer system which includes the Model 5901B Autoviscometer, the Model 5903A Programmer-Printer, and the Model 5910A constant temperature bath. Temperature control was maintained to within $\pm 0.01°C$ for the temperatures of measurement. Efflux times were measured to ± 0.01 sec by means of a photodetector system which consisted of an upper and lower photocell unit that detected meniscus movement. A drying tube containing activated 3 Å molecular sieves was placed between the external air pump and the autoviscometer to eliminate water absorption by the THF and TFE solvents.

Polymer concentrations were chosen such that the highest concentration had an efflux time between 200 and 300 sec. Six solution concentrations were made up volumetrically from a stock solution on a g solute/100 g solution basis and converted to g/dl via the solvent density at the temperature of measurement. The solvent densities used were $d_{TFE}^{50°C} = 1.3429$ obtained from pycnometric measurements (16) and $d_{THF}^{25°C} = 0.8811$ (17). In order to produce dust-free solutions, the stock solution and solvent were first drawn through a coarse fritted glass disk filter into the pipet before being delivered to the viscometer. Solution efflux times generally had an average deviation of ± 0.02 sec. All efflux times were of sufficient duration to justify neglect of kinetic energy corrections.

The intrinsic viscosity, $[\eta]$, was determined from an equivalent form of the Schulz-Blaschke equation (18) derived by Heller (19) and Ibrahim (20). The intrinsic viscosity was the reciprocal of the intercept obtained from a linear-least square fit of $\{C, \eta_{sp}/C\}$ where η_{sp} is the specific viscosity and C (g/dl) is the concentration and the abscissa of the parameter set. In most cases the experimental error in $[\eta]$ was less than $\pm 0.5\%$.

Light Scattering

The weight-average molecular weights of some of the polycaprolactam polymer fractions were determined from light-scattering measurements carried out with a S.O.F.I.C.A. light-scattering photometer. Measurements were made at room temperature in TFE with unpolarized light of 546 mμ wavelength. Solutions were filtered by gravity through Metricel Alpha-8 0.2 μ filters directly into the measuring cell in order to produce dust free solutions. The instrument was calibrated with benzene (21). The average specific refractive-index increment was determined with a Brice-Phoenix differential refractometer was 0.220 ± 0.002 cm³/g.

The light-scattering data were analyzed by the dissymmetry method (22), assuming the polymer molecules in solution could be described as polydisperse random coils, and by the Zimm-plot method (23). The experimental error in $\bar{M}_w$ was on the order of ± 5 to $\pm 10\%$. A more detailed report on the light-scattering measurements in TFE will be published later (24).

USE OF THE HYDRODYNAMIC VOLUME CONCEPT IN THE GENERATION OF MOLECULAR WEIGHT CALIBRATION CURVES

Benoit et al. (25, 26) and LePage, Beau, and DeVries (27) have shown that narrow MWD fractions of a variety of polymer types (PMMA, PS, PVC, as well as branched polymers and graft copolymers), which ordinarily have distinct molecular weight-elution volume calibration curves, have a common calibration curve when $\{[\eta]\, \bar{M}_w\}$, the effective HDV, is plotted against elution volume in THF. Similar solution behavior has been observed by other workers for linear and branched polyethylene and linear polystyrene in trichlorobenzene (28, 29) and in o-dichlorobenzene (30). Two methods will be discussed below whereby the HDV concept can be used to generate molecular weight calibration curves for a variety of polymer types.

Method I. Mark-Houwink Parameters Supplied

When the relationship between the intrinsic viscosity and molecular weight can be described adequately over the molecular weight range of interest by the functional form of Mark-Houwink equation,

$$[\eta] = KM^\epsilon \tag{2}$$

the relationship between the calibration curve for the polymer of interest, $f_x(v) = \log_{10} M_x$, and the molecular weight calibration curve for the polymer standards, $f_s(v) = \log_{10} M_s$, can be expressed according to the formalism of Coll and Prusinowski (31) as

$$\log_{10} M_x = \left(\frac{1}{1 + \epsilon_x}\right) \log_{10}\left(\frac{K_s}{K_x}\right) + \left(\frac{1 + \epsilon_s}{1 + \epsilon_x}\right) f_s(v) \tag{3}$$

The Mark-Houwink parameters ϵ_s, K_s and ϵ_x, K_x refer to the standard polymer and to the polymer of interest, respectively. For the special case of a linear calibration curve (1),

$$f_s(v) = \log_{10} M_S = \log_{10} D_1(s) - \{D_2(s)/2.303\}v \tag{4}$$

the calibration constants for the polymer of interest, $D_1(x)$, $D_2(x)$, can be expressed, with the aid of Eqs. (3) and (4), in terms of the calibration constants for the standard polymer, $D_1(s)$ and $D_2(s)$, as

$$D_1(x) = (K_s/K_x)^{1/(1+\epsilon_x)} D_1(s)^{[(1+\epsilon_s)/(1+\epsilon_x)]} \tag{5}$$

$$D_2(s) = \left(\frac{1 + \epsilon_s}{1 + \epsilon_x}\right) D_2(s) \tag{6}$$

Method II. Fit for Effective Mark-Houwink Parameters

If the Mark-Houwink parameters are unknown and there is insufficient data available for the direct generation of these parameters, effective values of ϵ and K can be obtained provided at least two out of the three experimental observables $\bar{M}_n$, $\bar{M}_w$, and $[\eta]$ are known for the polymer sample of interest. Pickett, Cantow, and Johnson (12) have obtained expressions for $\bar{M}_n$, $\bar{M}_w$, and $[\eta]$ in terms of the differential molecular weight distribution (DMWD) function da/dM as follows:

$$\bar{M}_n = \left[\int_{M_L}^{M_H} \frac{1}{M} \left(\frac{da}{dM}\right) dM\right]^{-1} \tag{7}$$

$$\bar{M}_w = \int_{M_L}^{M_H} M \left(\frac{da}{dM}\right) dM \tag{8}$$

$$[\eta] = K \int_{M_L}^{M_H} M^\epsilon \left(\frac{da}{dM}\right) dM \tag{9}$$

where

$$\frac{da}{dM} = F(v_M) \frac{1}{\left(\dfrac{df}{dv}\right)_{v_M}} \frac{\log_{10} e}{M} \tag{10}$$

The limits of integration M_L and M_H correspond to the lowest and highest molecular weight species, respectively, in the sample. The parameter a in the DMWD function is the weight fraction of polymer having molecular weights between M_L and M. The first factor on the right of Eq. (10), $F(v_M)$, is the normalized (i.e., area of the chromatogram is unity) baseline-adjusted chromatogram height at elution volume v_M, and the second factor is the reciprocal of the slope of the molecular weight calibration curve at v_M. The HDV can be expressed as

$$Z = \{[\eta]M\} = KM^{\epsilon+1} \tag{11}$$

and substituted into Eqs. (7), (8), (9), and (10) to yield the expressions

$$\bar{M}_n = \left[\int_{Z_L}^{Z_H} \left(\frac{Z}{K} \right)^{-1/(\epsilon+1)} \left(\frac{da}{dZ} \right) dZ \right]^{-1} \tag{12}$$

$$\bar{M}_w = \int_{Z_L}^{Z_H} \left(\frac{Z}{K} \right)^{1/(\epsilon+1)} \left(\frac{da}{dZ} \right) dZ \tag{13}$$

$$[\eta] = K \int_{Z_L}^{Z_H} \left(\frac{Z}{K} \right)^{\epsilon/(\epsilon+1)} \left(\frac{da}{dZ} \right) dZ \tag{14}$$

$$\frac{da}{dZ} = F(v_Z) \frac{1}{\left(\dfrac{df_H}{dv} \right)_{v_Z}} \frac{\log_{10} e}{Z} \tag{15}$$

where now the HDV calibration curve is expressed as $f_H(v) = \log_{10} Z$. The limits of integration Z_L and Z_H correspond to the lowest and highest HDV species, respectively, in the sample. The parameter a now represents the weight fraction of polymer having HDV's between Z_L and Z. By fitting to one of the parameter sets $\{\bar{M}_n, \bar{M}_w\}$ or $\{\bar{M}_n, [\eta]\}$ in a least square sense (32, 33), effective values of ϵ_x and K_x can be obtained. Then, the effective molecular weight calibration curve can be obtained from Eq. (3) or from Eqs. (4), (5), and (6) if the calibration curve is linear.

The values of ϵ_x and K_x obtained in this manner are called effective values because these parameters include the effects of (a) instrument spreading (1, 2) on the chromatogram due to axial dispersion, skewing, and flattening; (b) experimental errors in $\bar{M}_n$, $\bar{M}_w$, $[\eta]$, and in the chromatogram baseline; and (c) uncertainties associated with the degree to which the polymer of interest and polymer standard lie on a common HDV calibration curve. When experimental errors and instrument spreading effects are minimized, the parameters ϵ_x and K_x should be reasonably close to the true values.

By using two to three characterized polymer samples, calibration curve segments can be obtained that span the entire elution volume range of interest. Then a smoothed calibration curve can be constructed that spans the entire elution volume range. Some smoothing of the calibration curve segments may be necessary in the regions of overlap because of the variability in instrument spreading effects due to axial dispersion, skewing, and flattening as a function of elution volume (1, 2). Molecular-weight averages calculated from this smoothed curve should be closer to the absolute values than the infinite resolution values that would be obtained from a primary calibration curve of $\bar{M}_w$ vs. PEV. However, the smoothed calibration curve would retain some effects of instrument spreading at very low and very high elution volumes where the calibration curve tails up and down, respectively, due to a loss in resolution in these regions.

There are several distinct advantages of this new calibration method. It is not necessary to use very narrow MWD samples in contast to the primary calibration curve method. The entire GPC trace is used in constructing the calibration curve as opposed to one point in the primary calibration curve method. Errors associated with choosing the appropriate molecular-weight average to associate with PEV are eliminated. The measurement of $\bar{M}_w$ by light-scattering techniques is a time consuming and often experimentally difficult task and is subject to larger experimental errors than the determination of $\bar{M}_n$ by membrane or vapor pressure osmometry and $[\eta]$ by viscometry. Within the fitting technique $\bar{M}_w$ values are not required, whereas in the primary calibration curve method $\bar{M}_w$ values are needed for the construction of the $\bar{M}_w$ vs. PEV curve. As mentioned previously, instrument spreading effects are minimized by the fitting procedure. Calibration curves for both linear and branched polymers can be constructed by this method.

METHOD FOR THE GENERATION OF MOLECULAR WEIGHT AND HYDODYNAMIC VOLUME CALIBRATION CURVES IN TFE

Since the readily available well-characterized polystyrene samples are not soluble in TFE, it is not possible to construct directly a HDV calibration curve and subsequently construct molecular weight calibration curves according to Methods I and II for polymers of interest that are soluble in TFE. This difficulty can be circumvented with the aid of several samples of a given type of "test" polymer which are not necessarily narrow in MWD, but are soluble in both THF and TFE and cover the elution volume ranges of interest in both solvents.

By applying a sequence of transformations to the chromatograms of the test polymers run in THF and TFE, molecular weight and HDV calibration curves be generated in TFE. Integral distribution curves of elution volume (IDEV) for the test polymer are first constructed from the raw chromatograms. The IDEV and the wt-% polymer at elution volume v, $A(v)$, can be obtained from a transformation I on the normalized chromatogram $F(v)$ for the test polymer in THF and in TFE.

$$A_{\text{THF}}(v) = I[F_{\text{THF}}(v)] = -\int_{v_{M_L}}^{v} F_{\text{THF}}(v)\, dv \qquad (16)$$

$$A_{\text{TFE}}(v) = I[F_{\text{TFE}}(v)] = -\int_{v_{M_L}}^{v} F_{\text{TFE}}(v)\, dv \qquad (17)$$

where v_{M_L} is the elution volume in a particular solvent corresponding to the lowest molecular weight species of the sample. At equal wt-% polymer, a one-to-one correspondence can be made between the elution volume in THF, v_{THF}, and the elution volume in TFE, v_{TFE}. Thus, when

$$A_{\text{THF}}(v) = A_{\text{TFE}}(v) \qquad (18)$$

the elution volumes in THF and TFE are related by the equations

$$v_{\text{THF}} = A_{\text{THF}}^{-1}\{I[F_{\text{TFE}}(v)]\} \qquad (19)$$
$$v_{\text{TFE}} = A_{\text{TFE}}^{-1}\{I[F_{\text{THF}}(v)]\} \qquad (20)$$

where A^{-1} is the inverse function to A.

Once the relationship between v_{THF} and v_{TFE} is established, HDV and molecular weight calibration curves in TFE can be generated from a HDV curve based on PS standards in THF. Recalling that

$$f_H(v_{\text{THF}}) = \log_{10} Z \qquad (21)$$

a HDV curve in TFE, $g_H(v_{\text{TFE}})$, can be constructed by the use of Eqs. (18), (19), (20), and (21) and is formally given by

$$g_H(v_{\text{TFE}}) = \log_{10} Z = f_H[A_{\text{THF}}^{-1}\{I[F_{\text{TFE}}(v)]\}] \qquad (22)$$

In practice the construction of the elution volume calibration (v_{THF} vs. v_{TFE}) and HDV curves in TFE is best done graphically and is illustrated in Fig. 7. By use of several samples of the test polymer the entire elution volume range of interest in both solvents can be covered. Then, molecular weight calibration curves can be constructed by Methods I and II for polymers soluble in TFE. This approach only requires that the test polymer samples be completely soluble in both

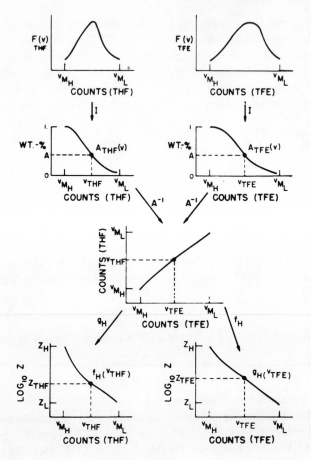

FIG. 7. Illustrative method for the generation of a hydrodynamic volume calibration curve in TFE.

solvents and that the response of the differential refractometer be linear for the sample concentrations used in both solvents.

Characterization of the test polymer samples by $\bar{M}_n$, $\bar{M}_w$, and $[\eta]$ determinations are not necessary. However, if such information is available on some of the test polymer samples, a molecular weight calibration curve can be constructed in THF by Method II. Then, a molecular weight calibration curve can be constructed in TFE through the elution volume calibration curve by the procedure discussed above. Subsequently, a HDV calibration curve can be constructed in THF if the Mark-Houwink relation for the test polymer in TFE is known.

HDV curves will be constructed for PMMA in TFE by both approaches.

RESULTS AND DISCUSSION

Generation of Molecular Weight Calibration Curves for PMMA in THF

Molecular weight calibration curves for PMMA in THF were generated by both Methods I and II and are shown in Fig. 8. The symbol $\times$ denotes the curve generated via Method I by use of Eqs. (1) and (3), Fig. 6, and the Mark-Houwink constants for PMMA in THF given in Table 3 (3, 4). The calibration curve denoted by the solid line is a smoothed curve constructed from calibration curve segments for fractions D, F, and H generated by Method II. The two curves are reasonably coincident over most of the elution volume range of the fractions, diverging above 36 counts ($M < 5000$). This divergence

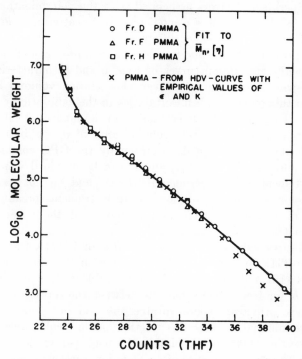

FIG. 8. PMMA molecular weight calibration curves in THF generated by Methods I and II.

can be attributed to experimental errors in the determination of $\bar{M}_n$ for fraction D and in the determination of the Mark-Houwink constants for PMMA in THF and for PS in THF for $M < 5000$. The values of $\bar{M}_n$, $\bar{M}_w$, $[\eta]$ and P calculated from these calibration curves best reflect the coincidence of the two curves. These values for fractions D, F, and H in THF are shown in Table 2 along with the corresponding true values and values obtained by directly fitting for ϵ and K. The values of $\bar{M}_n$, $\bar{M}_w$, $[\eta]$, and P calculated from these calibration curves best reflect the coincidence of the two curves. These values for fractions D, F, and H in THF are shown in Table 2 along with the corresponding true values and values obtained by directly fitting for ϵ and K. The values of $\bar{M}_n$, $\bar{M}_w$, and $[\eta]$ calculated by the various calibration procedures compare very favorably and are all closer to the true values than values normally obtained from the infinite resolution calibration curve (1) constructed by associating $\bar{M}_v$ or $\bar{M}_w$ with PEV. In fact, in most cases the values are within the normal experimental errors associated with the determination of $\bar{M}_n$, $\bar{M}_w$, and $[\eta]$ by membrane osmometry, light scattering, and viscometry, respectively. The larger differences in the $\bar{M}_n$, $\bar{M}_w$, $[\eta]$, and P values for fraction H are due to a loss of resolution at high molecular weights in the calibration curves of Figs. 6 and 8 characterized by a sharp upswing in the curves at very low elution volumes below 25 counts. In this region small uncertainties in the calibration curves are manifested by much larger uncertainties in the numerical calculations. Thus, over most of the elution volume range of the calibration curve above 25 counts, where good resolution by the GPC columns is attained, values of $\bar{M}_n$, $\bar{M}_w$, $[\eta]$, and P can be obtained which are in good agreement with experimental values, and do not have to be corrected for imperfect resolution due to instrument spreading effects such as dispersion, skewing, and flattening of the observer chromatogram (1, 2).

The effective values of ϵ and K obtained for PMMA fraction D, F, and H in THF, by fitting to $\{\bar{M}_n, [\eta]\}$ with the aid of Eqs. (12), (14), and (15) and Fig. 6, are shown in Table 3. The effective values of ϵ and K for fraction D should lie between the corresponding set of true values, provided experimental errors and instrument spreading effects are minimal, because this fraction has molecular weights above and below the 31,000 breakpoint in the $\log_{10} [\eta]$ vs. $\log_{10} M$ curve. Similarly, the effective values of ϵ and K for fractions F and H should lie close to the corresponding true values, provided experimental errors and instrument spreading effects are minimal. The increase in ϵ and

TABLE 2

Comparison of $\bar{M}_n$, $\bar{M}_w$, $[\eta]$, and P Obtained by Several Calibration Procedures in THF and TFE for PMMA Fractions D, F, and H

Calibration Methods	$\bar{M}_n \times 10^{-3}$	$\bar{M}_w \times 10^{-3}$	$[\eta]$	P
		Fraction D		
		THF		
True value	30.8	—	0.159	—
HDV-fit to $\{\bar{M}_n, [\eta]\}$	25.2	43.8	0.180	1.73
HDV-ϵ, K supplied	31.4	53.0	0.162	1.69
Smoothed MW-curve[a]	29.7	46.2	0.186	1.56
		TFE		
True value	30.8	—	0.225	—
HDV-fit to $\{\bar{M}_n, [\eta]\}$[b]	33.1	58.6	0.239	1.77
Smoothed MW-curve[c]	23.8	44.2	0.292	1.86
		Fraction F		
		THF		
True value	130	—	0.612	—
HDV-fit to $\{\bar{M}_n, [\eta]\}$	149	213	0.522	1.43
HDV-ϵ, K supplied	126	183	0.612	1.45
Smoothed MW-curve	139	227	0.562	1.63
		TFE		
True value	130	—	1.12	—
HDV-fit to $\{\bar{M}_n, [\eta]\}$	121	184	1.10	1.53
Smoothed MW-curve	143	214	0.942	1.49
		Fraction H		
		THF		
True value	513	—	1.46	—
HDV-fit to $\{\bar{M}_n, [\eta]\}$	469	1090	1.41	2.32
HDV-ϵ, K supplied	498	1050	1.42	2.10
Smoothed MW-curve	430	873	1.22	2.03
		TFE		
True value	513	—	3.03	—
HDV-fit to $\{\bar{M}_n, [\eta]\}$	463	692	2.67	1.50
Smoothed MW-curve	490	785	2.39	1.60

[a] The smoothed MW-curve in THF refers to the calibration curve constructed from the individual calibration curves obtained from Fractions D, F, and H by fitting to $\{\bar{M}_n, [\eta]\}$ according to Method II. The smoothed MW-curve is shown in Fig. 8 as a solid line.

[b] The HDV-curve in TFE used to fit to $\{\bar{M}_n, [\eta]\}$ data refers to the calibration curve obtained by transformation of the PS–HDV–THF-curve of Fig. 6 by the v_{THF}–v_{TFE} curve of Fig. 9. This HDV curve is designated by the solid line in Fig. 10.

[c] The smoothed MW-curve in TFE refers to the calibration curve obtained by transformation of the curve described in footnote a by the v_{THF}–v_{TFE} curve of Fig. 9. This calibration curve is designated by the solid line in Fig. 15.

TABLE 3

Comparison of Mark-Houwink Coefficients for PMMA Fractions in THF

	ϵ	$K \times 10^4$
True value $\begin{cases} \bar{M}_v < 31,000 \\ \bar{M}_v > 31,000 \end{cases}$	0.406 0.697	21.1 1.04
HDV, fit to $\{\bar{M}_n, [\eta]\}$		
Fraction D	0.675	1.06
Fraction F	0.762	0.953
Fraction H	0.774	0.359

decrease in K with increasing molecular weight of the fractions reflect the increased effects of instrument spreading on the chromatogram due to axial dispersion, skewing, and flattening (*1, 2*).

Construction of Hydrodynamic Volume Curves in TFE

The relationship between v_{THF} and v_{TFE} was obtained from the IDEV curves of the PMMA fractions C, D, F, G, H, I, and J in THF and in TFE according to the graphical method discussed above and illustrated in Fig. 7. A one-to-one correspondence between v_{THF} and v_{TFE} was made at the weight fractions 0.05, 0.1, 0.1, 0.3, 0.4, 0.5, 0.6, 0.7, 0.8, 0.9, and 0.95 for all the PMMA fractions. The resulting curve is shown in Fig. 9. The nonlinear regions of the curve below 25 counts (THF) and above 34 counts (THF) reflect the differences in resolving power and instrument spreading effects on the chromatogram at high and low molecular weights between the column set used with THF and the column set used with TFE.

A HDV calibration curve was constructed for TFE by using the PS-HDV curve of Fig. 6 in conjunction with the elution volume calibration curve of Fig. 9. This HDV curve is designated by the solid curve shown in Fig. 10. The construction of this curve did not require the determination of $\bar{M}_n$, $\bar{M}_w$, or $[\eta]$ for the PMMA polymer samples which are soluble in both THF and TFE. The only characterized polymer samples required were the readily available PS standards which were used for the construction of the HDV curve in THF.

For comparison purposes, a HDV curve was constructed in TFE from the smoothed PMMA molecular weight calibration curve in THF, designed at the solid line in Fig. 8, which was obtained by fitting to $\{\bar{M}_n, [\eta]\}$. This smooth PMMA–THF molecular weight calibration curve was used in conjunction with Fig. 9 to construct a

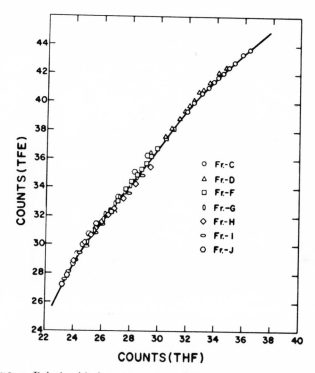

FIG. 9. Relationship between counts (THF) and counts (TFE).

PMMA molecular weight calibration curve in TFE, which is shown in Fig. 15 and designated by the solid line. Then, a HDV curve was constructed from the PMMA–TFE molecular weight calibration curve by use of Eq. (11) and the Mark-Houwink coefficients for PMMA in TFE (34), which are $\epsilon = 0.461$, $K = 1.81 \times 10^{-3}$ for $\bar{M}_v < 31{,}000$ and $\epsilon = 0.791$, $K = 5.95 \times 10^{-5}$ for $\bar{M}_v > 31{,}000$. This HDV curve is designated by the symbol + in Fig. 10. Within the experimental errors associated with the construction of the two HDV curves, the two curves are coincident.

Comparison of PMMA $\bar{M}_n$, $\bar{M}_w$, $[\eta]$, and P Values in TFE and in THF

The values of $\bar{M}_n$, $\bar{M}_w$, $[\eta]$, and P for the PMMA fractions D and F in THF and TFE, calculated by the various calibration procedures described in Table 2, compare favorably and agree with the corre-

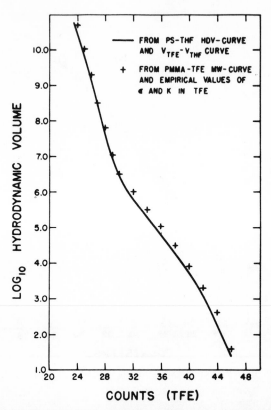

FIG. 10. Hydrodynamic volume calibration curves in TFE.

sponding true values within normal experimental errors. This is not true for fraction H. The differences in polydispersity values in THF and in TFE are too large to be attributable solely to differences in the experimental errors associated with the various procedures used to construct the calibration curves described in Table 2. The GPC columns set used with THF had nominal porosity designations, 10^6, 10^5, 10^4, 10^3, 250 Å, while the column set used with TFE had nominal porosity designations 10^7, 10^7, 10^6, 1.5×10^5, 1.5×10^4 Å. Thus, at high molecular weights the resolving power of the column set used with TFE was superior to that of the column set used with THF. Therefore, instrument spreading effects on the chromatograms run in THF were more severe than those run in TFE, resulting in much higher calculated values of $\bar{M}_w$ and consequently much higher polydispersity values.

Generation of Molecular Weight Calibration Curves for Polycaprolactams in TFE

Molecular weight calibration curves were generated for the anionically polymerized caprolactam samples and the hydrolytically polymerized caprolactam Sample P-8205, according to Method II, by using the raw chromatograms of these samples which are shown in Figs. 2–5, the HDV curve denoted by the solid line in Fig. 10, the parameter set $\{\bar{M}_n, \bar{M}_w\}$ or $\{\bar{M}_n, [\eta]\}$ for these samples, and Eqs. (3) and (7) through (10). The resulting calibration curves are shown in Figs. 11–15. No attempt was made to generate a common molecular weight calibration curve for the PC samples due to possible differences in the distribution of short- and long-chain branching in these samples (*9, 35*). An overlay of the resulting calibration curves reveals that, indeed, there are definite molecular structural differences among

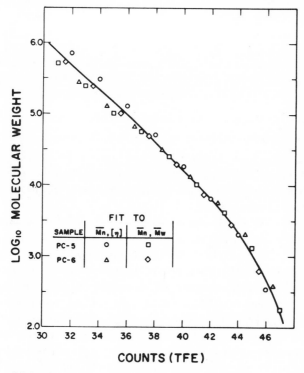

FIG. 11. Molecular weight calibration curve for Samples PC-5 and PC-6.

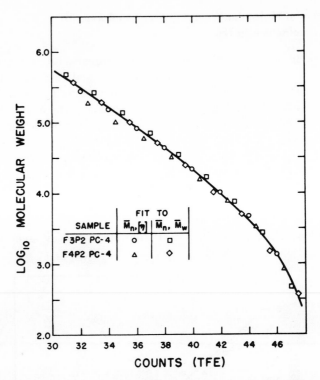

FIG. 12. Molecular weight calibration curve for Sample PC-4.

the PC samples because the curves are noncoincident. This point will
be discussed further later. However, the samples were separated into
classes according to the molecular weight ranges of the samples and
the methods of synthesis.

Samples PC-4, PC-5, and PC-6 essentially cover the same molecular weight range. However, as shown in Table 1, PC-4 was synthesized
under quite different experimental conditions than PC-5 and PC-6.
In Fig. 11 a common smoothed molecular weight calibration curve
was constructed from samples PC-5 and PC-6 using the calibration
segments obtained for both samples by fitting separately to both
parameter sets $\{\bar{M}_n, [\eta]\}$ and $\{\bar{M}_n, \bar{M}_w\}$. Over the common elution
volume range of the samples, $35 < v_{TFE} < 44$, the four calibration
curve segments represented by symbols in Fig. 11 are coincident within
experimental errors. Outside of the elution volume ranges of the
samples, the calibration curve segments tend to be divergent. However,

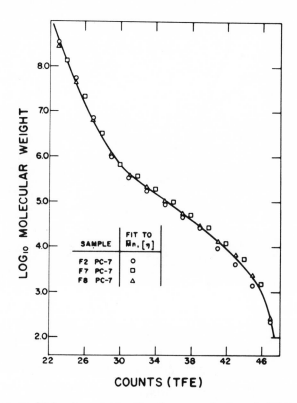

FIG. 13. Molecular weight calibration curve for Sample PC-7.

this behavior is to be expected in the extrapolated region of the count range due to the effects of experimental errors in $\bar{M}_n$, $\bar{M}_w$, $[\eta]$, and the chromatogram baseline, which are reflected in the fitted values of ϵ and K. Examination of Table 4 shows that the values of $\bar{M}_n$, $\bar{M}_w$, $[\eta]$, and P calculated by the various calibration procedures compare very favorably with the corresponding true values. The smoothed values generally do not agree as well with the true values as do the values obtained from fitting to $\{\bar{M}_n, [\eta]\}$ or to $\{\bar{M}_n, \bar{M}_w\}$. The smooth curve designated by the solid line in Fig. 11 was the result of subjective averaging by eye of the four calibration segments. A better smoothed curve would result if the data from the four calibration segments were smoothed in a least-square sense by fitting to the mathematical form of the Yau-Malone function (*32, 33, 36*). It is interesting to note that the $\bar{M}_w$ values resulting from the fit to $\{\bar{M}_n, [\eta]\}$ are in excellent agreement with the true values. Thus, the fitting

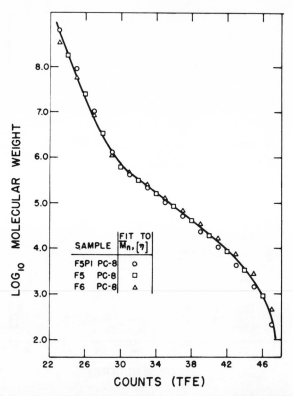

FIG. 14. Molecular weight calibration curve for Sample PC-8.

procedure via the HDV concept eliminates the necessity of empirically determining $\bar{M}_w$ values. The agreement of the calculated values of $\bar{M}_n$ and $[\eta]$, obtained from fitting to $\{\bar{M}_n, [\eta]\}$, with their respective true values; and the agreement of the calculated values of $\bar{M}_n$ and $\bar{M}_w$, obtained from fitting to the parameter set $\{\bar{M}_n, \bar{M}_w\}$, with their respective true values reflect the degree of fit obtained from the least-square process. The degree of fit reflects the experimental errors associated with $\bar{M}_n$, $\bar{M}_w$, $[\eta]$, the observed chromatogram, and the construction of the HDV curve.

Fractions F3P2 aid F4P2 of Sample PC-4 were used to construct a calibration curve for Sample PC-4. The calibration segments were generated by fitting to both parameter sets $\{\bar{M}_n, [\eta]\}$ and $\{\bar{M}_n, \bar{M}_w\}$, and are shown in Fig. 12 along with the smoothed curve designated by the solid line. The values of $\bar{M}_n$, $\bar{M}_w$, and $[\eta]$ in Table 5 calculated

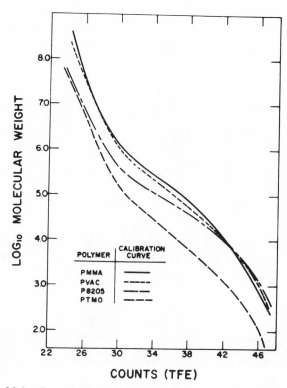

POLYMER | CALIBRATION CURVE

PMMA
PVAC
P8205
PTMO

FIG. 15. Molecular weight calibration curves for PMMA, PVAC, P8205, and PTMO polymers.

by the various calibration procedures compare favorably with their respective true values. The general quality of the results is the same as that for Samples PC-5 and PC-6. When the calibration curves of Figs. 11 and 12 are overlaid, it is seen that the two curves cross at $M \approx 40,000$. For $M < 40,000$ the calibration curve of PC-5 and PC-6 lies below that of PC-4. The converse is true for $M > 40,000$. These differences are greater than the experimental errors associated with the construction of these calibration curves and are believed to reflect molecular structural differences associated with the branching distributions of these samples. When these calibration curves are compared to that of the hydrolytically polymerized caprolactam Sample P8205 of Fig. 15, it is seen that they lie above the P8205 curve for $M > 10,000$. Based on the similar results of Drott and Mendelson (*28*) and Wild and Guliana (*29*) in their GPC studies of the effects of long-chain

TABLE 4

Comparison of $\bar{M}_n$, $\bar{M}_w$, $[\eta]$, and P Obtained by Several Calibration Procedures in TFE for PC-5 and PC-6

Calibration method	$\bar{M}_n \times 10^{-3}$	$\bar{M}_w \times 10^{-3}$ [a]	$[\eta]$	P	Fitted values	
					ϵ	$K \times 10^3$
		PC-5				
True value	8.52	(17.7)	0.45	2.08	—	—
HDV-fit to $\{\bar{M}_n, [\eta]\}$	8.52	22.0	0.45	2.59	0.335	16.0
HDV-fit to $\{\bar{M}_n, \bar{M}_w\}$	8.52	17.7	0.52	2.08	0.573	1.96
Smoothed MW-curve[b]	8.05	15.8	(0.53)[k]	1.97	—	—
		PC-6				
True value	18.2	43.0 (35.5)	0.85	2.36 (1.95)	—	—
HDV-fit to $\{\bar{M}_n, [\eta]\}$	18.8	38.3	0.89	2.03	0.635	1.05
HDV-fit to $\{\bar{M}_n, \bar{M}_w\}$[c]	18.2	39.2	0.86	2.15	0.445	7.32
Smoothed MW-Curve[b]	22.3	34.7	(0.88)[d]	1.55	—	—

[a] The true $\bar{M}_w$ values enclosed in parenthesis were determined by the dissymmetry method while those not enclosed in parenthesis were determined by the Zimm-plot method.

[b] The smoothed MW-curve was constructed from calibration curves for the samples generated by fitting to $\{\bar{M}_n, [\eta]\}$ and $\{\bar{M}_n, \bar{M}_w\}$ using the HDV-curve denoted by a solid line in Fig. 10.

[c] The average of the dissymmetry method and Zimm-plot method values of $\bar{M}_w$ was used in fitting to $\{\bar{M}_n, \bar{M}_w\}$.

[d] The values of $[\eta]$ in parenthesis were those determined using Mark-Houwink coefficients for linear PC in TFE. These coefficients were determined using experimental data on samples of PC, $[\eta]_{TFE}^{50}$, $[\eta]_{m\text{-cresol}}^{25°}$, and literature values (37) of Mark-Houwink coefficients for linear PC in m-cresol-at 25°C. This work will be published later (10).

branching in polyethylene, it can be concluded that the hydrolytically polymerized P8205 sample is much more linear than the anionically polymerized PC-4, PC-5, and PC-6 samples.

The calibration curves constructed from Fractions F2, F7, and F8 of Sample PC-7 and Fractions F5P1, F5, and F6 of Sample PC-8 are shown in Figs. 13 and 14, respectively. The calculated values of $\bar{M}_n$, $\bar{M}_w$, $[\eta]$, and P for the fractions of Samples PC-7 and PC-8, ob-

TABLE 5

Comparison of $\bar{M}_n$, $\bar{M}_w$, $[\eta]$, and P Obtained by Several Calibration Procedures
in TFE for Fractions of PC-4

Calibration method	$\bar{M}_n \times 10^{-3}$	$\bar{M}_w \times 10^{-3}$ [a]	$[\eta]$	P	Fitted values	
					ϵ	$K \times 10^3$
		F3P2 PC-4				
True value	31.5	60.5 (61.0)	1.15	1.92 (1.94)	—	—
HDV-fit to $\{\bar{M}_n, [\eta]\}$	23.6	59.8	0.90	2.54	0.906	0.0423
HDV-fit to $\{\bar{M}_n, \bar{M}_w\}$ [b]	25.1	69.2	0.78	2.75	0.801	0.103
Smoothed MW-curve [c]	22.9	51.3	$(1.14)^k$	2.24	—	—
		F4P2 PC-4				
True value	23.2	67.0 (62.0)	1.33	2.89 (2.67)	—	—
HDV-fit to $\{\bar{M}_n, [\eta]\}$	21.7	52.8	1.25	2.44	0.980	0.0294
HDV-fit to $\{\bar{M}_n, \bar{M}_w\}$ [b]	23.2	67.0	0.99	2.89	0.772	0.188
Smoothed MW-curve [c]	24.9	60.3	$(1.28)^d$	2.43	—	—

[a] The true $\bar{M}_w$ values enclosed in parenthesis were determined by the dissymmetry method while those not enclosed in parenthesis were determined by the Zimm-plot method.

[b] The smoothed MW-curve was constructed from calibration curves for the samples generated by fitting to $\{\bar{M}_n, [\eta]\}$ and $\{\bar{M}_n, \bar{M}_w\}$ using the HDV-curve denoted by a solid line in Fig. 10.

[c] The average of the dissymmetry method and Zimm-plot method values of $\bar{M}_w$ was used in fitting to $\{\bar{M}_n, \bar{M}_w\}$.

[d] The values of $[\eta]$ in parenthesis were those determined using Mark-Houwink coefficients for linear PC in TFE. These coefficients were determined using experimental data on samples of PC, $[\eta]^{50}_{\text{TFE}}$, $[\eta]^{25°}_{m\text{-cresol}}$, and literature values (37) of Mark-Houwink coefficients for linear PC in m-cresol-at 25°C. This work will be published later (10).

tained by fitting to $\{\bar{M}_n, [\eta]\}$ and obtained by using the smoothed calibration curves of Figs. 13 and 14, are shown in Tables 6 and 7, respectively, along with the true values of $\bar{M}_n$ and $[\eta]$. The general quality of the graphical and calculated results are similar to that obtained for Samples PC-4, PC-5, and PC-6. Upon overlaying the calibration curves for Samples PC-7, PC-8, and P8205, it is seen that the calibration curve for PC-7 lies above that for P8205 and the calibra-

TABLE 6

Comparison of $\bar{M}_n$, $\bar{M}_w$, $[\eta]$, and P Obtained by Several Calibration Procedures
in TFE for Fractions of PC-7

Calibration method	$\bar{M}_n \times 10^{-3}$	$\bar{M}_w \times 10^{-3}$	$[\eta]$	P	Fitted values	
					ϵ	$K \times 10^3$
F2 PC-7						
True value	7.69	—	1.04	—	—	—
HDV-fit to $\{\bar{M}_n, [\eta]\}$	7.69	35.2	1.04	4.58	0.703	0.725
Smoothed MW-curve[a]	10.7	45.6	$(1.03)^b$	4.27	—	—
F7 PC-7						
True value	165	—	5.97	—	—	—
HDV-fit to $\{\bar{M}_n, [\eta]\}$	166	1000	5.95	6.09	0.856	0.0502
Smoothed MW-curve[a]	118	769	$(7.32)^b$	6.50	—	—
F8 PC-7						
True value	140	—	5.29	—	—	—
HDV-fit to $\{\bar{M}_n, [\eta]\}$	140	563	5.30	4.03	0.831	0.105
Smoothed MW-curve[a]	132	540	$(5.62)^b$	4.09	—	—

[a] The smoothed MW-curve was constructed from calibration curves generated for the samples in the Table by fitting to $\{\bar{M}_n, [\eta]\}$ using the HDV curve denoted by the solid line in Fig. 10.

[b] The values of $[\eta]$ in parenthesis were those determined using Mark-Houwink coefficients for linear PC in TFE. These coefficients were determined using experimental data on samples of PC, $[\eta]_{TFE}^{50}$, $[\eta]_{m\text{-cresol}}^{25°}$, and literature values (37) of Mark-Houwink coefficients for linear PC in m-cresol-at 25°C. This work will be published later (10).

tion curve for PC-8 lies above that of PC-7 and P-8205. This would indicate that PC-8 is more branched than PC-7 and that Samples PC-7 and PC-8 are considerably more branched than Sample P8205.

Branching in Samples PC-7 and PC-8 also is reflected in the values of $[\eta](l)$, the intrinsic viscosity of a linear polymer having the same DMWD as the polymer in question, which may be branched. The values of $[\eta](l)$ for the fractions of PC-7 and PC-8 are computed from Eq. (9) using the Mark-Houwink coefficients for linear PC in TFE. These coefficients were obtained as described in footnote d in Table 4 and are $\epsilon = 0.736$ $K = 5.11 \times 10^{-4}$ for $\bar{M}_v > 29,000$. This work will be published later (10). The values of $[\eta](l)$ for fractions of PC-7 and PC-8 are shown in Tables 6 and 7, respectively, and designated by the superscript k. The DMWD function, da/dM, used

TABLE 7

Comparison of $\bar{M}_n$, $\bar{M}_w$, $[\eta]$, and P Obtained by Several Calibration Procedures in TFE for Fractions of PC-8

Calibration method	$\bar{M}_n \times 10^{-3}$	$\bar{M}_w \times 10^{-3}$	$[\eta]$	P	Fitted values	
					ϵ	$K \times 10^3$
		F5P1 PC-8				
True value	118	—	4.73	—	—	—
HDV-fit to $\{\bar{M}_n, [\eta]\}$	118	926	4.75	7.87	0.632	1.12
Smoothed MW-curve[a]	142	713	$(6.75)^b$	5.01	—	—
		F5 PC-8				
True value	202	—	7.41	—	—	—
HDV-fit to $\{\bar{M}_n, [\eta]\}$	202	1290	7.42	6.38	0.728	0.332
Smoothed MW-curve[a]	209	1080	$(9.50)^b$	5.18	—	—
		F6 PC-8				
True value	183	—	6.79	—	—	—
HDV-fit to $\{\bar{M}_n, [\eta]\}$	183	1160	6.79	6.34	0.819	0.0865
Smoothed MW-curve[a]	155	994	$(8.82)^b$	6.42	—	—

[a] The smoothed MW-curve was constructed from calibration curves generated for the samples in the Table by fitting to $\{\bar{M}_n, [\eta]\}$ using the HDV curve denoted by the solid line in Fig. 10.

[b] The values of $[\eta]$ in parenthesis were those determined using Mark-Houwink coefficients for linear PC in TFE. These coefficients were determined using experimental data on samples of PC, $[\eta]^{50}_{TFE}$, $[\eta]^{25°}_{m\text{-cresol}}$, and literature values (37) of Mark-Houwink coefficients for linear PC in m-cresol-at 25°C. This work will be published later (10).

in Eq. (9) is given by Eq. (10), where now $f(v) = \log_{10} M$ is the smoothed molecular weight calibration curves for Samples PC-7 and PC-8 shown in Figs. 13 and 14, respectively.

If the polymer in question is linear, the values of $[\eta](l)$ calculated in this manner will agree with the experimental value, $[\eta](t)$, within reasonable experimental errors. If the polymer in question is branched, then $[\eta](t) < [\eta](l)$. Therefore, the ratio $[\eta](t)/[\eta](l) = g^b$, which shall be designated as the branching factor, is a measure of the degree of long-chain branching in the sample. The parameter g is the classical branching index which is related to the weight-average number of branch points per molecule, n_w (38, 39) and depends upon the type of branching (i.e., random, comb, or star-type). The value of the parameter b depends upon the type of branching and usually lies in the range $\frac{1}{2} \leq b \leq \frac{3}{2}$ *(38)*. It is important to note that the branching factor calculated by the above method is a ratio of intrinsic viscosities of a linear and branched polymer having the same DMWD curve (i.e., the same $\bar{M}_n$, $\bar{M}_w$, $\bar{M}_z$, $\bar{M}_{z+1}$, etc.). Since the branched calibration curve has molecular weight greater than or equal to the linear calibration curve at corresponding elution volumes, the chromatogram of the linear polymer having the same DMWD as the branched polymer would have elution volumes less than or equal to that of the branched polymer at corresponding molecular weights. Thus the chromatograms of the linear and branched polymer having the same DMWD would be distinct.

This approach suggests a means of generating g^b values as a function of molecular weight for the linear and branched polymer having the same DMWD, independent of a specific molecular branching model relating g to n_w, by using characterized narrow MWD fractions of the branched polymer. Previous approaches have resorted to specific branching models and used g^b values where $[\eta](l)$ was calculated at the same $\bar{M}_w$ or $\bar{M}_r$ value as the branched polymer and not at the same DMWD. If only $\bar{M}_n$ and $[\eta]$ are readily available for these fractions, $\bar{M}_w$ values can be calculated for each fraction from the generated segmental calibration curve. Subsequently the $[\eta]$–$\bar{M}_w$ and g^b behavior can be obtained over the entire molecular weight range.

The value of $[\eta](l)$ obtained in this manner for F2 PC-7 is in good agreement with the true value indicating that this fraction has negligible amounts of long-chain branching that is detectable by intrinsic viscosity measurements. This is not unexpected because of the low molecular weights of this fraction. However, F7 PC-7 and F8 PC-7 have $[\eta](l)$ values greater than the true values indicating that these fractions, indeed, do have significant amounts of long-chain branching. Similarly, comparison of $[\eta](l)$ values for Fractions F5P1, F5, and F6 of Sample PC-8 with the corresponding true values shows these fractions to contain significant amounts of long-chain branching. The values of the branching factor g^b for the Fractions F7 PC-7 and F8 PC-7 are 0.82 and 0.94, respectively, while the values of g^b for the fractions F5P1 PC-8, F5 PC-8, and F6 PC-8 are 0.70, 0.78, and 0.77, respectively. Upon taking into consideration the large polydispersities of these fractions and the experimental errors contributing to the values of $[\eta](l)$, it can be seen that at high molecular weights Sample PC-8 has a greater degree of long-chain branching than does Sample PC-7. The results of this analysis are in qualitative accord with the earlier observation that Sample PC-8 should have a greater degree of long-chain branching than Sample PC-7 because of the nature of the molecular weight calibration curves.

Generation of Molecular Weight Calibration Curves for PVAC and PTMO

Molecular weight calibration curves for PVAC and PTMO were generated in TFE by fitting to the parameter sets $\{\bar{M}_n, [\eta]\}$ and

TABLE 8

Comparison of $\bar{M}_n$, $\bar{M}_w$, $[\eta]$, and P Obtained by Several Calibration Procedures in THF for PVAC, P8205, and PTMO

Calibration method	$\bar{M}_n \times 10^{-3}$	$\bar{M}_w \times 10^{-3}$	$[\eta]$	P	Fitted values	
					ϵ	$K \times 10^3$
		PVAC				
True value	58.6	—	1.40	—	—	—
HDV-fit to $\{\bar{M}_n, [\eta]\}$	58.4	223	1.40	3.81	0.681	0.380
		P8205				
True value	25.4	—	1.87	—	—	—
HDV-fit to $\{\bar{M}_n, [\eta]\}$	23.3	68.2	1.73	2.92	1.02	0.0199
		PTMO				
True value	107	800[a]	—	7.47	—	—
HDV-fit to $\{\bar{M}_n, \bar{M}_w\}$	107	799	—	7.47	0.643	16.3

[a] The $\bar{M}_w$ value was supplied by Dr. R. E. Wetton.

$\{\bar{M}_n, \bar{M}_w\}$, respectively; and using the raw chromatograms of these samples which are shown in Fig. 5, the HDV curve designated by the solid line in Fig. 10, and Eqs. (3) and (7) through (10). The calibration curves are shown in Fig. 15. The calculated values of $\bar{M}_n$, $\bar{M}_w$, and $[\eta]$ are shown in Table 8 and agree with the respective true values well within the usual experimental errors. Figure 15 is a composite plot of the molecular weight calibration curves for the linear polymers PMMA, PVAC, P8205, and PTMO. All but PTMO tend to coalesce to a common point at $M \approx 6000$. The divergence of the curves below this point probably is due to the effect of experimental errors on the numerical calculations in the construction of these curves. The divergence of the curves above this point reflects differences in the effective hydrodynamic molecular volumes among these polymer types. The molecular weight calibration curve for PTMO lies well below the curves for PMMA, PVAC, and P8205. From Fig. 5 it can be seen that the elution volumes of the PTMO sample cover the range $24.2 < v_{TFE} < 36.5$. The extrapolated curve above 36.5 counts, based on the fitted ϵ and K values, reflects experimental errors and instrument spreading effects on the chromatogram of this sample. At high molecular weights (low elution volumes) chromatogram spreading as measured by skewing, dispersion, and flattening parameters (1, 2) will significantly effect the magnitude of the fitted values of ϵ and K. Therefore, the use of these values to extrapolate the calibration curve to lower molecular weights may lead to an unrealistic calibration curve in this region.

CONCLUSIONS

The HDV concept has been used to generate molecular weight calibration curves by (I) using empirical Mark-Houwink parameters and (II) generating effective Mark-Houwink parameters by fitting to the parameter set $\{\bar{M}_n, \bar{M}_w\}$ or $\{\bar{M}_n, [\eta]\}$. Molecular weight calibration curves generated by these methods for PMMA in THF were shown to be coincident over the elution volume range of the samples, within experimental errors. This coincidence was reflected in a comparison of the calculated values of $\bar{M}_n$ and $[\eta]$ with the corresponding true values.

Some of the advantages in using Method II for the generation of molecular weight calibration curves are as follows:

(a) Only two to three characterized samples, which can have broad MWD's, are needed for the construction of the calibration curve to cover a wide elution volume range.

(b) The entire GPC trace of the sample is used in constructing the calibration curve as opposed to one point in the primary calibration curve method.

(c) Instrument spreading effects are minimized by the fitting procedure.

(d) By fitting to $\{\bar{M}_n, [\eta]\}$, the necessity for the measurement of $\bar{M}_w$ by light-scattering techniques is eliminated because $\bar{M}_w$ can be obtained from the generated calibration curve.

A method has been presented for the generation of a HDV calibration curve in TFE from which molecular weight calibration curves can be generated by Method II for a variety of polymers. This method makes use of the HDV curve in THF, generated from the readily available PS standards.

REFERENCES

1. T. Provder and E. M. Rosen, *Separ. Sci.,* **5,** 437 (1970).
2. E. M. Rosen and T. Provder, *Separ. Sci.,* **5,** 485 (1970).
3. C. V. Goebel, "Fourth International GPC Seminar," Miami Beach, Florida, Seminar Proceedings. 90 (1967).
4. J. D. Gouveia, L. A. Prince, and H. E. Staplefeldt, "Sixth International GPC Seminar," Miami Beach, Florida, Seminar Proceedings, 78 (1968).
5. W. A. Dark, R. F. Levangie, and K. S. Bombaugh, *Ibid.,* p. 414.
6. T. G Fox, J. B. Kinsingner, H. F. Mason, and E. M. Schuele, *Polymer,* **3,** 71 (1962).
7. T. Provder, J. H. Clark, and E. E. Drott, Unpublished Data.
8. G. B. Gechele and G. Stea, *Eur. Polym. J.,* **1,** 91 (1965).
9. R. Z. Greenley, J. C. Stauffer, and J. E. Kurz, *Macromolecules,* **2,** 561 (1969).
10. T. Provder, M. Ohta, and J. C. Woodbrey, Unpublished Data.
11. W. W. Yau, H. L. Suchan, and C. P. Malone, *J. Polym. Sci., Part A-2,* **6,** 1349 (1968).
12. H. E. Pickett, M. J. R. Cantow, and J. F. Johnson, *J. Appl. Polym. Sci.,* **10,** 917 (1966).
13. D. F. Alliet and J. M. Pacco, "Sixth International GPC Seminar," Miami Beach, Florida, Seminar Proceedings, 274 (1968).
14. ArRo Laboratories, Inc., *Membranes for Osmometry, Conditioning Procedures for Organic Solvents,* 1967.
15. W. R. Krigbaum and P. J. Flory, *J. Polym. Sci.,* **9,** 503 (1952).
16. F. Daniels, J. W. Williams, P. Bender, R. A. Alberty, and C. D. Cornwell, *Experimental Physical Chemistry,* McGraw-Hill, New York, 1962, p. 452.
17. J. Brandrup and E. H. Immergut, *Polymer Handbook,* Wiley (Interscience), New York, 1966, p. VIII-53.
18. G. V. Schulz and E. Blaschke, *J. Prakt. Chem.,* **158,** 130 (1941).
19. W. Heller, *J. Colloid Sci.,* **9,** 547 (1954).
20. F. W. Ibrahim, *J. Polym. Sci., Part A,* **3,** 469 (1965).
21. J. E. Kurz, *J. Polym. Sci., Part A,* **3,** 1895 (1965).
22. P. J. Debye, *J. Phys. Colloid Chem.,* **51,** 18 (1947).
23. B. H. Zimm, *J. Chem. Phys.,* **16,** 1093 (1948).
24. M. Ohta, T. Provder, and J. C. Woodbrey, Unpublished Data.
25. H. Benoit, Z. Grubisic, P. Rempp, D. Decker, and J. G. Zilliox, *J. Chem. Phys.,* **63,** 1507 (1966).
26. M. Benoit, Z. Grubisic, and P. Rempp, *J. Polym. Sci., Part B,* **5,** 753 (1967).
27. M. LePage, R. Beau, and A. J. DeVries, *J. Polym. Sci., Part C,* **21,** 119 (1968).
28. E. E. Drott and R. A. Mendelson, "Fifth International GPC Seminar," London, England, Seminar Proceedings, No. 15 (1968).
29. L. Wild and R. Guliana, *J. Polym. Sci., Part A-2,* **5,** 1087 (1967).
30. D. J. Pollock and R. F. Kratz, "Sixth International GPC Seminar," Miami Beach, Florida, Seminar Proceedings, 336 (1968).

31. H. Coll and L. R. Prusinowski, *J. Polym. Sci., Part B*, **5**, 1153 (1967).
32. D. W. Marquardt, *J. Soc. Ind. Appl. Math.*, **2**, 431 (1963).
33. E. J. Henley and E. M. Rosen, *Material and Energy Balance Computations*, Wiley, New York, 1969, p. 547f and pp. 560–566.
34. T. Provder, J. C. Woodbrey, and J. H. Clark, Unpublished Data.
35. O. Wichterle, J. Šebenda, and J. Káliček, *Fortschr. Hochpolym. Forsch.*, **2**, 578 (1961).
36. W. W. Yau and C. P. Malone, *J. Polym. Sci., Part B*, **5**, 663 (1967).
37. Z. Tuzar, P. Kratochvil, and M. Bohdanecky, *J. Polym. Sci., Part C*, **16**, 663 (1967).
38. M. L. Miller, *The Structure of Polymers*, Reinhold, New York, 1966, pp. 124–131.
39. R. A. Mendelson and E. E. Drott, *J. Polym. Sci., Part B*, **6**, 795 (1968).

Modification of a Gel Permeation Chromatograph for Automatic Sample Injection and On-Line Computer Data Recording*†

ANNIE R. GREGGES, BARRY F. DOWDEN,
EDWARD M. BARRALL II, and TERUO T. HORIKAWA

RESEARCH DIVISION
IBM CORPORATION
SAN JOSE, CALIFORNIA 95114

Summary

Gel permeation chromatograms are usually subjected to mathematical analyses of varying degrees of complexity to obtain molecular weight distribution and analytical data. Hand reduction of the recorder output data is difficult and time consuming even on an occasional basis. This paper describes in detail the modifications necessary to adapt the Waters Gel Permeation Chromatograph Model 200 for on-line data acquisition by a computer equipped with an analog to digital converter (IBM 1800). Since the operation is done in the background mode, time sharing with several other users or instruments is possible. Modifications to the auto injection unit are also described. This system has proven to remove all of the usual hand data reduction steps and to be extremely flexible for nonroutine analyses.

Gel phase or permeation chromatography (GPC) has become a routine analytical procedure for the qualitative and quantitative analysis of materials soluble in organic solvents. The technique has been applied to low, medium, and high molecular weight materials. The method is commonly used for the determination of molecular

---

*Presented at the ACS Symposium on Gel Permeation Chromatography sponsored by the Division of Petroleum Chemistry at the 159th National Meeting of the American Chemical Society. Houston, Texas, February, 1970.

† II in a series on the automation of analytical instruments.

weight distributions on a wide variety of soluble medium and high molecular weight polymers. Since the time scale of an individual chromatogram is long with respect to most other analytical methods, maximum use of a given instrument can be obtained only if samples are interleafed. That is, injected so that several samples are being chromatographed in a given column at one time; spaced so that elution peaks do not overlap. For most GPC systems this is best done by some form of automatic injection system. This permits loading of a large number of samples at a convenient time for injection as column space becomes available.

For molecular weight determinations extensive calculations must be performed on data taken from the GPC peak. Although two computer programs have been described for the necessary calculations, data reduction from the curves is done by hand (1, 2). This is a slow and exacting process subject to many errors. Some form of direct digital recording of the chromatographic data independent of the recorder chart would greatly improve the convenience and ultimate accuracy of the data processing and calculated molecular weight distributions.

The subject matter of this paper is the detailed discussion of the modifications necessary to a Waters Model 200 GPC for automatic operation with on-line, time share recording of data by an IBM 1800 computer. In addition, the computer programs for the calculation of M_w, M_n, M_z, M_v, M_w/M_n, graphical representation of recorded data, integral distribution, per cent distribution and per cent by volume eluted are outlined.

The problems of recording chromatographic type data where the independent variable is time on a time shared computer have been discussed before (3). For our particular monitor system no experiment is allowed to place undue demands on the computer (4). Data collection is accomplished using software timers and each instrument is serviced at the user's specified time interval, DT. There is no guarantee, however, that the requested time interval between data points will be realized. To overcome this, the real time at which the data point is collected is recorded along with the differential refractometer deflection. The differential refractometer is the detector in the apparatus. Since GPC is an extremely "slow" experiment, the nominal DT is 30.0 sec. However, interaction with the system and other users can result in a 5% variation. The actual interval, however, can be measured to an accuracy of 1 msec. A record at equal abscissa increments may be readily generated using a simple numerical linear interpolation.

During the course of a GPC run the measurement of the flow rate must be made. Here a demand/response mode of operation is utilized. The emptying of the siphon counter causes the computer to immediately respond and record the time at which the event occurred. This is a perfectly acceptable mode of operation since normally these "interrupts" occur at ~5.0 min intervals. The same method is used to record the infrequent auto-injection markers.

INSTRUMENTATION

Indexing Pen Modification

Under normal operating conditions a 15 mV signal is injected via a relay closure into the data recording each time the elution siphon counter empties. This causes the output of the differential refractometer to be displaced by 15 mV for approximately 15 sec. Should a computer reading be made of the differential refractometer output during a count marking cycle, an error of ~15 mV would be introduced. Thus, prior to any attempt at direct digital data recording of GPC curves, it is necessary to remove the siphon counter markers from the differential refractive index recording.

This requirement was met by the addition of a Perkin-Elmer event marker for a Speedomax W recorder, No. 124491. The pen coil was actuated by tapping a signal from Pins 13 and 14 of the relay plug-in, see Fig. 1. The relay is numbered KHP17A11. Since the output of Pins 13 and 14 is 110 V ac, a simple voltage dropping and rectifier arrangement is necessary to obtain 24 V dc to operate the pen coil. The 15 mV signal from the siphon counter is suppressed by the removal of the 1.5 V battery shown in Fig. 2.

In addition to the siphon counter marks, it is also necessary to record the injection point of the automatic sample injector. The necessary modifications to the auto inject system are shown in Fig. 3. With the above modifications, the siphon counter and refractometer signals are completely isolated from one another. These two signals are now in the correct form for direct connection to the 1800 computer.

Modifications to the Auto-Injection Unit

The Waters auto injector is an adequate solution to the problem of automatic sample injection. However, the attachment was found to have one flaw as received. The cycle of the loop, once started, goes

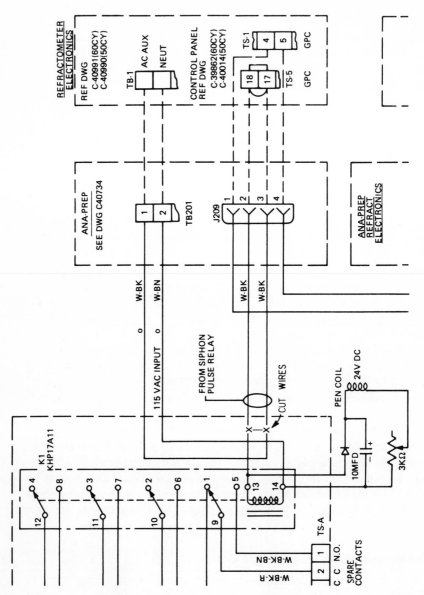

FIG. 1. Attachment of the event marker pen coil.

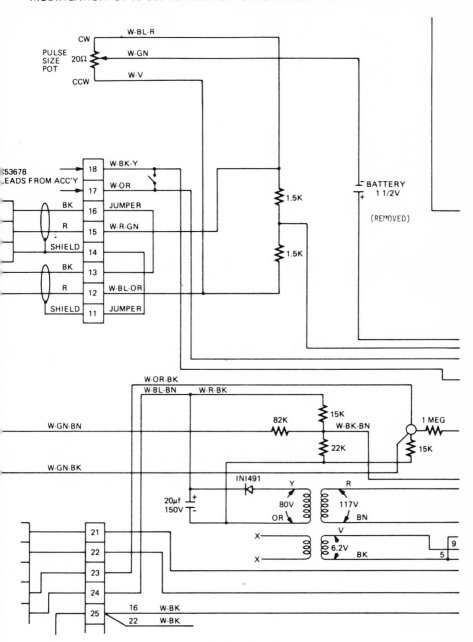

FIG. 2. Battery location.

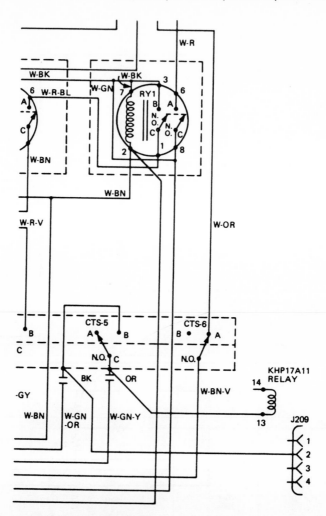

FIG. 3. Addition of the auto injector pulse to the volume count signal to the IBM 1800 and the event marker.

on until the main power switch of the auto injector is turned off manually. This operation results in empty (solvent filled) loops being injected after the last "loaded" sample. Were the refractometer less sensitive, this would be no problem. However, the solvent placed in the first loop when the first sample was injected is not identical to the machine solvent after the sixth loop is injected. This results in a series

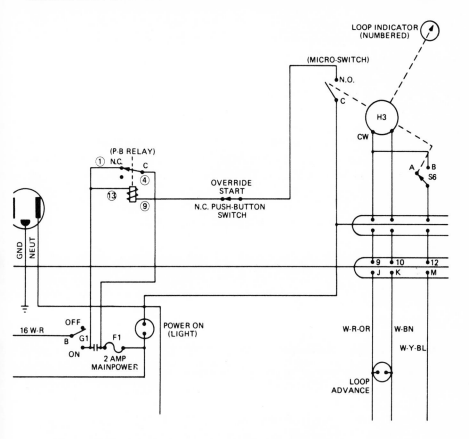

FIG. 4. Modification of the auto injector for automatic stop after six injections.

of baseline variations corresponding to each blank injection. Since with proper interleafing and column choice the average peak requires only 1.5 to 2 hr for actual elution, only 9 hr are required to go through the six sample cycle. This leaves, if the machine is run overnight, 7 hr for blank injection. The resultant unstable baseline ruins the next day's work.

To solve this problem, a microswitch was placed on the motor which rotates the loop. Corresponding to each numbered loop position is a mechanically attached cam. The microswitch is located between the first and sixth loop (cam) position. After the sixth injection, a relay operated by the switch causes the main power to the injector unit to shut off. The power to the auto injector remains off until the power override switch is reset. The circuit modifications are shown in Fig. 4. This requires a small change in the operating sequence as follows:

Loading
1. Turn on the main power switch and press the override switch. Wait for the loop assembly to advance to the 1-2 position.
2. Flush the loop with solvent and inject sample.
3. Press the loop advance switch (hold for approximately 5 sec) to advance to the next loop and repeat Step 2 until all loops have been filled. Each loop *must* be filled with either *sample* or *solvent*.
4. After loading the 6-1 position, press the loop advance switch and advance the loop to the cutoff switch.
5. Press the override switch to restart. The loop will advance to the 1-2 position which is the initial injection position.

Injection
1. Set timer to 3 counts and then to the desired injection interval.
2. Turn on pulse counter. The injection sequence will then begin after 3 counts.
3. After the last loop has been injected the microswitch will turn off the auto injector.

Data Collection—Hardware

Direct connection of a GPC to a digital computer is a relatively simple task. The interface bears many similarities to that recently described for gas–liquid chromatography (*3*). Only two direct connections between the IBM 1800 and the GPC are required. These are: 1) a shielded twisted pair conductor connected across the recorder input from the refractometer. This connection does not alter the operation of the GPC refractometer or the recorder. This pair carries the analog signal (± 100 mV) to the 1800 analog-digital converter. 2) A twisted pair conductor connected across Pins 3 and 7 of the siphon counter relay KHP17A11, Fig. 1. The other end of this pair is connected to a contact interrupt at the 1800. A count or pulse from the siphon which closes relay KHP17A11 produces a resistance drop (>10000 to $<4\Omega$) which is recognized as a contact interrupt at the computer. This latter connection causes the time of the siphon pulse to be automatically recorded at the computer.

The analog output of the refractometer is reduced at the computer by a factor of 3 (chosen so that the recorded signal is always on scale) using a simple voltage divider, and the reduced signal is connected to a low-level filtered analog input on the relay multiplexer of the 1800. No other treatment of the signal is required. The relay multiplexer input provides filtering with a time constant of 0.3 sec.

The remainder of the hardware consists of a panel at the GPC having a momentary closure switch which is connected to a second contact interrupt at the 1800. This switch is used to initiate action by the computer. In addition, toggle switches on the same panel at the GPC are connected to seven bits of a digital input group at the 1800. Of these switches, six are used to specify run times (i.e., how long the 1800 records the GPC output) of 0.5 to 9 hr, and one to determine if the switches are to be read or if data are to be collected. Two bits of digital output from the 1800 are used, one to switch on a light on the GPC panel indicating that data collection is in progress and the other to close a relay completing the siphon counter/contact interrupt circuit. This latter precaution is necessary to prevent the siphon counter from generating interrupts at the 1800 when analog data collection is not actually being carried out. The interface is completed by a 1053 typewriter near the GPC which prints out run status information.

No attempt has been made thus far to provide automatic operation of the various functions of the GPC by the computer. The programmable Waters auto injection system is efficient. All that is required of the 1800 is to record the "real time" at which an automatic injection has occurred. This is performed by the counter monitoring circuit. Automatic gain control on the refractometer output is unnecessary, for the dynamic range of the 1800 analog to digital converter is more than adequate to handle the entire linear range of the refractometer.

Data Collection—Software

Three programs are used for data collection from the GPC. Two, "GELIN" and "COUNT," are interrupt core leads. These are loaded from disk and executed each time an interrupt is generated at the instrument (e.g., by depression of the momentary closure switch discussed above).

> GEL-PERMEATION CHROMATOGRAPH
> NO. DATA POINTS = 720
> RUN TIME = 6.0   INTERVAL = 30.00 SECS.

FIG. 5. Printout of 1053 Typewriter run status information.

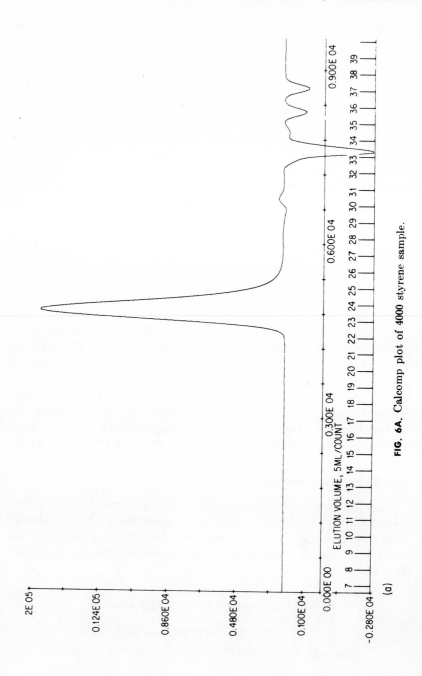

FIG. 6A. Calcomp plot of 4000 styrene sample.

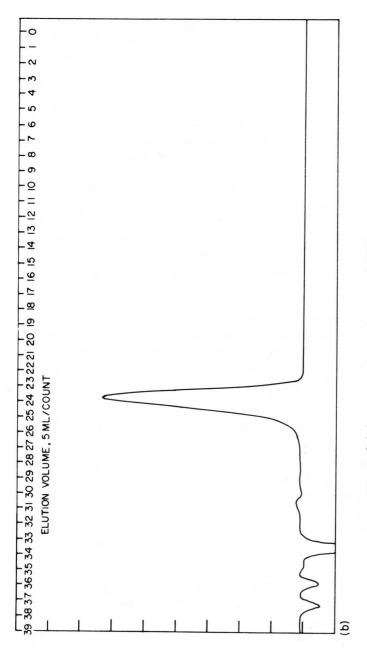

FIG. 6B. Gel chromatograph recorder output of **4000** styrene sample.

The function of "GELIN" is twofold: 1) to read the data switches, set up the run parameters [run time, time interval between points (DT), and number of data points (NPTS)], and to type these out for the operator to check at the 1053 typewriter; 2) to initiate data collection, switch on the "run" light, and activate a relay that starts the monitoring of count markers. The programming for the time-shared acquisition of data is a part of the TSX computer monitor system described previously (4). All that is required in the GPC program is a simple Fortran subroutine call with the appropriate arguments (DT, NPTS). See Fig. 5 for 1053 message.

The program "COUNT" is loaded and executed each time an interrupt is generated by the siphon emptying. The program's function is to read and record the computer "time-of-day" clock and to store these time values on disk. The first value in this data file contains the time at which data collection started. With this initial value, at the conclusion of a run all of the subsequent count markers can be related to the recorded differential refractometer readings. It should be emphasized that these interrupts are generated *only* while data collection is in progress.

At the conclusion of the run time the third core load, "GELEX," is queued and upon execution the time interval and refractometer output files are read. A record is provided at equal abscissa increments by interpolation of the digital record. The "count" file is then read and reconstructed in terms of real time in seconds from the run start. The interpolated record and the count markers, together with injection markers, are then plotted on the Calcomp plotter. Thus, the two sets of data (count interrupts and refractometer output) are reconstructed to give the conventional gel chromatogram at the plotter. This record is identical to the output of the GPC recorder with the exception that, irrespective of sample size, the chromatogram may be scaled to utilize the full width of the plotter. Such a plot is contrasted to the recorder plot in Fig. 6.

Punched cards suitable for off-line analysis of the above data are obtained by executing a simple nonprocess program "GPNCH."

Data Reduction—Software

Once a digitized record of a chromatogram or set of chromatograms has been collected, the type of reduction which may be applied is extensive. It is convenient to analyze our data off-line in the nonprocess

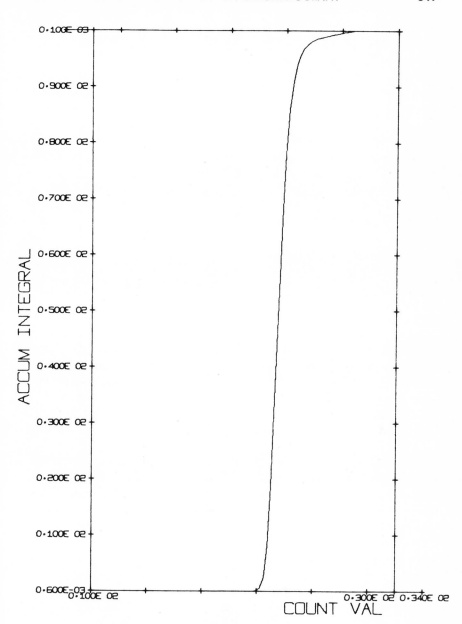

FIG. 7. Integral distribution curve from Program GPCl.

RUN IDENTIFICATION

RD-690181   POLYSTYRENE 4000   8/21/69

INTEGRATION LIMITS   START = 12   FINISH = 23   FROM RUN INITIATION

                     START = 18   FINISH = 29   FROM INJECTION POINT

BASE-LINE DETERMINED AT   12   AND   34   VOLUME COUNT

FRACTIONAL COUNT VALUE = 0.25

VOL. COUNT (5 ML.)	TIME (SECS)	TRUE COUNT (5 ML.)	AREA (MV. SECS.)	PERCENT AREA	ACCUM. PERCENT AREA
12.0000	3250.8006	18.0000	0.1173	0.0006	0.0006
12.2500	3321.9000	18.2500	0.2745	0.0016	0.0023
12.5000	3392.9999	18.5000	0.1437	0.0008	0.0031
12.7500	3464.0994	18.7500	0.0243	0.0001	0.0032
13.0000	3535.2002	19.0000	0.3502	0.0020	0.0053
13.2500	3606.3001	19.2500	0.8594	0.0050	0.0104
13.5000	3677.3996	19.5000	0.3088	0.0018	0.0122
13.7500	3748.4995	19.7500	0.1325	0.0007	0.0129
14.0000	3819.6003	20.0000	0.0000	0.0000	0.0129
14.2500	3891.6003	20.2500	0.0000	0.0000	0.0129
14.5000	3963.6003	20.5000	0.0000	0.0000	0.0129
14.7500	4035.6003	20.7500	0.0000	0.0000	0.0129
15.0000	4107.6013	21.0000	0.0000	0.0000	0.0129

15.2500	4177.8001	21.2500	0.0000	0.0000	0.0129
15.5000	4247.9999	21.5000	0.5179	0.0030	0.0160
15.7500	4318.1988	21.7500	2.6519	0.0155	0.0316
16.0000	4388.4005	22.0000	14.1645	0.0832	0.1149
16.2500	4459.5009	22.2500	80.9777	0.4760	0.5909
16.5000	4530.6003	22.5000	332.0036	1.9518	2.5427
16.7500	4601.6997	22.7500	942.5332	5.5410	8.0838
17.0000	4672.8011	23.0000	1827.6154	10.7444	18.8282
17.2500	4743.0009	23.2500	2617.4763	15.3879	34.2161
17.5000	4813.1997	23.5000	2883.4424	16.9515	51.1677
17.7500	4883.3986	23.7500	2579.3120	15.1635	66.3312
18.0000	4953.6013	24.0000	1989.3811	11.6954	78.0266
18.2500	5024.7007	24.2500	1349.6815	7.9346	85.9613
18.5000	5095.8001	24.5000	850.8946	5.0023	90.9636
18.7500	5166.8995	24.7500	517.2921	3.0411	94.0047
19.0000	5238.0009	25.0000	310.2628	1.8240	95.8288
19.2500	5309.1003	25.2500	189.5566	1.1143	96.9431
19.5000	5380.1997	25.5000	119.4729	0.7023	97.6455
19.7500	5451.2992	25.7500	78.9908	0.4643	98.1099
20.0000	5522.4005	26.0000	53.9512	0.3171	98.4270
20.2500	5593.5009	26.2500	40.3798	0.2373	98.6644
20.5000	5664.6003	26.5000	31.3992	0.1845	98.8490
20.7500	5735.6997	26.7500	26.2657	0.1544	99.0034
21.0000	5806.8011	27.0000	22.6152	0.1329	99.1364
21.2500	5877.9005	27.2500	19.9101	0.1170	99.2534
21.5000	5949.0009	27.5000	22.2061	0.1305	99.3839
21.7500	6020.1003	27.7500	24.7020	0.1452	99.5291
22.0000	6091.2007	28.0000	24.6913	0.1451	99.6743
22.2500	6161.3995	28.2500	22.7965	0.1340	99.8083
22.5000	6231.5993	28.5000	18.8729	0.1109	99.9192
22.7500	6301.7982	28.7500	13.7392	0.0807	100.0000

GEL-PERMEATION DATA REDUCTION

STD-700042　PS STD 4000　HI-4　1/19/70

COUNT	AREA	LOG. MOL. WT.	MOL. WT.
24.0000	0.1890	4.3791	23940.8
24.2500	0.0000	4.2813	19112.1
24.5000	0.0000	4.1906	15510.9
24.7500	0.3836	4.1070	12796.4
25.0000	17.1669	4.0302	10720.8
25.2500	94.0825	3.9592	9104.9
25.5000	337.5266	3.8928	7814.2
25.7500	909.2407	3.8299	6760.6
26.0000	1850.3046	3.7690	5875.8
26.2500	2871.1673	3.7086	5112.9
26.5000	3536.1746	3.6473	4439.9
26.7500	3709.3164	3.5839	3836.7
27.0000	3345.0807	3.5173	3291.0
27.2500	2765.7160	3.4468	2797.7
27.5000	2029.7957	3.7204	5252.9

NO. AVERAGE MOL. WT.　　=　　　4412.899

WT. AVERAGE MOL. WT.　　=　　　4742.558

Z-AVERAGE MOL. WT.　　=　　　5096.742

DISPERSIVITY　　　　=　　　1.074

VISCOSITY AV. MOL. WT. =　　　4567.6401

FIG. 9. Table of molecular weight from GPC2.

job stream of the 1800 using two programs that give the user some flexibility of operation. The data may be supplied to these programs either from magnetic disk or from data cards (GPNCH).

The program "GPC1" performs simple tasks such as background reduction, linear base line drift correction, digital smoothing (five point quadratic), and integration between fractional count values. The record, which was collected by the computer as integers ranging between ±32767 (5), is converted to millivolts of true output from the instrument. As an option the user may request a Calcomp plot of the integral distribution curve, see Fig. 7. Also, an interpolated printout of the refractometer deflection at 0.25 count values is produced. The same program produces a set of punched cards in a format suitable for use with the Chevron Research Programs (1, 2). A printout is shown in Fig. 8.

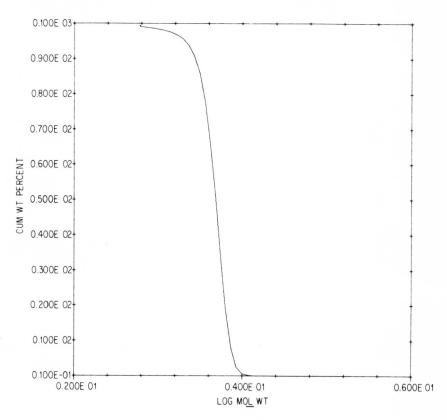

FIG. 10. Molecular weight integral.

The second program, "GPC2," uses the table of fractional count vs. areas generated by GPC1 to give a molecular weight distribution. The molecular weight (M_i) corresponding to a particular elution volume is determined by interpolation into a table of calibration data. The distributions are obtained using the conventional equation

$$\bar{M} = \frac{\Sigma n_i M_i^j}{\Sigma n_i M_i^{j-1}}$$

where n_i is the number of molecules of weight M_i.

$$M = \text{number-average mol wt if } j = 1$$
$$= \text{weight-average mol wt if } j = 2$$
$$= Z \text{ average mol wt if } j = 3$$

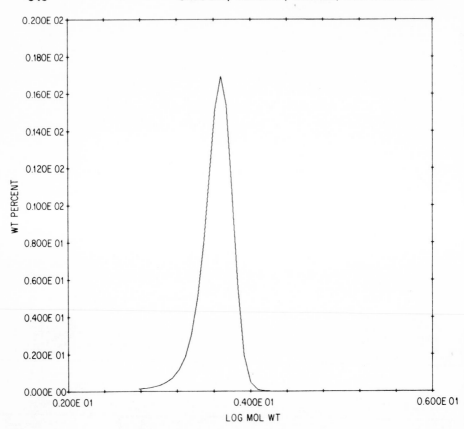

FIG. 11. Molecular weight differential.

Viscosity-average molecular weight may also be calculated from GPC2. The relationship

$$\bar{M}_v = \left\{ \frac{\Sigma n_i M_i^{(1+a)}}{\Sigma n_i M_i} \right\}^{1/a}$$

where a = polymer–solvent interaction parameter. This a must be determined by an alternate method.

The GPC2 output is in the form of a table, Fig. 9, and Calcomp plots of accumulated area vs. log molecular weight and area vs. log molecular weight (a differential plot with 0.25 count as the step). Figures 10 and 11 show these plots. It is planned that several calibration curves shall be included in GPC2 and the user permitted to specify the desired curve.

CONCLUSION

With the data collection system described in this study it is possible to obtain useful molecular weight distribution data and liquid chromatographic analyses in a semi-automatic manner. The programs "GELIN," "COUNT," and "GPNCH," remove most of the work from data reduction. Although programs "GPC1" and "GPC2" are very simple, they provide useful data given adequate calibration. The punched cards furnish data in an ideal form for more sophisticated treatment.

Acknowledgments

The helpful discussions with D. C. Clarke concerning the programming and the suggestions made by T. Kuga about the electronic modifications are gratefully acknowledged.

REFERENCES

1. H. E. Pickett, M. J. R. Cantow, and J. F. Johnson, *J. Appl. Polym. Sci.,* **10,** 917 (1966).
2. H. E. Pickett, M. J. R. Cantow, and J. F. Johnson, *J. Polym. Sci., Part C,* **21,** 67 (1968).
3. H. M. Gladney, B. F. Dowden, and J. D. Swalen, *Anal. Chem.,* **41,** 883 (1969).
4. H. M. Gladney, *J. Comp. Phys.,* **2,** 255 (1968).
5. The word size of the IBM 1800 Computer is fifteen bits plus a sign bit, as a consequence the absolute value of the largest integer number which can be represented in a single word is $2^{15} - 1$.

Characterization of Crude Oils
by Gel Permeation Chromatography*

H. H. OELERT,† D. R. LATHAM, and W. E. HAINES

LARAMIE PETROLEUM RESEARCH CENTER
BUREAU OF MINES
U.S. DEPARTMENT OF THE INTERIOR
LARAMIE, WYOMING 82070

Summary

The classical method of dividing crude oil into narrow molecular-weight cuts—distillation—is not practical in the high-boiling region. Thus, gel permeation chromatography was evaluated as a technique for characterizing the high-molecular-weight portion of crude oils. Crude oils, stripped to 180°C, were separated in a dual-column GPC system. The fractions obtained provided information concerning the amounts of material with molecular weights between 400 and 3000. Gross estimates were made of the proportion of ring and nonring carbons in these fractions. The latter estimates were obtained by relating molecular weights determined on the fractions to a calibration plot made up from data on model compounds.

INTRODUCTION

The first step in the characterization of crude oils usually involves separation into molecular-weight ranges by distillation. The cuts thus obtained are then analyzed either superficially or in depth, depending upon capabilities and needs. Considerable effort has been expended to develop analytical techniques to analyze hydrocarbon compound types in fractions boiling below 400°C, and this knowledge has materially aided in the efficient use of the lower-boiling fractions of

*Presented at the ACS Symposium on Gel Permeation Chromatography, sponsored by the Division of Petroleum Chemistry at the 159th National Meeting of the American Chemical Society, Houston, Texas, February, 1970.

† Max Kade Foundation Fellow 1968–1969. Permanent address: University of Clausthal, Clausthal-Zellerfeld, Germany.

crude oils. The increasing demand for gasoline and jet fuel requires the petroleum industry to convert the higher-boiling residual material into lower-boiling fractions. Thus, a knowledge of the composition of the high-boiling cuts of petroleum—the heavy ends—is of growing importance.

Distillation-based methods for the analysis of crude oils do not separate the heavy portion of the oil; they stop at about 400°C and lump the high-boiling materials in the crude oil (often more than 50%) together as residue. Although some further tests may be performed on these residues, the material is largely uncharacterized. The distillation temperature limit can be extended somewhat by use of special high-vacuum equipment, but the practical limit is not much above 400°C. High temperatures must be avoided because of the thermally sensitive materials that are often present in crude oils. The efficiency of actual distillation usually decreases as boiling point increases; but, even if ideal distillations were possible, the molecular-weight range of the expected cut increases dramatically with increase in temperature. For example, an ideal cut at 200°C would have a spread of 6 carbon numbers, while at 400°C the spread would increase to 18.

Because of the limitations of distillation, it seemed worthwhile to evaluate the capabilities of a relatively new chromatographic technique, gel permeation chromatography (GPC), to characterize crude oils, especially that portion of the oil boiling above 400°C. The development of gels for use in organic solvent systems (1) provided an opportunity for work with petroleum samples. Several workers (2–7) have applied GPC to distillate fractions, to asphalts and asphaltenes (8–12), and to porphyrins (13, 14); but only one attempt has been made to characterize crude oils with GPC (6).

GPC has the advantages of high reproducibility of separation runs, short run times resulting in low labor costs, and the ability to fractionate thermally sensitive compounds without exposing them to high temperatures. However, as pointed out by Giddings (15), the expected separability is not great because the effective working range is limited by the internal volume of the gel. Therefore, for a highly complex mixture such as crude oil, a gross separation is expected.

This paper describes the separation of crude oils in a calibrated GPC system. The relationship between fraction elution volume and molecular weight provides information concerning the molecular weight spread in the heavy portion of the oil and allows an estimation of the proportion of ring and chain carbons in the fractions.

EXPERIMENTAL PROCEDURES

Materials

The polystyrene cross-linked gels used were Poragel A-1 and Poragel A-3 (Waters Associates, Framingham, Mass.). The molecular-weight exclusion limits for these two gels are 1000 and 3000, respectively; each gel has a particle-size range of 36–75 μ.

Reagent-grade methylene chloride, flash distilled to remove high-boiling impurities and water, was used as the solvent. The gels were swollen for 10 hr the methylene chloride before packing in the columns.

Five crude oils, Wilmington (Calif.), Wasson (Tex.), Red Wash (Utah), Ponca City (Okla.), and Recluse (Mont.), were separated in this study. Before GPC, the material boiling up to an equivalent temperature of 180°C was vacuum stripped from each oil in a Büche Rotovapor apparatus.

The model compounds used in calibration studies were reagent grade or better.

Apparatus and Procedure

The GPC separations were carried out using two water-jacketed glass columns 0.5 in.$\times$5.0 ft, in series. The first was packed with Poragel A-3 and the second with A-1. Solvent flow was upward under a constant pressure of 15 psig of N_2.

The system had a total volume of 480 ml, an exclusion volume of 100 ml, and an internal volume of 260 ml. About 300 mg of oil was dissolved in 0.5 ml methylene chloride and charged to the column. Three to five charges of each oil were separated to provide sufficient sample for molecular-weight characterization. For model compound studies, about 50 mg of sample was used.

Fractions of 5.2 ml were collected by means of a siphon. Solvent was evaporated, and the sample was weighed. Molecular weights were determined by vapor-phase osmometry, with benzene as the solvent.

RESULTS AND DISCUSSION

Calibration of System

The dual-column system used for the separation was studied using model compounds representative of various types. These calibration data were used to construct the curves shown in Fig. 1, in which the relative elution volume (V_R) (benzene = 1.0) is plotted against the log of the molecular volume (V_m). Line I includes n- and isoparaffins,

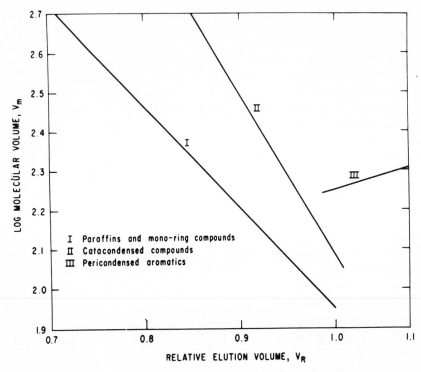

FIG. 1. Model compound data for dual-column system.

mono-ring cycloparaffins, benzenes, and polyaryls; Line II includes catacondensed cycloparaffins and aromatics; and Line III includes pericondensed aromatics.

Figure 1 shows that the elution volume is related to the molecular volume but that the relationship is different for different types of compounds. The difference in types of compounds that are described by the three lines is the amount of condensation. Line I includes branched chain paraffins, mono-ring cycloparaffins, and noncondensed aromatics—no condensed ring compounds are included in this group. All of the catacondensed aromatics and their saturated analogs fall on Line II. The two-ring, three-ring, etc., compounds fall on a straight line as long as condensation is cata (i.e., four carbons are added to two already in the system to form a new ring). If, however, the new ring is formed by pericondensation (i.e., the new rings is formed by adding three carbons to three already in the system), an extreme effect is noted, as shown by Line III. Because types of molecules in the I and II lines differ, not in polarity but in condensation, the difference in slopes is probably not due to sorption effects. Rather, this difference must be due primarily to a variation between the effective size of the molecules in a GPC column and the size of the molecule as determined by the molecular weight/density relationship.

Line III is tentative because it represents only a few aromatic compounds. However, it can be partly justified by the same argument. If the variation in slope of Lines I and II is due to catacondensation, then pericondensation should produce at least an equal change in slope so that Line III might assume an almost vertical position. However, the extreme shift of Line III suggests that factors such as sorption are overruling size effects for these compounds. These types of compounds would emerge much later than the others of the same molecular weight. However, if other compounds having V_R's near or greater than 1.0 are absent from the sample, one can assume that materials eluting in this range are pericondensed aromatics or other, as yet undefined, adsorbed types.

Comparison of GPC and Distillation Separation Data

Because distillation is the classical separation procedure for crude oils and because both distillation and GPC provide separations on a gross-molecular-weight basis, the temptation exists to equate the fractions from the two techniques. To compare the separation parameters of distillation and gel permeation chromatography, the molecular volume data in Fig. 1 have been translated to a carbon-number basis and superimposed on a similar plot for distillation, as shown in Fig. 2.

TABLE 1

Crude Oils Studied

Crude oil	Stripped crude (>180°C) wt% of crude	Residue (>400°C) wt% of crude	Residue (>400°C) wt% of stripped crude
Ponca City	64	12	19
Recluse	67	18	27
Red Wash	99	57	58
Wasson	72	27	38
Wilmington	91	47	52

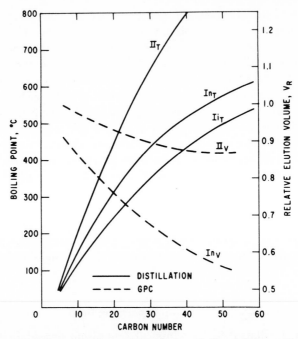

FIG. 2. Correlation of boiling point, relative elution volume, and carbon number for hydrocarbons. $Ii =$ iso-paraffins; $In = n$-paraffins; $II =$ condensed rings.

The dotted lines represent the limits of GPC and the solid lines the limits for boiling point. The translation from a molecular volume basis to a carbon number basis produces a family of curves or bands. The isoparaffins of increasing carbon number form a band above the n-paraffin line shown. The one-ring compounds lie above this line, then the two-ring, and on up until the limiting line for the most catacondensed systems. The intermediate lines are not shown, simply the limits. Just as Ii_T and II_T are limits for distillation, I_V and II_V are limiting lines for GPC.

If the objective of the separation procedure is to produce fractions with a narrow molecular-weight or carbon-number range, either procedure can be used. Figure 2 shows that for both methods of separation a larger carbon-number spread must be expected as the carbon number increases. The spread for distillation is less than that for GPC so that distillation would be preferable if usable. However, Fig. 2 also suggests that GPC would give some discrimination in the high-molecular-weight region where distillation is impractical or impossible.

Figure 2 demonstrates two differences between the separation techniques—one obvious and irritating, the other less obvious but more important. The data in Fig. 2 are arranged so that, for both techniques, the separation proceeds from bottom to top. Thus in distillation the small molecules emerge first, but in GPC the large molecules emerge first. This reversal in order of emergence is initially troublesome to the distillation-oriented petroleum chemist because he must reverse his thinking. Of considerably more importance, when attempting to equate distillation fractions and GPC fractions, is the difference in the effect of structural variations on the two separation schemes. Examination of Fig. 2 shows that a distillation cut containing the n-C_{20} paraffin might also contain isoparaffins up to C_{27} and aromatics down to C_{15}. The n-C_{20} paraffin would be approximately in the middle of the

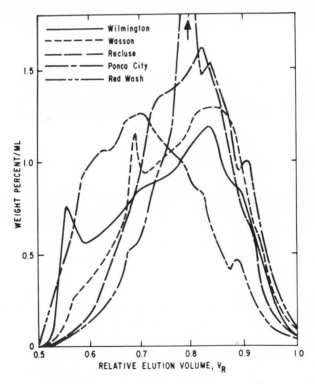

FIG. 3. GPC chromatograms of five crude oils.

carbon-number spread; and one might expect that the average molecular weight of this cut would be near that of the n-C_{20}, or about 280. If condensed molecules predominate in the mixture, the average molecular weight would decrease, with about 200 as the limiting value. The GPC cut in which the n-C_{20} paraffin would be expected would contain no molecules of lower molecular weight but would have isoparaffins and aromatics of higher carbon number; condensed aromatics with carbon numbers over 50 might be included. The average molecular weight of this cut would be well above 300 and, if the condensed rings predominated, could approach 1000. Thus, the types of molecules associated with each other are quite different in fractions from the two procedures.

The difference in the effect of structural variations frustrates a one-to-one comparison of fractions from the two techniques. For example, Fig. 2 shows that a narrow fraction at 180°C (the temperature to which our crude oils were stripped) could contain a C_9 ring compound, a C_{11} paraffin, and a C_{15} isoparaffin. Transferring these carbon numbers to the GPC scale suggests relative elution volumes of .98, .85, and .81. A similar inspection for a fraction at 400°C (the end temperature for a temperature for a practical assay by distillation) shows a C_{18} ring compound, a C_{27} n-paraffin, and a C_{35} isoparaffin. These materials should elute at V_R of .94, .70, and .64. Thus, materials from the 180°C cut will elute from .81 to .98 and will overlap those from the 400° cut, which will elute from .64 to .94. Similar fractions should not be expected from the two procedures.

Crude Oils

The lack of similarity of the GPC and distillate fractions does not negate the promise of GPC for producing usable fractions. Figure 2 suggests that GPC would provide some molecular-weight discrimination in the high-molecular-weight region where distillation is impractical. Five crude oils were stripped of material boiling up to 180°C and separated by GPC. Table 1 shows the relationship that these stripped crudes had to the whole crude oil and also the percentage of each crude that was classified as residue by the Bureau of Mines crude oil analysis.

The GPC chromatograms are shown in Fig. 3, in which weight percent of material per milliliter of eluant is plotted against relative elution volume. These curves show some similarities and some differences among the oils. Four of the oils are quite similar at high-relative

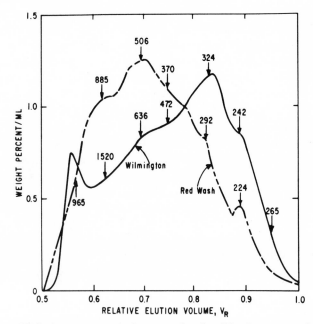

FIG. 4. GPC chromatograms of two crude oils. (Numbers are molecular weights of fractions.)

elution volumes (low-molecular-weight region) and show the maximum concentration of material between the relative elution volumes of .80 and .85. The fifth oil, Red Wash, is deficient in low-molecular-weight material and shows a maximum at .71.

A comparison of the amount of material in the high-molecular-weight portion of the GPC chromatograms (V_R's lower than about .7) and the amount of material in the residues from distillation (Table 1) shows that for these oils each technique arranges the crudes in the same order of increasing residual material—Ponca City, Recluse, Wasson, Wilmington, and Red Wash. However, the magnitude of the variations is different for the oils. For example, the chromatograms show a much greater variation between the Wilmington and Red Wash oils than the difference in their distillation residue contents (52 and 58%, respectively) indicates. This is probably due to major compositional differences between the two oils, which, as discussed earlier, influence the separations by GPC and distillation. Therefore, the correlation observed may not be found when another group of oils is compared.

In Fig. 4 the chromatograms of Wilmington and Red Wash crude oils are repeated, and the molecular weights of several fractions are added to the curves. These oils were chosen for this comparison because they are quite different and may suggest limits within which most crude oils should fall. Wilmington is an asphaltic oil, and Red Wash is very paraffinic. A comparison of the molecular weights of fractions with the same elution volume shows gross differences, particularly in the first part of the separation. The early peak at V_R .55 in the Wilmington oil must be due to large asphaltenic molecules that exceed the exclusion limit of the gel (3000) and emerge as an unseparated peak. Following these large asphaltenic molecules, the separated molecules begin to emerge; from elution volumes of .6 to .85 the amounts increase as the molecular weight decreases. Both the amount and molecular weight then decrease until a relative elution volume of .95. An increase in molecular weight at V_R .95 suggests the presence of adsorbed species such as pericondensed aromatics.

The Red Wash oil begins to elute slightly before the Wilmington, but the molecular weights are much lower. No maximum is observed for the large excluded molecules. The molecular weights decrease in a slow, orderly fashion across the entire curve; while the amounts show a similar distribution on both sides of the V_R .7.

Thus, these two stripped crude oils, which have similar amounts of distillation residue and similar average molecular weights (Wilmington 381, Red Wash 433), are shown by GPC to be quite different in the high-molecular-weight region. If amounts were plotted vs molecular weight, Wilmington would show 20% of the oil in a long tail, with molecular weights from 1500 to above 3000. The Red Wash has only 5% of material with molecular weights above 1000; 70% of its molecules have molecular weights from 300 to 900.

Compositional Estimations

The differences in average molecular weight at similar elution volumes are caused by differences in composition. An understanding of these differences should provide compositional information. Some insight can be gained by examination of the calibration plot for the effect of different types of molecules. In Fig. 5 the data for the Wilmington and Red Wash oils are plotted on the calibration plot for our GPC system. For the purpose of this plot, the molecular weights were divided by 14 to obtain an estimated average carbon number for each fraction. A gross interpretation of the calibration data suggests

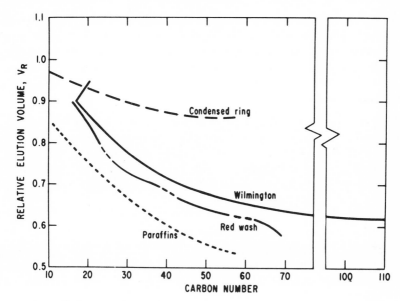

FIG. 5. Carbon number data for two oils as compared with GPC calibration data.

that the position of the fraction between the two calibration lines is related to the amount of ring vs nonring carbons. The position of the Wilmington curve is somewhat above the Red Wash curve and nearer to the limiting curve for completely condensed compounds. This shows that the Wilmington fractions are rich in ring carbons, as would be expected in this asphaltic oil. The position of the Red Wash curve— near the I or nonring-carbon line—shows that the Red Wash fractions contain predominantly nonring carbons. Although not shown in Fig. 5, the curves of Ponca City, Wasson, and Recluse fall between the Wilmington and Red Wash curves.

The above data show that by using molecular weights and a GPC calibration plot, one can obtain some information on the size and type of the molecules present in GPC fractions.

CONCLUSIONS

GPC was used to separate several crude oils, especially the high-molecular-weight portion of those oils. The relationship of fraction elution volume and molecular weight provides information on the proportion of ring and nonring compounds in the fractions. Comparison of GPC and distillation separation data shows the difference between the two techniques and that similar fractions will not be obtained using these two procedures. GPC provides discrete fractions in the molecular-weight range from 400 to 3000 using the system described.

Acknowledgments

Work presented in this paper was done as part of API RP 60 under a cooperative agreement between the American Petroleum Institute and the Bureau of Mines, U.S. Department of the Interior.

Mention of specific models of equipment or brand names of materials is made to facilitate understanding and does not imply endorsement by the Bureau of Mines.

REFERENCES

1. J. C. Moore, *J. Polym. Sci.*, **A2**, 835 (1964).
2. B. J. Mair, P. R. T. Hwang, and R. G. Ruberto, *Anal. Chem.*, **39**, 838 (1967).
3. P. C. Talarico, E. W. Albaugh, and R. W. Snyder, *Anal. Chem.*, **40**, 2192 (1968).
4. H. J. Coleman, D. E. Hirsh, and J. E. Dooley, *Anal. Chem.*, **41**, 800 (1969).
5. H. H. Oelert, *Erdoel Kohle*, **22**, 19 (1969).
6. H. H. Oelert, *Erdoel Kohle*, **22**, 536 (1969).
7. H. H. Oelert and J. H. Weber, *Separ. Sci.*, **5**, 669 (1970).
8. K. H. Altgelt, *Amer. Chem. Soc. Div. Petrol. Chem. Preprints*, **10**(3), 29 (1965).
9. K. H. Altgelt, *Makromol. Chem.*, **88**, 75 (1968).
10. K. H. Altgelt, *J. Appl. Polym. Sci.*, **9**, 3389 (1965).
11. W. B. Richman, *Proc. Assoc. Asphalt Paving Technol.*, **36**, 106 (1967).
12. L. R. Snyder, *Anal. Chem.*, **41**, 1223 (1969).
13. H. Blumer and W. D. Snyder, *Chem. Geol.*, **2**, 35 (1967).
14. H. P. Pohlmann and R. J. Rosscup, *Amer. Chem. Soc. Div. of Petrol. Chem. Preprints*, **12**(2), A103 (1967).
15. J. C. Giddings, *Anal. Chem.*, **39**, 1027 (1967).

Separation and Characterization of High-Molecular-Weight Saturate Fractions by Gel Permeation Chromatography*

J. H. WEBER and H. H. OELERT†

LARAMIE PETROLEUM RESEARCH CENTER
BUREAU OF MINES
U.S. DEPARTMENT OF THE INTERIOR
LARAMIE, WYOMING 82070

Summary

Gel permeation chromatography was applied to the saturate portion of a 450–475°C cut of Wilmington crude oil prepared by American Petroleum Institute Research Project 60. Selected fractions were analyzed by mass spectrometry, which demonstrated that a separation based on the number of rings was achieved. An early fraction contained 95.5 wt% no-, one-, and two-ring compounds; and a late fraction contained 95.9% wt% three-, four-, five-, and six-ring compounds. The relationship of the experimental results to those predicted from calibration curves relating carbon number to both distillation and GPC is discussed.

INTRODUCTION

The characterization of selected petroleum oils boiling above 400°C is an objective of American Petroleum Institute Research Project 60. To achieve this objective, the specimen oil must be divided into subfractions which are amenable to suitable analyses. Any separation scheme must be selective and chemically nondestructive. One such

* Presented at the ACS Symposium on Gel Permeation Chromatography, sponsored by the Division of Petroleum Chemistry at the 159th National Meeting of the American Chemical Society, Houston, Texas, February, 1970.

† Max Kade Foundation Fellow 1968–1969. Permanent address: University of Clausthal, Clausthal-Zellerfeld, Germany.

technique is gel permeation chromatography (GPC). This technique has been used previously on petroleum fractions (1–6). We report here preliminary results from the GPC of a saturate concentrate isolated from a distillate fraction of a Wilmington crude oil.

The separation of the high-boiling saturates into types is important not only for the characterization of the particular oil but to provide standards for the calibration of analytical techniques which may be applied in this molecular-weight range. A satisfactory method for dividing the saturates into types has not been available. Adsorption chromatography has not been useful because the saturate molecules do not possess the varying polar characteristics that are prerequisite for adsorption chromatography. Although thermal diffusion has been successfully applied (7, 8), this technique is not always practical because it requires a large sample and long separation periods. The lengthy exposure to elevated temperatures may alter the composition of some heat-sensitive samples. GPC minimizes these difficulties; it is thermally gentle to the sample, and the separation may be completed in a few hours on samples of less than 1 g.

EXPERIMENTAL PROCEDURES

A cut equivalent to 450–475°C at 760 torr was prepared by molecular distillation of a Wilmington crude oil and was separated into saturate and aromatic fractions by adsorption chromatography on alumina. The saturate fraction thus prepared (33% of the cut) was further separated by GPC, using Poragel A-1 preswollen in methylene chloride. A jacketed glass column 5 ft × 0.5 in. i.d. was filled with the preswollen gel, and the column was sealed with plugs of porous glass wool. The upward flow of the solvent, methylene chloride, through the column was maintained from a pressure reservoir (15 psig N_2) and held to a constant flow of 0.8 ml/min by a needle valve at the discharge side of the column. Samples were injected from a syringe into a valve connected directly with the lower end of the column. By means of a siphon, constant volume fractions (3.4 ml) were collected. Solvent was stripped from these fractions under nitrogen, and the weights of the fractions were determined. The refractive indices were determined on selected fractions, and molecular weights were measured for these fractions by vapor pressure osmometry in benzene. Mass spectra were obtained on a CEC 103 mass spectrometer and the data processed by Hood and O'Neal's saturate method (9).

RESULTS AND DISCUSSION

Separation of materials by GPC is a function of several variables; however, one may hold many of these variables constant within a given experiment so that the elution volume for a compound will depend primarily on its molecular volume. The sorption effects noted for many types of compounds should be negligible for the saturate compounds.

In Fig. 1 a correlation diagram is presented which relates boiling point, carbon number, and relative elution volume. The solid lines relate carbon number with literature values for normal boiling points. Curve In_T represents the relationship between boiling point and carbon number for the normal paraffins. Curve Ii_T represents the lowest boiling branched paraffins of each carbon number, and II_T approximates

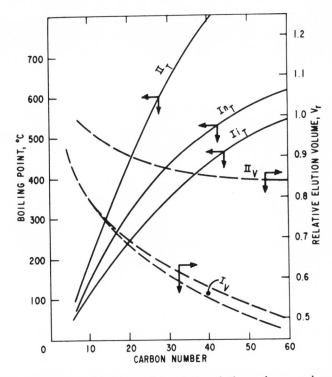

FIG. 1. Correlation of boiling point, relative elution volume, and carbon number for petroleum compounds. (——) Boiling point; (- -) GPC (Poragel/methylene chloride. Ii = branched paraffins; In = n-paraffins; II = condensed rings.

the most highly condensed compounds. Thus, Curves Ii_T and II$_T$ define
the carbon-number limits for a given distillation cut. Between Curves
In_T and II$_T$ a family of curves should exist which define the limits for
saturates with one ring, two condensed rings, etc. The band of one-
ring compounds lies immediately above the In_T line with its lower
limits extending somewhat below this line. The two-ring band lies
above the one-ring, etc. The dashed lines represent similar limits for
the GPC that we have established for our system, using model compounds. The relative elution volume, V_r, is plotted against carbon
number. The V_r is the ratio of the elution volume of the material to
the elution volume of benzene. The curve for the n-hydrocarbons,
shown at the bottom of the Fig. 1, has been well established by calibration compounds for our system. Based on a few points, we believe
the branched paraffins will be included in the band labelled I$_V$. The
II$_V$ curve is an estimated limiting curve for the most condensed
paraffins; it is based on only a few points. Between Curve I$_V$ and
Curve II$_V$ a family of curves should exist that describe limits for
saturates with one ring, two rings, etc. These should lie successively
above the I$_V$ curve and approach the II$_V$ curve as condensation
increases.

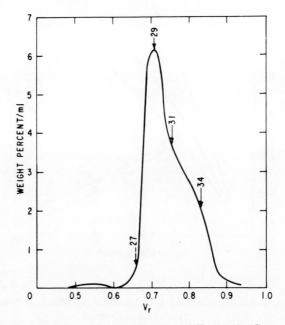

FIG. 2. Amount of material as a function of V_r 450–475°C saturate.

According to Fig. 1 the 450–475°C cut, if ideal, should contain normal and branched paraffins (no-ring compounds) between lines In_T and Ii_T or those with carbon numbers of 32 to 46. The ring compounds should have carbon numbers somewhat lower, ranging from about 20 to 35. Moving these limits to the GPC curve suggests that the GPC separation of this cut could be represented by a line from the intersection of the C_{46} line and the I_V line and moving upward and to the left. Thus the highest molecular weight no-ring compounds should emerge first, followed by lower molecular weight no-rings which would be followed by, and undoubtedly mixed with, the highest molecular weight one-ring compounds. These in turn would give way successively to condensed two-ring, three-ring, and four-ring compounds. Thus, in successive fractions the number of rings should increase while the molecular weight decreases. These data suggest that GPC, when applied to a narrow-boiling saturate cut, will effect a separation according to number of rings.

Data from GPC of the saturate cut are shown in Fig. 2. The weight per cent of material per milliliter of eluent is plotted as a function of the V_r. Most of the material emerges between the V_r's of .65 and .90 as predicted from Fig. 1. The compositional changes across this curve were checked by a detailed examination of four fractions noted by arrows. Data on these fractions are shown in Table 1. The increasing refractive index strongly suggests increasing condensation. Mass spectral analyses confirm this suggestion. The no-ring compounds predominate in Fraction 27, one-ring compounds in Fraction 29, two-ring compounds in Fraction 31, and finally four-ring compounds in Fraction 34. (The skipping of the three-ring compounds is probably due to fraction selection—two fractions intervened—and to the predominance of four-ring compounds in the saturate cut.) The change in composition from the first to the last fraction is striking. The early Fraction, 27, contains 95.5 wt% no-, one-, and two-ring compounds; and a late Fraction, 34, contains 95.9 wt% three-, four-, five-, and six-ring compounds.

Inspection of Fig. 1 predicts a decrease in molecular weight in successive fractions. The average molecular weights of Fractions 29, 31, and 34 suggest carbon numbers of 38, 36, and 29, respectively, which are consistent with the increasing condensation and complexity predicted. The low molecular weight of Fraction 27 appears to be anomalous. More detailed examination of Fig. 1 suggests that the no-ring compounds that emerge at the V_r corresponding to Fraction 27 should have carbon numbers between 20 and 30. Although most of

TABLE 1

GPC-MS Data on 450–475°C Saturates

Mass spectral analyses, wt%

	No-ring	One-ring	Two-ring	Three-ring	Four-ring	Five-ring	Six-ring	Mono-aromatics	Mol wt	Refractive index
Charge	12.3	9.5	19.3	20.3	27.6	7.4	3.0	0.6		
Fraction 27	67.6	20.1	7.8	3.1	0	0	0	1.4	427	1.451
Fraction 29	18.5	40.6	22.6	10.4	6.3	0	0	1.6	535	1.460
Fraction 31	1.0	18.9	27.2	17.6	17.2	9.2	3.7	5.1	510	1.481
Fraction 34	1.5	0	2.8	23.7	51.6	13.0	7.6	0	400	1.503

these compounds would not be present in an idealized 450–475°C cut, they are inevitable in a real cut; thus these low-molecular-weight no-ring compounds appear with the higher-molecular-weight one-ring compounds and lower the average molecular weight of this fraction. We have observed this same phenomenon in other cuts from this oil.

The data from this separation suggest the positions of the families of GPC curves that might be added to Fig. 1. Similar data on additional fractions from this and other crude oils should provide sufficient information to estimate the elution parameters for various condensed systems. Such data will aid in the design of the separation system and in the interpretation of the results.

CONCLUSIONS

Gel permeation chromatography is a thermally gentle, rapid, and effective method for separating the saturates from high-boiling distillate fractions according to number of rings. The technique provides concentrates of condensed ring compounds that are needed as calibration standards for other methods of separation and of analysis.

Acknowledgments

Work presented in this paper was conducted under a cooperative agreement between the American Petroleum Institute and the Bureau of Mines, U.S. Department of the Interior.

Mention of specific models of equipment or brand names of materials is made to facilitate understanding and does not imply endorsement by the Bureau of Mines.

REFERENCES

1. K. H. Altgelt, *Makromol. Chem.*, **88**, 75 (1965).
2. K. H. Altgelt, *J. Appl. Polym. Sci.*, **9**, 3389 (1965).
3. R. J. Rosscup and H. P. Pohlmann, *Amer. Chem. Soc. Div. Petrol. Chem. Preprints*, **12**(2), A103 (1967).
4. B. J. Mair, P. T. R. Hwang, and R. G. Ruberto, *Anal. Chem.*, **39**, 838 (1967).
5. P. C. Talarico, E. W. Albaugh, and R. E. Snyder, *Anal. Chem.*, **40**, 2192 (1968).
6. H. J. Coleman, D. E. Hirsch, and J. E. Dooley, *Anal. Chem.*, **41**, 800 (1969).
7. L. Linderman, *Amer. Chem. Soc. Div. Petrol. Chem. Preprints*, **14**(3), B186 (1969).
8. F. W. Melpolder, R. W. Sayer, and T. A. Washall, *Amer. Soc. Testing Mater. Spec. Tech. Publ.* **224**, 94–110 (1957).
9. A. Hood and M. J. O'Neal, *Advances in Mass Spectrometry* (J. D. Waldron. ed.), Pergamon, New York. 1959, p. 175.

Fractionation of Residuals by Gel Permeation Chromatography*

E. W. ALBAUGH, P. C. TALARICO, B. E. DAVIS,
and R. A. WIRKKALA

GULF RESEARCH & DEVELOPMENT COMPANY
PITTSBURGH, PENNSYLVANIA 15230

Summary

Gel permeation chromatography has been effectively used for the fractionation of residuals. The preparative-scale gel permeation chromatograph
employed for these studies is described along with selected operating
parameters. The instrument has been used to fractionate a shale oil
residue, 50% reduced crude, and 25% reduced crude, into fairly narrow
molecular weight fractions. The individual cuts from each residual have
been analyzed and the elemental distribution established. The trends of
heteroatom distribution, i.e., nitrogen, sulfur, nickel, and vanadium, as a
function of molecular weight for the various residuals are discussed.

The heavy ends of petroleum are a complex mixture of hydrocarbons
that vary widely in molecular weight and composition. A serious
problem in characterizing these heavy ends has been the lack of a
good and simple method for separating the molecules according to
molecular weight. Gel permeation chromatography (GPC) has been
shown to produce a molecular weight separation for many materials.
The mechanism of the separation appears to be primarily a function
of the size of the molecules in solution. However, with certain types
of compounds, such as highly condensed aromatics, secondary effects
become important and produce anomalous behavior. This work was
undertaken to determine if GPC could separate residuals into mo-

*Presented at the ACS Symposium on Gel Permeation Chromatography,
sponsored by the Division of Petroleum Chemistry at the 159th National Meeting of the American Chemical Society, Houston, Texas, February, 1970.

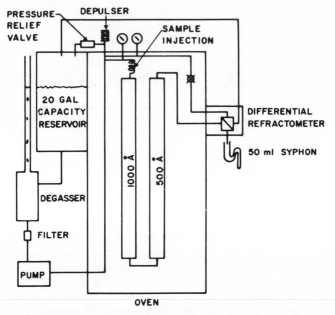

FIG. 1. Preparative scale gel permeation chromatograph.

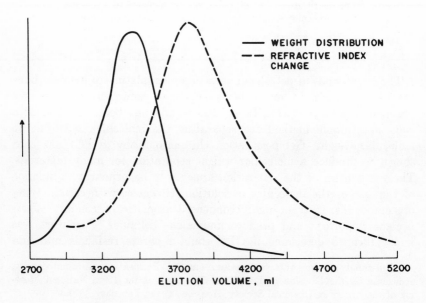

FIG. 2. Chromatogram of 25% reduced Kuwait VTB.

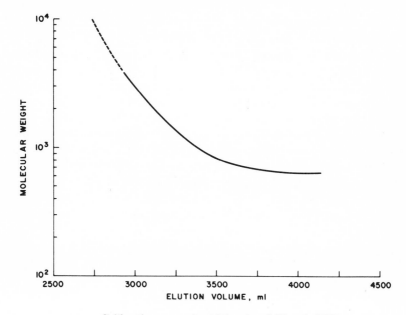

FIG. 3. Calibration curve for 25% reduced Kuwait VTB.

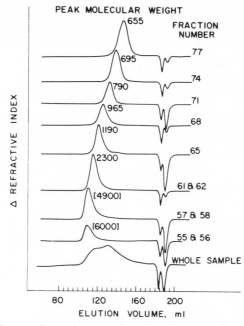

FIG. 4. Preparative scale separation of 25% reduced Kuwait VTB.

lecular weight fractions that could be used to determine the hetero-
atomic distribution as a function of molecular weight.

Preliminary separations were applied to a 25% reduced Kuwait
vacuum tower bottom (VTB) on an analytical scale using a unit of
our design with four Waters Associates Styragel columns. A satisfactory molecular weight separation was achieved with approximately
99% recovery.

In order to obtain sufficient material for characterization, a preparative-scale gel permeation chromatograph was built. A schematic
diagram of the instrument is shown in Fig. 1. Benzene was used as
solvent and was stored in a 20-gal reservoir. The degasser consisted of
a 2-liter stainless steel cylinder wound with resistance heating wire
and fitted with a glass standpipe. A 10-μ porous metal filter was placed
between the degasser and pump. A piston-type pump with a capacity
of 80 ml/min was used. For reducing the pump surges, a stainless steel
bellows was placed in the line ahead of the columns. A pressure relief
valve was also incorporated ahead of the columns. The sample valve
was air-actuated and contained a 50-cc sample loop. The columns,
4 ft $\times$ 2-$\frac{1}{2}$ in., were purchased from Waters Associates and contain
Styragel with the designation of 1000 and 500 Å. Both the sample
valve and columns were housed in an oven maintained at 40°C. A
Waters R-4 differential refractometer was used for detection. The
volume of eluant was measured with a 50-cc siphon. When the siphon
empties, the effluent flows over a thermistor that is part of a relay
circuit. The heating of the thermistor by the eluant actuates the relay,
placing a mark on the recorder chart and activating the fraction
collector.

The chromatogram of a 1-g charge is shown in Fig. 2. The dashed
curve represents the change in refractive index during elution and the
solid line the weight distribution. The refractive index curve is displaced toward lower molecular weights. The molecular weight calibration curve is shown in Fig. 3. The molecular weights were determined by vapor pressure osmometry on selected fractions. The slope
of the curve shows a molecular weight separation up to an elution
volume of approximately 3750 ml. Over 90% of the sample falls in the
size separation range.

To determine the effectiveness of the separation, selected fractions
were analyzed using five analytical columns (10^5, 10^3, 250, 100, and
60 Å). The results are shown in Fig. 4. The bottom curve is the whole
sample, and the curves above represent selected fractions from the

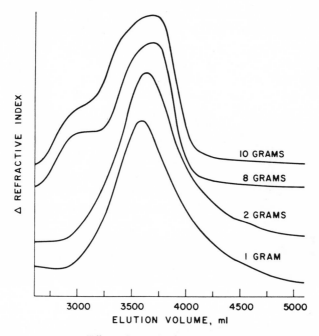

FIG. 5. Effect of sample size on separation.

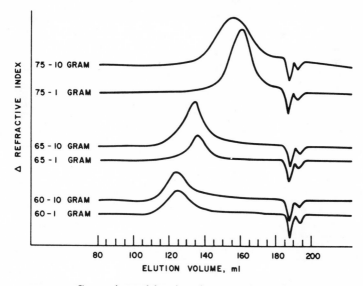

FIG. 6. Comparison of fractions from 1 and 10 g charges.

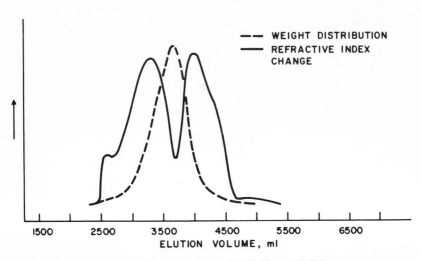

FIG. 7. Chromatogram of 50% Kuwait VTB.

preparative-scale separation. The molecular weight at the peak was obtained from the calibration curve. The numbers in brackets are extrapolated values. For the purposes of our study, these fractions were sufficiently narrow. If more narrow fractions are desired, the

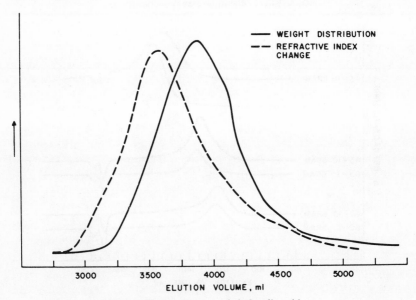

FIG. 8. Chromatogram of shale oil residue.

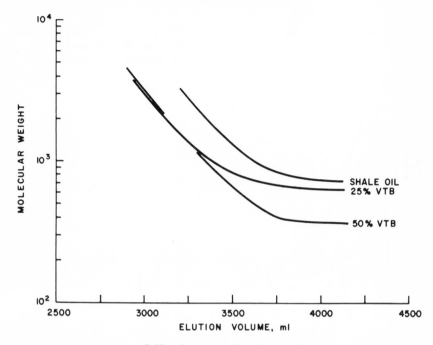

FIG. 9. Calibration curves for three residuals.

individual fractions can be rechromatographed and only the center cuts retained.

To determine the optimum size sample, a series of runs was made in which the sample size was varied from 1 to 10 g. In Fig. 5 are shown the curves from these runs. They are displaced vertically for clarity. The 1- and 2-g samples have nearly identical contours, while the 8- and 10-g samples show some distortion, possibly due to overloading. Several of the fractions from the 1- and 10-g runs were analyzed using the 5-analytical column set. The chromatograms of these fractions are shown in Fig. 6. In all cases, the 10-g fractions are nearly as narrow as the 1-g. The effects of overloading should be most noticeable in Cut 60. However, no appreciable difference is seen.

Two other residues were also fractionated: a 50% Kuwait VTB and a shale oil residue. The chromatogram of the 50% reduced material is shown in Fig. 7. The solid curve is the change in refractive index during elution, while the dashed curve represents the actual weight distribution. This is a very good example of how a refractive index curve can be misleading for complex materials. In Fig. 8 is shown

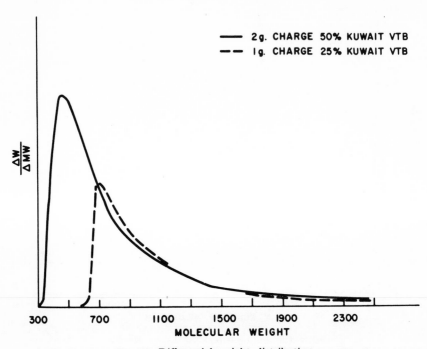

FIG. 10. Differential weight distribution.

the chromatogram for the shale oil. Here, the contours of the curves are similar, but the refractive index curve is displaced toward higher molecular weights, the converse of the 25% reduced material shown earlier. The differences between the weight and refractive index curves for the three materials are due to the variation in compositions. Figure 9 shows the comparison of the calibration curves for the three residues. The slopes of the curves show that a nearly-equivalent separation has been achieved for all three materials. The differential weight distribution of the 25 and 50% Kuwait VTB is shown in Fig. 10. These curves clearly show the weight change during distillation and demonstrate the usefulness of GPC in monitoring the distillation of petroleum.

The fractions from the 25 and 50% reduced Kuwait VTB were analyzed for nitrogen, sulfur, vanadium, and nickel. The differential weight distribution for the 50% reduced Kuwait VTB is shown in Fig. 11. The curves show that the heteroatom distribution is not uniform and the heaviest concentration is in the 350–450 molecular weight

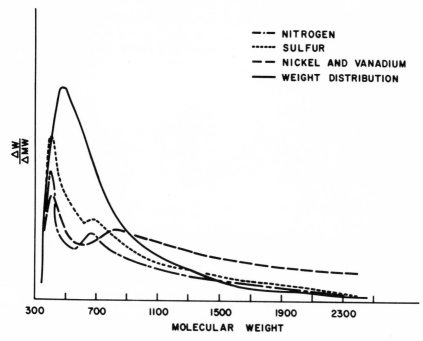

FIG. 11. Heteroatom distribution for 50% Kuwait VTB.

range. While the sample differential weight distribution decreases uniformly with increasing molecular weight, the heteroatom differential weight curve shows a second maximum at 600–700 molecular weight for nitrogen and sulfur and 750–850 for nickel and vanadium.

The heteroatom differential weight distribution curves for the 25% reduced Kuwait VTB are shown in Fig. 12. The sulfur differential distribution shows a maximum in the 700–800 molecular weight range and decreases uniformly with increasing molecular weight. This is the same distribution shown by the VTB differential weight curve, indicating the sulfur is fairly uniformly distributed throughout the sample. The nitrogen shows a similar maximum in the 700–800 molecular weight range and, in addition, a second maximum in the 1000–1100 molecular weight range. The nickel and vanadium distribution curves likewise have a maximum in the 700–800 molecular weight range with a second maximum centered around 1800 molecular weight.

Elemental analyses were obtained on the shale oil. The oxygen contents of the individual fractions were all higher than the original

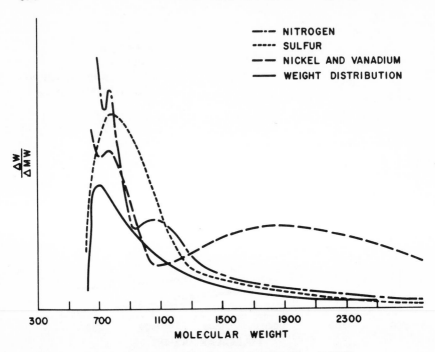

FIG. 12. Heteroatom distribution for 25% Kuwait VTB.

sample. This suggests oxidation during solvent removal, although precautions were taken to minimize this. The elemental data do indicate,
however, that the heteroatom distribution is uniform throughout the
entire molecular weight range.

In conclusion, we have found that large-scale separations of
residuals into narrow molecular weight fractions can be accomplished
by gel permeation chromatography. As a result of these separations,
the study of heteroatom distribution as a function of molecular weight
can be more easily accomplished.

Combined Gel Permeation Chromatography—NMR
Techniques in the Characterization of Petroleum Residuals*

F. E. DICKSON, R. A. WIRKKALA, and B. E. DAVIS

GULF RESEARCH & DEVELOPMENT COMPANY
PITTSBURGH, PENNSYLVANIA 15230

Summary

It has been shown that proton nuclear magnetic resonance spectra of fractions obtained by gel permeation chromatography techniques can yield significant information in the characterization of petroleum residuals. Application of empirical relationships developed for the analysis of NMR spectra is shown to be useful when applied to gel permeation chromatography fractions. Whole residues and a deasphaltened fraction have been separated and studied. Unit weights and aromaticities have been obtained by NMR and compared with molecular weights from vapor pressure osmometry measurements. In addition, supplemental information from infrared and mass spectrometry on GPC fractions has been used to characterize a "neutral" portion of the residue.

Adsorption effects have caused an apparent shift in aliphatic and aromatic components as observed by NMR spectra suggesting that information relating to the mechanism of GPC separations, particularly in regard to factors other than molecular size which influence elution rates, might be obtained from petroleum residual characterizations.

INTRODUCTION

The composition of petroleum residuals has been of interest for a number of years, and several analytical methods have been applied to the problem. Over the past 10 years, various spectroscopic techniques have been used in residual characterization as they have become available or have been improved. Of particular importance has been the

---

* Presented at the ACS Symposium on Gel Permeation Chromatography, sponsored by the Division of Petroleum Chemistry at the 159th National Meeting of the American Chemical Society, Houston, Texas, February, 1970.

application of proton nuclear magnetic (NMR) resonance spectroscopy to these problems. In 1958 Williams (1) used proton NMR in the characterization of saturate, aromatic, and olefinic petroleum fractions. Although the spectra used in this early work were of poor quality and low resolution, the empirical relationships derived were valid and have proven to be extremely useful in NMR petroleum characterization studies. Similar empirical relationships based on the observable proton resonances were reported by Brown and Ladner (2) for vacuum carbonization products of coal. Both of these studies were performed on low frequency (30–40 MHz) and relatively low resolution instruments.

The first application of 60 MHz, high resolution NMR techniques to petroleum characterizations was reported by Yen and Erdman (3) in 1962 in characterization studies of several asphaltene samples from various sources. A typical high resolution NMR spectrum of a petroleum residual is given in Fig. 1, showing the resonances of interest. Empirical relationships have been developed based on the integrated areas of the proton resonances which are then used to calculate structural parameters.

All of the above studies were concerned with unfractionated petroleum or coal residuals. In 1967, Ramsey, McDonald, and Petersen (4) first reported NMR structural studies on various asphalts and asphalts fractionated by elution techniques. Ferris, Black, and Clelland (5) studied a series of 22 asphaltene fractions separated according to solubility. NMR analysis of the series revealed a systematic increase in parameters such as aromaticity and aromatic ring size. Parameters measured by other techniques correlated well with the NMR data.

Altgelt (6), in 1965, was the first to report the fractionation of asphaltenes by gel permeation chromatography (GPC) techniques. It was clearly shown that GPC was to become an invaluable tool in residual characterization. It was further revealed that a definite molecular weight separation was achieved, and in addition, preliminary spectroscopic studies by infrared (IR) suggested that systematic changes in structural characteristics could be monitored. Dickie and Yen (7) later reported the GPC separation of a Baxterville crude petroleum which offered additional evidence for a fractionation by molecular weight (size). More recently, Altgelt (8) studied molecular association in asphaltenes through vapor pressure osmometry (VPO) studies on GPC fractions, pointing out that such fractionation was essential in that it strongly reduced the effects of heterogeneity on

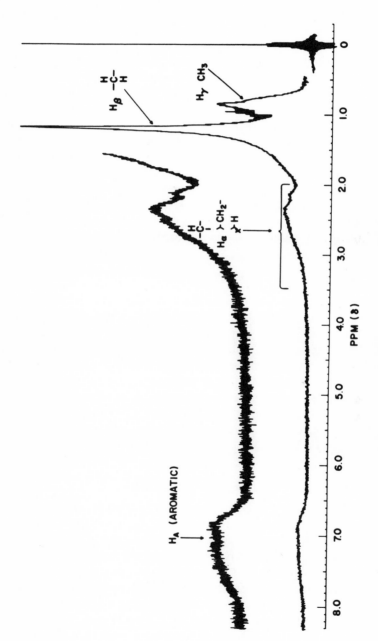

FIG. 1. Proton NMR spectra of Kuwait residual.

the average molecular weight values obtained and it also allowed higher molecular weight samples to be studied.

The application of NMR to the analysis of a series of GPC fractions of petroleum residuals should provide an opportunity to utilize the methods and empirical relationships developed over the past 10 years. Indeed, we have observed systematic changes in aromaticity, f_a, and unit weight (7) with GPC fractions. In addition, through NMR analysis, we concur with recent reports (9, 10) that factors in addition to molecular size influence the fractionation process. A recent publication by Coleman, Hirsch, and Dooley (9) reported significant molecular separation by GPC as elucidated by mass spectral data. NMR spectra revealed that two of their fractions showed a large difference in basic compound types. We have pursued similar studies on a series of GPC fractions of Kuwait residuals and have shown NMR to be a potentially valuable tool in this application.

EXPERIMENTAL

Samples of Kuwait residual dissolved in benzene were fractionated on a gel permeation unit packed with Styragel of 500 and 1000 Å pore size. This system has been previously described by Albaugh et al. (12).

All NMR spectra were obtained on a Varian HA-60 spectrometer system. Samples were dissolved as received from the GPC (solvent removed) in CS_2. Tetramethylsilane was used as the internal reference. Spectra were obtained over a sweep width of 500 Hz at a rate of 2 Hz/sec.

Molecular weights by vapor pressure osmometry were obtained on a Mechrolab VPO unit in benzene.

RESULTS AND DISCUSSION

The effectiveness of GPC separations is usually described by distribution plots of material recovered as well as elemental distributions. Typical GPC distribution plots obtained for a whole Kuwait residual and a deasphaltened Kuwait residual are shown in Fig. 2. Note that an obvious decrease in the amount of material (as indicated by Δ refractive index) eluted in the earlier fractions has occurred with the removal of the asphaltenes. In general, the distribution of hetero elements (e.g., sulfur) and the carbon-to-hydrogen atomic ratio of the fraction series takes the form of a "V"-shaped plot (Fig. 3). Similar

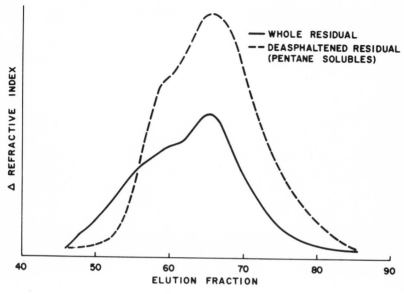

FIG. 2. Kuwait residual.

distributions have been observed for most of the residual fractions studied to date.

Two descriptive parameters developed from NMR data reveal unusual distributions as a function of elution volume. Using the empirical relationships developed for NMR data most recently reported by Ramsey et al. (4) and others (1–3), one can calculate the aromaticity, f_a, which expresses in percentage the number of aromatic carbon atoms in the material and a second parameter, the unit weight, described by Dickie and Yen (7) as the basic "building block" of asphaltic material and as being the weight of a "unit sheet" of pericondensed aromatic rings with aliphatic side chains of varying lengths.

To obtain the unit weight, one must first calculate a "hypothetical unsubstituted unit ring" (4) present in the residual material by using the NMR integrated areas. Additional calculations using the aromaticity value, f_a, described above will yield a total unit weight. Since NMR is capable of observing only the "average" structure present in a given GPC fraction, we have suggested (11) that the weight thus calculated represents the weight of the unit sheet as proposed by Dickie and Yen.

The distribution or change in these NMR descriptive parameters as a function of GPC fraction should supply meaningful information in

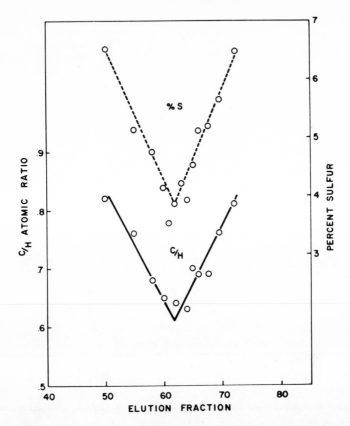

FIG. 3. Kuwait residual. Elemental data vs. GPC elution fractions.

the characterization of the residual material. For example, the distribution of aromaticity, f_a, takes the form of a "V" distribution similar to that observed for sulfur content in the whole residue, suggesting a relation between sulfur and aromaticity.

Of even greater interest, however, is the distribution of unit weights calculated from the NMR data (Fig. 4). These weights agree with VPO measurements in the molecular weight region below ~1000. Above ~1000 the VPO weights increase rapidly as a function of GPC fraction while the NMR unit weights are seen to increase somewhat erratically to a maximum of ~1200 at elution count #50. This behavior of the NMR unit weight distribution plot has been interpreted (*11*) as resulting from the combination of two effects. First, the fact that NMR unit weights are the weights of the *average* unit sheets, or

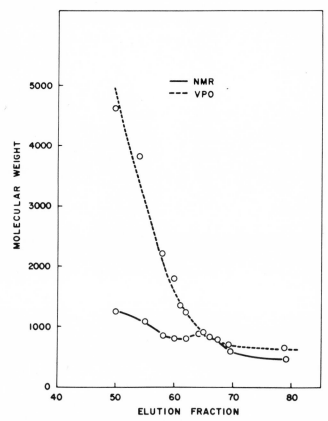

FIG. 4. Molecular weight of Kuwait residual vs. elution fraction from GPC.

the *average* monomer weight of any associated (or coordinated) species in the material (while the higher VPO weights, >1000, represent the weight of coordinated species), and second that the GPC is separating not only by molecular size, but, in addition, adsorption effects have caused a separation by molecular type (e.g., asphaltenes, oils, etc.). Several recent GPC studies concur in this latter observation (*9, 10*). For example, the material above elution count $\sim$64 has been shown to have a relatively high concentration of aromatic material while the material at earlier elution counts is apparently a mixture of overlapping paraffinics, naphthenics, and aromatics. The change in concentration (and molecular size) of these various molecular types with

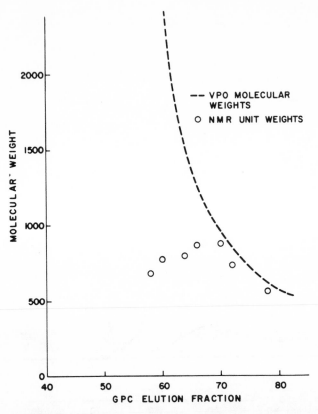

FIG. 5. Kuwait residual "neutral" fraction. Molecular weight vs. elution fraction.

elution volume contributes to the behavior of the plot at early elution volumes. The plot suggests, however, that a maximum unit weight of 1000–1200 is present below elution count $\sim \#62$.

This interpretation has been supported by a more rigorous analysis of a "neutral" fraction of the Kuwait residual obtained by fractionation of the pentane soluble portion. Figure 5 shows the behavior of the NMR "unit weights" and VPO weights as a function of GPC fraction. Note that the behavior is very similar to that observed for the whole residue with excellent agreement below molecular weight ~ 900. Aromaticities (Fig. 6) calculated by NMR show a "V"-shaped distribution similar to the distribution of heteroatoms and aromaticities observed for the whole residue. These data suggest an interpretation

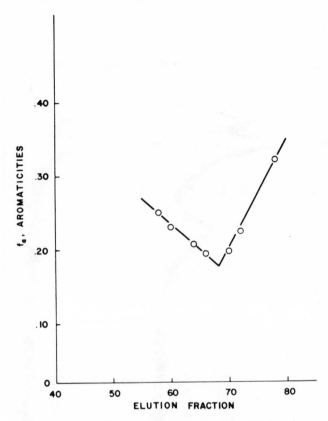

FIG. 6. Kuwait residual "neutral" fraction. Aromaticities, f_a, vs. GPC elution fraction.

similar to that proposed for the whole residue outlined above. However, NMR and IR data clearly show that the early GPC cuts contain no aromatic material while the late GPC fractions (i.e., #80, etc.) are highly aromatic (Fig. 7). This apparent contradiction may be resolved by an examination of the empirical equation reported in the literature for the calculation of aromaticities (Fig. 8) and reiterates the value of GPC in these studies. A serious dependence on the presence of naphthenic material in the sample is evidenced in the equation. If the concentration of condensed ring saturates is significant, then the calculation will give positive aromaticity values in the absence of aromatic material. This is due to the fact that the calculation is dependent on the assumption that the average number of aliphatic

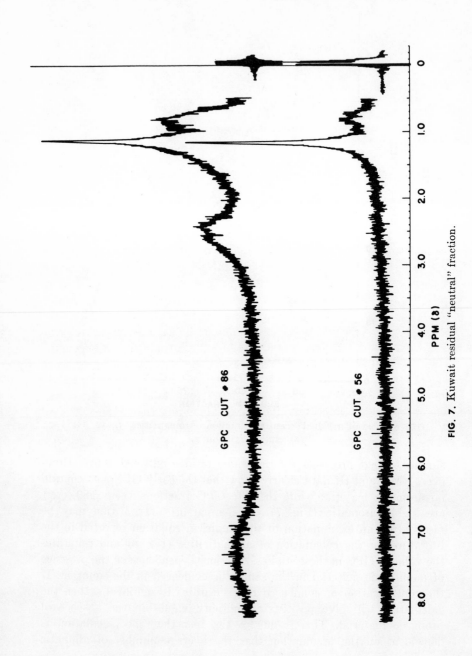

FIG. 7. Kuwait residual "neutral" fraction.

$$\text{Aromaticity, } f_a = \frac{\dfrac{C}{H} - \dfrac{H_\alpha^*}{x} - \dfrac{(H_\beta^* + H_\gamma^*)}{y}}{\dfrac{C}{H}}$$

where C/H = atomic ratio of carbon to hydrogen,
$H_\alpha^*, H_\beta^*, H_\gamma^*$ = normalized NMR integrated areas,
x = av number of protons/α-carbons
y = av number of protons/β- and γ-carbons.
usual assumption, $x = y = 2$
for naphthenic material present, $x = y < 2$

FIG. 8. Equation for calculation of aromaticity from NMR data (2).

protons per aliphatic carbon atom in the system is equal to two ($x = y = 2$). This assumption is necessary to properly interpret the methylene and methine (H_β and H_α) absorptions and is usually valid for large aromatic ring systems with relatively long aliphatic side chains such as asphaltenes and whole residues. However, if the material should contain a high concentration of condensed naphthenic ring systems, as is the case in our "neutral" material, then the assumption is invalid ($x = y < 2$) and introduces errors in the aromaticity calculations and ultimately into the unit weight calculations which are dependent on f_a. It has already been shown that the unit weights determined thus far for all the residue fractions studied agree with vapor pressure osmometry measurements and mass spectral measurements below $\sim$1000 ($\sim$elution cut #62) (Table 1). This indicates that the NMR calculations are valid in this region and presumably little or no naphthenic material is present. The reason NMR unit weights for the earlier eluted material tend not to agree with VPO data is, in the case of the "neutral" fraction, a result of the presence of condensed naphthenic rings yielding abnormally high aromaticity values. When introduced into the unit weight calculations, these aromaticity values will yield low unit weights.

Mass spectral analysis of the earlier eluted material of the "neutral" fraction supports our NMR interpretation, showing that the material is composed of a high concentration of small ring condensed and non-condensed saturates. Furthermore, ultimate separation of the saturate material from the aromatic material in the bulk "neutral" fraction indicated that 20% of the original material was saturate.

Even though the possibility of error in the aromaticity calculations due to the presence of naphthenic material exists in the earlier GPC fractions, the NMR average unit sheet weights in this region still

TABLE 1

Kuwait Residual Fractions Molecular and Unit Weights

Fractions	NMR	VPO	Mass spectroscopy
		Whole Residual	
74–84	460	650	400
69–70	590	675	546
66	820	840	>700
		Deasphaltened Residual	
78	325	—	350
74	460	—	—
70	670	765	620
68	760	760	—
66	925	920	—
64	1064	1180	980–1000
		Neutral Fraction	
78	550	660	—
72	740	825	—
70	880	850	—
66	860	800	—
64	800	~1000	—

tend to be grossly below those reported by VPO measurements. The agreement between mass spectral molecular weights, VPO number-average molecular weights, and our calculated NMR unit weights in the later GPC fractions (>elution count #62) suggests the NMR empirical relationship to be valid and useful in characterizing GPC fractions with respect to unit weight. In addition, the anomalies caused by naphthenic material suggest that the technique may offer a method of detection for naphthenic material (e.g., comparison of calculated aromaticities with actual aromatic content by NMR).

CONCLUSIONS

Because of the ability of gel permeation chromatography to separate materials by molecular size as well as possibly other inherent properties, its usefulness in the characterization of petroleum residuals has been recognized. The application of proton NMR to analysis of the separated fractions can yield supplementary information.

We have suggested that the empirical relations developed using NMR integrated areas can yield meaningful trends in fundamental characterizing parameters as a function of gel permeation fraction. Aromaticity and unit weight have been shown to vary with elution volume. Other more esoteric but perhaps useful NMR parameters which have been described in the literature (1–4) may also show the same trends. Future studies will explore these possibilities. In addition, it appears that the application of NMR techniques to GPC fractions can yield data not only on the characterization of petroleum residuals but also information related to the mechanism of GPC separations, possibly showing trends in the fractionation of molecular types.

Acknowledgments

The authors wish to express their appreciation to Dr. E. W. Albaugh and Mr. P. C. Talarico for helpful consultations during the course of this research and to Dr. A. B. King for several valuable discussions relating to interpretation of the data.

REFERENCES

1. R. B. Williams, *Symposium on Composition of Petroleum Oils* (ASTM Special Technical Publication 224), 1958.
2. J. K. Brown and W. R. Ladner, *Fuel* (London), **39**, 87 (1960).
3. T. F. Yen and J. G. Erdman, *Amer. Chem. Soc. Petrol. Chem. Preprints,* **7**(4). (September, 1962).
4. J. W. Ramsey, F. R. McDonald, and J. C. Petersen, *Ind. Eng. Chem., Prod. Res. Develop.,* **6**(4), 231 (1967).
5. S. W. Ferris, E. P. Black, and J. D. Clelland, *Ind. Eng. Chem., Prod. Res. Develop.,* **6**(2), 127 (1967).
6. K. H. Altgelt, *Amer. Chem. Soc. Div. Petrol. Chem. Preprints,* **10**(4), (September, 1965).
7. J. P. Dickie and T. F. Yen. *Anal. Chem.,* **39**(14), 1847 (1967).
8. K. H. Altgelt, *Amer. Chem. Soc. Div. Petrol. Chem. Preprints,* **13**(4), (September, 1968).
9. H. J. Coleman, D. E. Hirsch, and J. E. Dooley, *Anal. Chem.,* **41**(6), 800 (1969).
10. B. C. B. Hsieh, R. E. Wood, L. L. Anderson, and G. R. Hill, *Anal. Chem.,* **41**(8), 1067 (1969).
11. F. E. Dickson, B. E. Davis, and R. A. Wirkkala, *Anal. Chem.,* **41**(10). 1335 (1969).
12. E. W. Albaugh, P. C. Talarico, B. E. Davis, and R. A. Wirkkala, *Separ. Sci.,* **5**, 801 (1970).

A Rapid Method of Identification and Assessment of Total Crude Oils and Crude Oil Fractions by Gel Permeation Chromatography*

J. N. DONE and W. K. REID

BP RESEARCH CENTRE
THE BRITISH PETROLEUM COMPANY LIMITED
SUNBURY-ON-THAMES, MIDDLESEX, ENGLAND

Summary

"Fingerprint" chromatograms of crude oils have been obtained using a simple gel permeation chromatographic system with a differential refractometer as detector and tetrahydrofuran as solvent. The elution time is less than 1 hr using a sample size of about 6 mg and the repeatability is good. When toluene is used as solvent, a different profile is obtained which is again unique for each crude oil.

The applications of this technique to crude oil assessment and composition of crude oil mixtures will be discussed since the "fingerprints" obtained have been shown to be unique for the 50 crude oils examined. Various crude oil fractions, e.g., gas oils and waxy distillates, have also been investigated. Chromatograms of oil pollution samples from the UK have indicated that certain types of pollution can be identified although present experience with a wide range of sample types is limited.

INTRODUCTION

The separation achieved by gel permeation chromatography (GPC), mainly based on molecular size, has given polymer and petroleum chemists a new method of attack on some of their more complex problems. Originally GPC using organic solvents was developed by Moore (1) who prepared cross-linked polystyrene gels of known pore

---

*Presented at the ACS Symposium on Gel Permeation Chromatography, sponsored by the Division of Petroleum Chemistry of the 159th National Meeting of the American Chemical Society, Houston, Texas, February, 1970.

size. The application of this technique to polymers (2–4) has been widespread using commercial apparatus (5) which produces molecular weight distribution analyses.

Although commercial instrumentation has been used almost exclusively in the polymer field, the petroleum chemist has been aware of the advantage of the gel permeation process. His usual approach, however, has been to conduct preparative scale separations and then characterize the fractions obtained by using such techniques as mass spectrometry, high temperature programmed gas liquid chromatography, NMR, etc. (6–8). Nevertheless, some molecular weight distributions of heavy petroleum materials have been reported using GPC (9). Bombaugh et al. have also shown (10) that it is possible to obtain chromatograms of total crude oils and have published the GPC chromatogram of a Kuwait crude oil before and after removal of the light ends by distillation. Other contributions have also been made by workers in the petroleum field to the fundamental studies of the GPC of both large and small molecules (11, 12).

This paper describes the use of a liquid chromatography detector (a Waters Associates R4 Differential Refractometer) in a simple GPC system to obtain chromatograms of total crude oils and crude oil fractions. By this method "fingerprint" chromatograms can be obtained in under 1 hr and the profiles have been shown to be unique for each of the crude oils examined.

The applications of the technique to crude oil identification, classification, and characterization and to crude oil fractions are discussed and so too is its application to the examination of oil pollution samples. Essentially the technique provides a simple, analytical scale separation procedure which produces results useful to the petroleum chemist.

EXPERIMENTAL

Columns of 60 Å Styragel (24 in. × ⅜ in. i.d.) were packed in both tetrahydrofuran (THF) and toluene using the high pressure packing apparatus previously described by Waters Associates (13). Both the solvents were redistilled before use and, during the chromatography, a nominal solvent flow of 70 ml/hr was maintained using a micropump (Type M, Distillers Company Limited) at a fixing setting. The efficiencies of the columns were estimated using samples of toluene and THF and the columns were found to have theoretical plate counts in excess of 2000/ft.

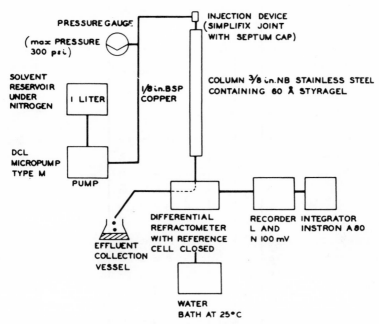

FIG. 1. Block diagram of system used to study crude oils and crude oil fractions by GPC.

The detector used in this work was a differential refractometer (Waters Associates), the inlet connections of which had been modified to reduce the dead volume. The reference cell, which was used with pure solvent trapped inside it, and the sample cell of the refractometer were kept at 25°C by a constant temperature water bath. A block diagram of the system is shown in Fig. 1.

The response of the detector will depend on the components being eluted and this complicates interpretation of the chromatograms. The use of the two separate solvents, THF and toluene, gives different chromatograms because the base refractive index (RI) is changed from 1.4040 to 1.4969. These effects will be discussed later in the paper.

Samples of crude oils, crude oil fractions, pollutants, and blends of crude oils and pure materials were run. The less viscous crude oil fractions and blends were injected without dilution but the other samples were first diluted in the ratio of 1:2 with solvent. The samples were injected with a microsyringe through a septum onto the column and each was run three times. The sample size chosen was between

TABLE 1

Typical Measurements of a Crude Oil Chromatogram to Give Figures for Profile Definition

	Distance of ordinate from exclusion peak (mm)																
	-8.5	-2	0	3	11	15.5	25	33	38	44.5	51	57	64	67.5	72	76	82
Height of ordinate (mm)																	
Run 1	3.5	166.1	226.1	155.7	94.2	93.0	83.8	74.2	64.3	50.3	40.6	30.5	20.3	22.1	15.2	6.1	0.5
Run 2	.3	162.3	231.9	158.8	92.2	90.9	82.0	72.6	63.5	49.8	39.4	32.0	21.1	22.1	14.7	5.1	0.8
Run 3	2.8	160.0	230.0	152.9	91.7	90.4	81.3	71.9	61.5	47.5	37.8	30.2	19.3	19.8	12.2	3.8	0.0
Average value	3.2	162.8	229.3	155.8	92.7	91.4	82.4	72.9	63.1	49.2	39.3	30.9	20.2	21.3	14.0	5.0	0.4

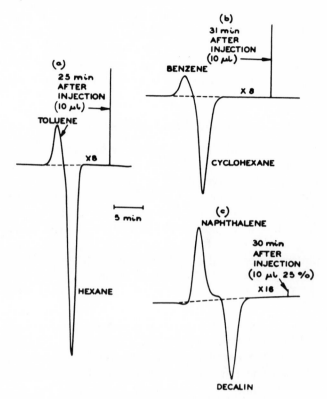

FIG. 2. GPC chromatograms of blends of pure hydrocarbons. Detector: Waters Differential Refractometer. Column: 2 ft × ⅜ in. NB stainless steel, 60 Å Styragel. Solvents: (a) tetrahydrofuran; (b) and (c) toluene. Temperature: 25°C.

10 and 30 μl depending on the response of the detector to the sample.

The chromatograms obtained using THF were treated in two ways. All of them were integrated to give the area of the exclusion peak and the total area using an Instron A80 integrator. Three chromatograms of each crude oil were run and the average values calculated. Using the average figures, the per cent area corresponding to the exclusion peak and the response of the sample, measured in counts per milligram, were calculated. The chromatograms obtained for 16 crudes were defined by measuring the height of the profile above the baseline at a number of fixed distances from the exclusion peak. The three repeat chromatograms produced for each of the crudes were treated in this way and the average values calculated. A typical set of figures for the Kuwait crude oil is shown in Table 1.

All the crude oil samples were also chromatographed using toluene as solvent. Kuwait crude oil fractions were run using both solvents while the blends of pure materials were run under the column conditions given in Fig. 2.

RESULTS AND DISCUSSION

The authors consider that this technique of examining crude oils by GPC is novel and that it may have a wide range of applications in petroleum and refining technology.

The chromatograms obtained in this work from crude oils and crude oil fractions are essentially plots of change of RI compared to a reference against molecular weight or, more correctly, molecular size. The two parameters of RI and molecular weight are important in petroleum technology and have been utilized previously, for example in the n-D-M method (14), for characterizing and analyzing petroleum materials. The peaks and valleys produced in the GPC "fingerprints" of crude oils and crude oil fractions are obviously related to types and amounts of hydrocarbons present, although the present work has not considered this in detail.

Important features of this work are the production of chromatograms in only 40 min and the requirement for less than 10 mg of crude oil. It is considered that the technique could easily be put on a routine basis and, furthermore, as the residence time of the components of the oil is between 20 and 40 min, samples can be injected about every 20 min without risk of overlap. Thus, in practice, 3 chromatograms can be produced in each hour. The cost of the equipment is approximately £1700 ($4000), including refractometer, recorder, integrator, and pump.

The use of efficient columns with different pore sizes from those considered here may well reveal more detail in the exclusion peak of the chromatogram. However, the aim of the present system has been to keep the technique as simple and as rapid as possible.

Pure Compound Blends

It is well known that adsorption effects in GPC cannot be ignored and some blends of pure hydrocarbons have been examined on the present GPC system to determine the magnitude of this effect in toluene and THF. The chromatograms obtained from blends of aro-

matic hydrocarbons with saturated hydrocarbons of approximately
the same molecular weight (Fig. 2) show that adsorption of the aromatics occurs both in toluene and THF. This is evident from the
increased elution time of the aromatics compared to saturate compounds of the same average molecular size. It can also be seen that
this adsorption effect is more pronounced in toluene and that the use
of toluene as solvent (and thus as reference) can indicate the aromatic or saturate nature of crude oils or crude oil fractions from the
positive or negative peaks.

Crude Oil Fractions

Crude oil fractions from Kuwait crude oil, gasoline (C_5, 149°C),
kerosene (149–232°C), light gas oil (232–343°C), heavy gas oil (343–
371°C), waxy distillate (371–525°C), and residue (>525°C), have all

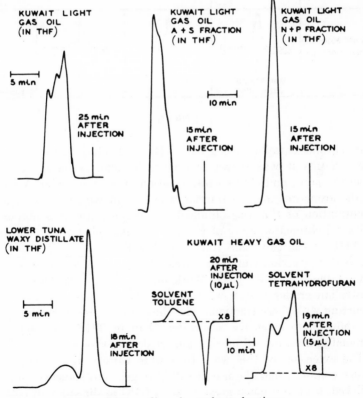

FIG. 3. GPC profiles of petroleum fractions.

Group	"Fingerprint"	Approximate Grouping according to "Fingerprint" Type in THF		
A		Amna, Khafji, Kuwait, Ratawl, Arabian Heavy, Sarir, Tia Juana 102LP, Bachaquero (Venezuela), Trinidad Soldado, Marguerite Lake (Canada)		
B		Iranian Light, Arabian Light, IMEG "A", IMEG "B", Iranian Heavy Middle Ground Shoal, Soldatno Creek Swanson River (both Alaska) Fyzabad (Trinidad), Oficina (Venezuela), Kern River (USA)		
C		Arzew Blend, Zarzaitine (Algeria), ADLEG, Qatar, ADMEG (Umm Shaif), ADMEG (Zakum)		
D		Ebocha, Imo River, Nigerian Light, Obigbo North (all Nigeria)		
E		Flounder, Halibut, Kingfish, Lower Tuna (all Australia) Maui (New Zealand)		
Profiles not occurring in the above groups	 Moonie (Australia)	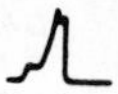 Cori (Libya)	 Minas (Sumatra)	 Puri (Papua)
	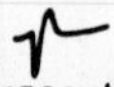 Barracouta (Australia)			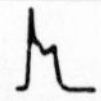 Kapuni (New Zealand)

FIG. 4.

been chromatographed using both THF and toluene as solvent. From these results it was shown that 60 Å Styragel excluded the residue from the pores and effected fractionation of the remainder. Adsorption of the aromatics in each fraction was evident with both solvents and confirmation of this was obtained by studying fractions obtained by silica gel chromatography of Kuwait light gas oil. The chromatograms in THF of the two main fractions (A + S and N + P) showed that the N + P fraction eluted first followed by the A + S fraction which gave a much broader peak. Work on the waxy distillates from two Australian crudes has shown that this separation by polarity is also occurring with these fractions. The chromatograms obtained in toluene all show more extensive adsorption than in THF. The appropriate chromatograms, not all on the same scale, are shown in Fig. 3.

The extension of this work to a wide variety of fractions, use of longer columns, and the use of different gels and solvents could, it is believed, lead to a much greater use of GPC in the study of petroleum

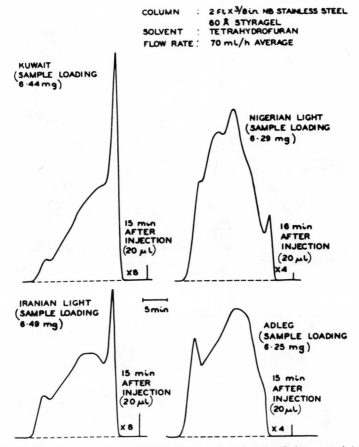

FIG. 5. Examples of GPC chromatograms of crude oils. Column: 2 ft ×
⅜ in. NB stainless steel, 60 Å Styragel. Solvent: tetrahydrofuran. Flow
rate: 70 ml/hr average.

fractions. The interpretation of these chromatograms would, however,
not be straightforward as the response of the detector to different
A + S fractions would not necessarily be the same due to compositional
differences.

Crude Oils

By using the experimental conditions discussed above, over 40 crude
oils listed in Fig. 4 have been chromatographed. Each chromatogram
obtained has been shown to be unique and may be considered as a

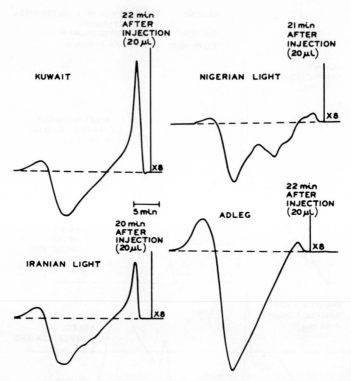

FIG. 6. GPC chromatograms of crude oils using toluene as solvent (compare Fig. 5). Column: 2 ft × ⅜ in. NB stainless steel, 60 Å Styragel.

"fingerprint" of the crude oil. Four examples of such chromatograms are shown in Fig. 5 for Kuwait, Nigerian Light, Iranian Light, and ADLEG crude oils in THF. When the solvent is changed to toluene, different, but once again "fingerprint" chromatograms are obtained for these crudes as shown in Fig. 6.

Although unique, the crude oil chromatograms fall into definite classes depending mainly on the light or heavy nature of the crudes. The four Gippsland crudes, Kingfish, Flounder, Lower Tuna, and Halibut, give similar chromatograms in THF with little or no exclusion peak visible. However, each chromatogram has a characteristic pattern as is shown in Fig. 7. In addition, profiles in THF of two North American crudes, a South American crude, and one Russian crude are shown in Fig. 8.

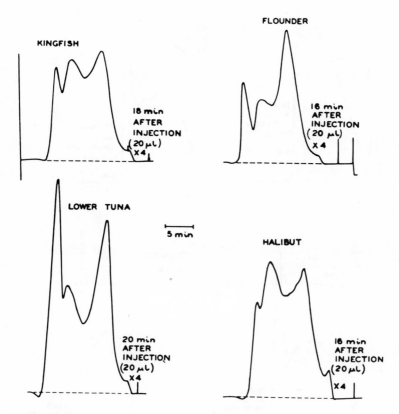

FIG. 7. GPC chromatograms of four Gippsland crudes using tetrahydrofuran as solvent.

The definition of a crude oil by a unique "fingerprint" is obviously valuable, and this technique may prove a quick method for examining a new crude or crude blend as information on the general character of the material can be established very quickly.

The fingerprints, although unique for each crude, appear to fall into definite groups as shown in Fig. 4. The significance of these similarities is, however, not yet understood. Obviously they do not correspond to the traditional methods of classification and, apart from the Nigerian and Australian crudes, there is no geographical or geological correlation.

Some of the crude profiles obtained do not fall into these groups and these are noted also in Fig. 4. These in general tend to be the profiles

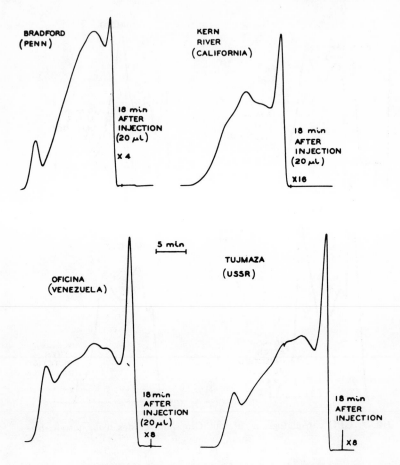

FIG. 8. Further examples of crude oil profiles on tetrahydrofuran.

from the lighter crudes. The "fingerprint" obtained from Barracouta crude is worthy of mention. From the chromatogram, which has a large negative peak in the low molecular region and a very small exclusion peak, it can be concluded that this is a light crude with a large amount of paraffinic material in the low molecular weight portion and having practically no residue above 525°C. This has been confirmed by GLC analysis and distillation.

Chromatograms in THF have been obtained from different wells in the same field. In the two fields examined, Imo River and Zakum, the different well samples give different profiles, although those from the Zakum wells were very similar to each other.

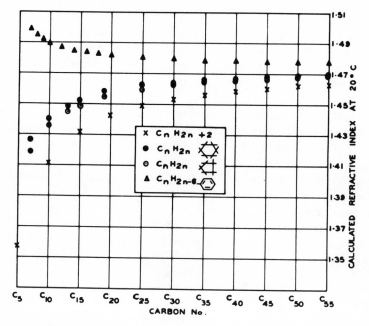

FIG. 9. Graph of refractive index vs. carbon number of four series of hydrocarbons.

Interpretation of Profiles

The interpretation of the profiles is dependent on knowledge of the refractive index of the components eluted. The calculated refractive indices (*15*) of four series of hydrocarbons are shown in Fig. 9. It can be seen that the values appear to approach a common figure as the molecular weight increases, and that in the low molecular weight region the differences in refractive index are greatest. This means that interpretation of the profile with respect to hydrocarbon type will have most meaning in the low molecular weight region of the profile. A qualitative estimate of the aromatic nature of the light end of the crude can thus be obtained by inspection of the final peak which is evident in many of the profiles. These estimates are, however, complicated by the fact that adsorption of polycyclic aromatics occurs and di- and tricyclic aromatics elute in the same region as monocyclic aromatics.

Another feature of this region of the profile is the accentuation of the final aromatic peak by the low molecular weight saturated and cyclic hydrocarbons. These compounds elute just before the aromatic peak and give a negative contribution to the response. Consequently the over-all effect observed in THF is a fairly sharp terminal peak. This effect is, however, most evident in the chromatograms using toluene when a large negative peak is nearly always present just before the final positive aromatic peak.

General interpretation is complicated by the overlap of aromatic hydrocarbons with nonaromatic hydrocarbons of lower molecular weight. The presence of polycyclic aromatics in certain crude oil fractions may also have considerable effect on the response since these compounds have appreciably higher refractive indices than nonaromatic and monocyclic aromatic hydrocarbons.

It can thus be seen that the interpretation of the profiles with respect to hydrocarbon composition of the crude is extremely difficult although it is possible to obtain some information of this nature from the profile.

Repeatability

In order for this technique to be useful in the applications discussed in this paper, the repeatability of the method must be good. It has been found that consecutive injections give good repeatability but poor injections or slight variations in flow rate do alter the peak height and the total peak width. However, with repeat injections of samples made up at different times, up to 2 weeks apart, slight differences have occurred in the low molecular weight region of the chromatogram. This could mean that the crude has lost some of the light ends or that water has been absorbed during storage. It is important therefore to ensure representative sampling of the crude and that water contamination be avoided if repeatable chromatograms are to be obtained. However, both short-term and long-term repeatability should be improved if the chromatographic conditions are carefully controlled.

Corrections can be applied to the chromatograms if the flow rate has differed from a standard value. Using the ratio of elution time of the whole profile to elution time of a standard chromatogram obtained at a fixed flow rate of 1.1 ml/min, the total area can be corrected to this standard elution time. The position of the ordinates used for the definition of the crude can also be corrected using this ratio.

Examples of the repeatability of consecutive runs of some of the crude oils are shown in Tables 1 and 2. The repeatability of the total

TABLE 2

Repeatability of Exclusion Peak Areas and Total Areas

Crude oil	Country of origin	Exclusion peak				Total			
		Run 1	Run 2	Run 3	Mean	Run 1	Run 2	Run 3	Mean
Kuwait Export	Kuwait	6516	6406	6220	6381	24143	23450	23455	23683
Amna	Libya	7380	7558	7648	7529	20127	19821	20635	20194
Khafji	Neutral Zone	6724	7110	7738	7191	17551	17902	17908	17787
Iranian Light	Iran	3802	3852	3814	3823	22271	21846	22482	22200
Arabian Light	Saudi Arabia	3734	3374	3668	3592	20966	20842	20575	20794
Ratawi	Neutral Zone	7305	6958	5978	6747	17638	17604	17415	17552
Sarir (Tobruk)	Libya	6186	5824	6254	6088	17575	17828	18086	17833
IMEG "A"	Iraq	5078	5228	5148	5151	26589	26666	26233	26495
IMEG "B"	Iraq	4418	4758	4650	4609	20989	21340	20788	21039
Tia Juana 102LP	Venezuela West	6176	6194	6340	6237	19792	20074	19929	19931
Arabian Heavy	Saudi Arabia	5126	6116	6440	5894	13220	16626	16517	16422
Nigerian Light	Nigeria	1376	1468	1488	1444	34670	34321	34090	34360
ADMEG (Zakum)	Abu Dhabi	3732	3718	3614	3688	40386	38922	38527	39193
Arzew Blend	Algeria	1614	1624		1619	26687	26358		26522
ADILEG	Abu Dhabi	2076	2112	2124	2106	30576	31354	31027	30986
Iranian Heavy	Iran	5856	5814	5716	5795	25182	24780	24262	24741

area is better than that of the exclusion peak. With some of the samples, such as Zakum crude, the exclusion peak measurement is difficult due to incomplete resolution.

Crude Oil Properties

Since the chromatographic profile of each crude oil examined to date appears to be unique, the possibility of determining crude oil properties by this method is being examined. Already it has been possible to estimate fraction yields from the chromatograms with reasonable accuracy.

For example, from investigations on Kuwait crude oil fractions it was shown that 60 Å Styragel excluded completely the crude residue (>525°C). Therefore the area of the exclusion peak should be related to actual per cent of residue and this has been shown to be reasonably correct.

The possibility of the method being used to estimate rapidly the amount of each crude in a crude oil blend has also been examined. Results to date are encouraging, although the accuracy of the estimates for a wide range of crudes and complexity of blends has not yet been established.

Oil Pollution

Work on oil pollution has so far been somewhat limited. However, whenever samples have been available they have been examined using the GPC system here described. The GPC chromatograms of two types of pollutants in the two solvents are shown in Fig. 10. Examples of the three main types of pollution, weathered crude oil, crude oil sludge, and fuel oil, have been examined and it is considered that with further experience, identification of the type of pollution will be possible.

As would be expected, the profile from the weathered crude (see Fig. 10) is similar to its parent crude (see Fig. 5) except that loss of light material is evident. Crude oil sludges, since they are crudes with extra waxy material, should give a profile similar to the weathered crude with an extra peak due to the wax just after the exclusion peak. This has been shown to be so for two crude oil sludge samples. Fuel oil pollution is the most difficult to deal with due to the wide variety of fuel oils used. It has been found, however, that the sharp cutoff in the chromatogram in the low molecular weight region is, so far, the best indication that the pollutant is a fuel oil.

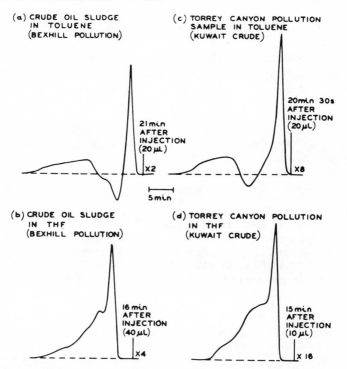

FIG. 10. GPC chromatograms of two pollution samples in toluene and tetrahydrofuran.

In the study of pollution, GPC has the distinct advantage that a profile of the *total* sample can be obtained rapidly. Present methods of pollution identification (*16*) are based on sulfur, vanadium, and nickel contents and GLC analysis, the latter being restricted by the relative involatility of the sample. GPC may well prove to be a valuable additional method for pollution analysis.

CONCLUSIONS

A rapid, relatively inexpensive method of producing GPC profiles of total crude oils has been developed. Each crude oil gives a unique chromatogram which, it is considered, opens up many possibilities for the application of the technique in the petroleum industry. The identification and partial characterization of crude oils and to a lesser extent crude oil fractions is considered to be possible, and further applications to crude oil blends and oil pollution problems have been indicated.

Further development of the technique, by using different size gels, different solvents, thermostatic control of columns, and possibly longer columns is obviously important in order to realize its full potential.

Acknowledgment

Permission to publish this paper has been given by The British Petroleum Company Limited.

REFERENCES

1. J. C. Moore, *J. Polym. Sci., Part A,* **2**, 835 (1964).
2. J. Cazes, *J. Chem. Educ.,* **43**, A567 (1966).
3. M. Determann, *Angew. Chem. Internat. Ed.,* **3**, 608 (1964).
4. R. L. Pecsok and D. Saunders, *Separ. Sci.,* **1**(5), 613 (1966).
5. J. Cazes, *J. Chem. Educ.,* **43**, A625 (1966).
6. K. H. Altgelt, American Chemical Society, Division of Petroleum Chemistry, Atlantic City Meeting, September, 1965.
7. K. H. Altgelt, *J. Appl. Polym. Sci.,* **9**, 3389 (1965).
8. H. J. Coleman, D. E. Hirsch, and J. E. Dooley, *Anal. Chem.,* **41**, 800 (1969).
9. W. B. Richman, *Molecular Weight Distribution of Asphalts,* Waters Associates, Framingham, Massachusetts.
10. K. J. Bombaugh, W. A. Dark, and R. F. Levangie, *Separ. Sci.,* **3**, 375 (1968).
11. P. C. Talarico, E. W. Albaugh, and R. E. Snyder, *Anal. Chem.,* **40**, 2192 (1968).
12. B. J. Mair, P. T. R. Hwang, and R. G. Ruberto, *Anal. Chem.,* **39**, 838 (1967).
13. Waters Associates Inc., *Polystyrene Gel Column Packing,* Framingham, Massachusetts.
14. K. Van Nes and H. A. van Westen, *Aspects of the Constitution of Mineral Oils,* Elsevier, Amsterdam, 1951.
15. Ref. *14,* p. 81.
16. J. V. Brunnock, D. F. Duckworth, and G. G. Stevens, *J. Inst. Petrol.,* **54**, 310 (1968).

Gel Permeation Analysis of Asphaltenes from Steam Stimulated Oil Wells*

C. A. STOUT and S. W. NICKSIC

CHEVRON OIL FIELD RESEARCH COMPANY
LA HABRA, CALIFORNIA 90631

Summary

Gel permeation chromatography was used to obtain molecular weight distributions of asphaltenes from 5 wells before and after steam injection. These wells all produce from the same horizon in one Kern River, California, field. A standard Q-mode correlation analysis showed no significant relationship of stimulation response to any of the molecular weight data, i.e., molecular weight distribution before, immediately after, or 200 days after treatment. A standard R-mode analysis showed a significant relationship of the slope of the molecular weight curve before stimulation with the slope just after stimulation, but not with the slope 200 days after stimulation. The wells that returned the least amount of oil per barrel of water injected as steam produced asphaltenes after stimulation very deficient in high molecular weight compounds. It is inferred from this that these materials precipitated within the formation and are restricting flow. At the other extreme, the change in asphaltene molecular weight profile from the most successful stimulation job can best be explained by the introduction of new production sources into the oil flowing towards the well. This is consistent with this well's having received the greatest quantity of steam.

In summary, GPC provides a valuable increase in analytical sensitivity to changes in the fraction of crude most responsible for determining its production flow rate.

INTRODUCTION

The high molecular weight and heterogeneous composition of crude oil heavy ends have presented formidable barriers to fractionation and

* Presented at the ACS Symposium on Gel Permeation Chromatography, sponsored by the Division of Petroleum Chemistry at the 159th National Meeting of the American Chemical Society, Houston, Texas, February, 1970.

611

subsequent characterization. Column chromatography, thermal diffusion, molecular distillation, and solvent separation have been the major separation techniques used on petroleum residues in the past. These tools have now been supplemented by gel permeation chromatography (GPC) which may become, if it has not already done so, the most generally satisfactory procedure for obtaining reasonably narrow cuts of crude oil heavy ends for detailed analytical studies. In the present study gel permeation chromatography is applied to asphaltenes from steam stimulated oil wells in the search for a better understanding of the role asphaltenes play in oil production before and after stimulation (1).

Steam is commonly injected into a producing oil well to stimulate production. Characteristically, wells which produce low gravity (API) crude from a relatively shallow depth respond most favorably to steam stimulation. The principal beneficial action of steam is the reduction in fluid viscosity brought about by local heating of the formation rock. Several additional processes of lesser importance undoubtedly take place at the same time. One is the clean-up of the well bore and of the production hardware area by the hot steam. Such action increases surface area and exposes additional cation exchange capacity for subsequent interaction with asphaltenes and other basic nitrogen compounds of crude oil. Another action of the steam is the partial degradation of rock and the attendant separation and dispersion of silt and clay. This is shown by increased silica, alumina, boron, iron, and other rock constituents in the produced water after steaming. Also, sanding problems appear or become more severe after steaming. Still another process occurring during steam stimulation is steam distillation of the more volatile crude oil components as steam flows through the volume of partially oil-saturated rock. The oil that remains in place is subjected to higher temperatures for a longer time the closer it lies to the well bore. The effect of this treatment on the composition of heavy ends may be predicted by laboratory studies but several questions remain unanswered. Two of these are the adhesion of the heat-treated residue to the formation rock surface and the effect this coating has on the distribution of the returning crude oil and brine.

The combined chemical and physical effects of steam on the rock, oil, and brine will be reflected in the composition of the fluids produced from the well following the treatment. Careful analysis of these fluids before and after stimulation is the only means available to study these effects. The validity of conclusions from such a study depends on the sensitivity and comprehensiveness of the analytical techniques. Prior to GPC, characterization of the heavy ends of crude oil was a major, inadequacy in the analysis of oil field produced fluids. In the present study of steam-stimulated Kern River wells, the samples were a part of a group for which a large amount of analytical data had already been obtained. It was desired to use GPC to obtain fractions so that molecular weight distributions of the asphaltenes could be determined. The molecular weight information, while interesting in its own right, was expected to be a worthwhile complement to the customary data already obtained on the samples. In a sense, a principal objective is to see if the molecular weight distribution adds materially to an understanding of the stimulation process and of the factors that cause success or failure of the treatment.

EXPERIMENTAL SECTION

Crude oil samples were collected at the wellhead and were stored in glass until used. Asphaltenes were separated from the crude by a procedure starting with 20 volumes of pentane added to 1 volume of water-free sample. After overnight standing, the asphaltenes were removed by filtration, washed briefly with pentane, and then exhaustively (12 hr or more) extracted with pentane in a Soxhlet-type apparatus. At the end of this treatment no colored extract was being removed from the asphaltenes. The product was a black free-flowing, finely divided powder.

Gel permeation was carried out by a procedure previously described (2) except that a somewhat shorter column (70 vs. 120 cm) was used. Also, the sample was placed on the column in benzene instead of a mixed solvent.

The eluting solvent was 5 volume % reagent grade methanol in reagent grade benzene. The methanol diminishes but does not entirely eliminate absorption of asphaltenes on the styrene–divinylbenzene copolymer gel. The gel was prepared for maximum separation in the 300–30,000 molecular weight range. A manual system of sample collection was used to obtain 18 fractions for molecular weight determination.

Molecular weights were obtained using a Mechrolab vapor pressure osmometer Model 302 operated at 37°C. Chloroform solvent was used and asphaltene concentrations were kept below 10 g/1 to minimize association effect (3). Other details are given in the previously cited Ref. 2.

The first fraction, and at times the second, were of lower molecular weight than the subsequent fractions. This anomalous behavior of asphaltene samples has been observed many times in this laboratory. No systematic investigation of this phenomenon has been made or reported in the literature. It has been suggested that this forerun passes through the column in highly associated form because of the 5% methanol added to the eluting solvent. When molecular weights are determined methanol is absent and the aggregates dissociate into lower molecular weight species. This anomalous forerun represents usually less than 5% of the total sample.

Several GPC runs showed some increase in molecular weights in the final few cuts. This probably arises from absorption of part of the sample on the gel surface. Increasing the methanol content of the solvent diminishes this but at the cost of a considerable increase in

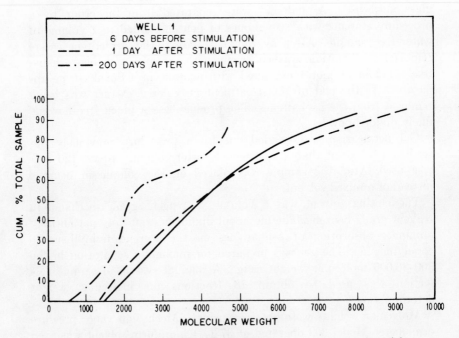

FIG. 1. Molecular weight profile. Cumulative per cent total sample weight vs. molecular weight.

the quantity of low molecular weight forerun. A volume of solvent equal to a complete gel permeation run was passed through the column between samples to minimize cross-contamination due to this tailing effect. Figure 6 shows a typical elution curve with molecular weight plotted against elution volume. Note the low molecular rate forerun in this figure.

RESULTS AND DISCUSSION

The molecular weight distribution of asphaltenes from 5 Kern River wells in this survey are shown in Figs. 1–5. Each Figure shows the profile for the asphaltenes from an oil sample taken before stimulation, 1 day after stimulation, and about 200 days after stimulation. In each Figure the vertical axis is the cumulative weight % of asphaltenes with molecular weight equal to, or less than, a particular value. The horizontal axis is molecular weight. These data will be discussed against a background of production, stimulation, and composition data given in Table 1 (4).

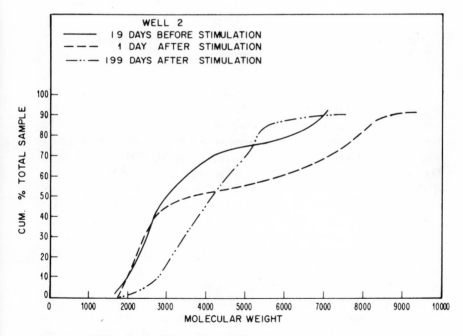

FIG. 2. Molecular weight profile. Cumulative per cent total sample vs. molecular weight.

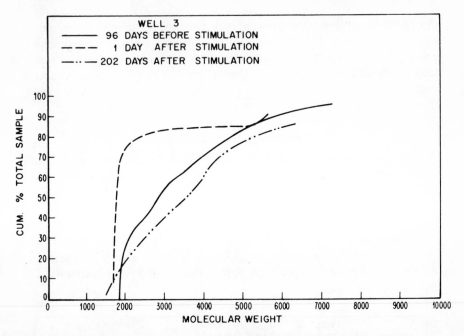

FIG. 3. Molecular weight profile. Cumulative per cent total sample vs. molecular weight.

The mass of data is difficult to assimilate by visual comparison. To help find molecular weight effects that relate to stimulation efficiency, a standard Q-mode correlation analysis was programmed and run by computer. Input information included slopes of the plots of cumulative weight fraction versus molecular weight, taken at 1000 unit intervals. The results of this analysis showed no significant correlation of any of the input variables with the efficiency of stimulation. This does not imply that asphaltenes do not play a role in determining stimulation response, but that molecular weight profiles alone cannot predict successful steam stimulation. An R-mode correlation analysis showed a statistically significant relationship between the slope of the molecular weight profile curve prior to stimulation and immediately after stimulation. No other significant relationships were found.

Note from the molecular weight profile that despite the fact that all 5 wells produced from the same geologic formation, there is a large variation in asphaltene composition in samples taken before stimulation. There is no way to tell whether this is due to differences in crude

TABLE 1

Treatment and Performance Data

Line	Treatment and results	Well				
		1	2	3	4	5
1	Steam (expressed in bbl of H_2O) injected per vertical foot of oil-bearing sands	26.48	70.08	81.51	77.55	17.31
2	Bbl of oil produced in 200 days per bbl of H_2O injected as steam	0.3304	0.3828	0.6152	0.4770	0.5595
3	Bbl of oil produced in 200 days per vertical foot of oil-bearing sands	8.75	26.83	50.15	36.99	9.69
4	Per cent asphaltenes in oil before stimulation	5.3	4.2	4.7	4.3	4.3
5	Per cent asphaltenes in oil 1 day after stimulation	3.8	4.2	4.7	4.5	4.3
6	Per cent asphaltenes in oil 200 days after stimulation	4.8	5.8	5.0	4.8	5.0
7	Short-term change in asphaltene content (Line 5 minus Line 4)	−1.5	0	0	+0.2	0
8	Long-term change in asphaltene content (Line 6 minus Line 4)	−0.5	+1.6	+0.3	+0.5	+0.7
9	Additional oil gained by stimulation, bbl/200 days	−165	6040	9680	4690	1350

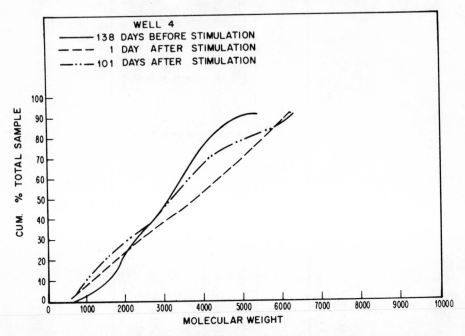

FIG. 4. Molecular weight profile. Cumulative per cent total sample vs. molecular weight.

oil composition as a function of the reservoir region being drained or to natural separation processes, such as absorption and deposition. Since the prestimulation asphaltenes differ widely from well to well, it is necessary to compare changes in composition caused by the stimulation process well by well rather than on an absolute basis. For example, the asphaltene molecular weight profile 199 days after stimulation of Well 2 shows an increased quantity of all components below molecular weight 5200 relative to the prestimulation curve with the same well. Compared with the prestimulation curve for Well 5 however, it would show a much greater increase in quantity of higher molecular weight components across the entire composition range.

Added to the variation in the composition in the prestimulation asphaltene is considerable variation in the stimulation process itself. Line 1 of Table 1 shows that the quantity of steam injected into the wells ranged from 17.31 to 81.51 barrel equivalent of water per vertical foot of oil bearing sand, a factor of 4.7. This variation in quantity of steam injected is included in the steam efficiency calculation, Line 2

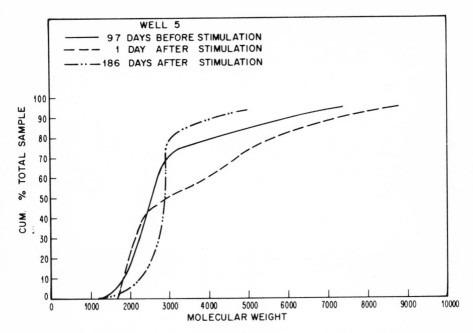

FIG. 5. Molecular weight profile. Cumulative per cent total sample vs. molecular weight.

of Table 1. It is recognized in oil-field technology that the volume of steam injected bears strongly but not rigidly on stimulation response.

Both the molecular weight profile curves and the stimulation efficiency data show clearly that Wells 1 and 3 stand at the extremes of behavior for the 5 wells. In terms of productivity per barrel of injected water, Well 1 is least satisfactory and Well 3 the most. Figure 1 shows that 200 days after stimulation the asphaltenes from this well are unique in the loss of fractions above 4600 molecular weight. This contrasts with the very close correspondence between the curves for prestimulation and 1 day following stimulation asphaltenes. Figure 3 shows the exact opposite behavior for Well 3. The molecular weight profile for asphaltene produced 202 days after stimulation is richer in higher molecular weight species from 1900 up to 6300 molecular weight but is not markedly different from the prestimulation sample. The asphaltenes produced 1 day after stimulation show a unique molecular weight profile, fully 76% of the sample lies at or below molecular weight 2200. In two more fractions amounting to 16% of the

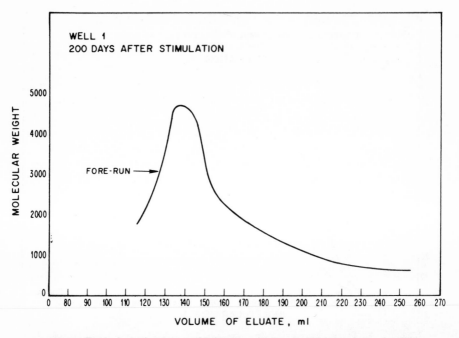

FIG. 6. Typical elution curve. Molecular weight vs. volume of eluate, ml.

sample, the curves rises to 5700, indicating either a selective removal of the intervening molecular weight species from the crude oil before it flowed into the well, the admixture of new production rich in low molecular weight asphaltene, or thermal degradation of heavy components. Wells 2, 4, and 5 represent intermediate cases between Wells 1 and 3 from the standpoint of reproductivity and alteration of asphaltene composition following stimulation.

Well 1, in addition to producing the least oil based on injected steam (Lines 2 and 9 of Table 1), is the only well to show a decrease in asphaltene 200 days after stimulation (Lines 7 and 8, Table 1). From Fig. 1 it will be seen that asphaltene reduction was entirely non-selective 1 day after stimulation but chiefly in the 4700 plus molecular weight range by the end of the 200 day poststimulation period. These results indicate that there is some deposition of asphaltenes from the oil onto the rock surfaces following steam stimulation and that this deposition changes from nonselective to very selective over a 200-day interval. Two important consequences of this process are a reduction of permeability resulting from restriction of flow channels and a change in the distribution in water and oil phases resulting from the development of a new solid phase surface. The quantity of steam injected cannot be the controlling factor in asphaltene composition and content as this variable for Well 1 is bracketed by Wells 5 and 2, neither of which show similar effects (Line 1, Table 1). The mechanism therefore depends on an interaction of steam with locally situated oil, rock, and brine.

For Well 3 the unique shape of the 1 day after stimulation molecular weight profile deserves additional comment. Increased production may be due to the admixture of a second oil stream of different composition. Note that there was no reduction in asphaltenes (Line 7, Table 1). From Fig. 3 it is seen that the 2200–5300 molecular weight range represents 57% of the prestimulation asphaltene sample but only 10% of the sample 1 day after stimulation. Such a reduction in this fraction would be expected to produce a reduction in per cent asphaltene found in the oil. Asphaltene content increased slightly during poststimulation period (Line 8, Table 1). Well 3 received the largest quantity of steam per vertical foot of oil bearing rock; therefore the probability of stimulating a wider distribution of adjacent rock is greatest in this well. A marked change in asphaltene composition soon after stimulation would be a result of the presence of oil from these heretofore unproductive areas, rather than selective precipitation of asphaltenes as suggested for Well 1.

Regarding the possibility of thermal degradation of asphaltenes by the prolonged intensive steam treatment, the results shown do not generally support such a hypothesis. In most cases a slight increase of molecular weights is noted compared with prestimulation values. This is consistent with our concept of asphaltenes stability and structure which favor polymerization rather than degradation. Thermogravimetric studies show that decomposition occurs by cleaving of alkyl groups, generally short ones from an aromatic nucleus, and that this occurs at an appreciable rate only when the temperature approaches 430°C or about twice steam temperature.

CONCLUSIONS

The 15 molecular weight profiles from 5 wells in 1 field show that substantial differences in asphaltene characteristics occur even prior to stimulation. This reinforces the notion of great local heterogeneity of geologic formations, oil brine distribution, and oil composition suggested by almost every detailed study of these topics. Such variation together with a lack of correct information on subsurface processes makes correlation of experiments or analyses very uncertain and limits the elaboration of mechanisms to explain field observations. Nevertheless GPC is of considerable assistance by allowing molecular weight profiles to be constructed for whatever inferences can be drawn within the framework of limitations just described. By confining the discussion to two apparently exceptional examples, a case is made for selective precipitation of asphaltene fractions by steam treating of a crude oil. An equally valid suggestion would be selective absorption of asphaltene by a steam-altered surface. In a second well the most likely explanation of productivity increase in the stimulation of a hitherto unproductive portion of the reservoir. Molecular weight profiles alone cannot predict steam stimulation results, perhaps because in this case the asphaltene content (about 5%) is lower than for many California groups (up to 20%). However, major differences in asphaltene molecular weight profiles during the steam cycle do occur and additional interpretations may become possible when additional subsurface information develops.

REFERENCES

1. For a review of gel permeation chromatography, techniques, and applications, see J. Cazes, *J. Chem. Educ.,* **43**, A567, A625 (1966).

2. C. A. Stout and S. W. Nicksic, "Separation and Gel Permeation Analysis of Natural Emulsion Stabilizers," *J. Soc. Petrol. Eng.,* **8**(3), 253 (1968).

3. Private communication, K. H. Altgelt, Chevron Research Company, Richmond, California.

4. This information was taken from an engineering study of steam stimulation made by Messrs. K. C. Hong and C. P. Coppel of Chevron Oil Field Research Company, La Habra, California.

GPC Separation and Integrated Structural Analysis of Petroleum Heavy Ends*

K. H. ALTGELT and E. HIRSCH

CHEVRON RESEARCH COMPANY
RICHMOND, CALIFORNIA 94802

Summary

Three hundred grams of a Venezuelan asphalt was separated on a 70-liter GPC column in one run. The 30 cuts collected were consolidated to 7 fractions. After separating the insoluble asphaltenes from the soluble maltenes by n-pentane extraction, the maltenes were chromatographed on deactivated alumina into oil and three different resin fractions. Selected fractions were analyzed by NMR, IR, density, and elemental analysis. The experimental results were evaluated by a novel method making use of an extensive scheme of structure relations. Percentages of aromatic, naphthenic, and paraffinic carbon; number of ring systems per molecule, their average size, compactness, composition, and substitution are given explicitly or implicitly.

The structural analysis of crude oil residua, coal, and similar complex organic mixtures has been the target of research for many years. The main difficulties of this endeavor are the great variety of chemical structures and the large range of molecular weights encountered in these materials. Because of this variety, any measurements on the whole mixture will generally give insufficient information. More detailed data are often needed for optimizing processing conditions, for blending specifications, etc. It is then mandatory to separate the mixtures into distinct subgroups and to analyze these.

*Presented at the ACS Symposium on Gel Permeation Chromatography, sponsored by the Division of Petroleum Chemistry at the 159th National Meeting of the American Chemical Society, Houston, Texas, February, 1970.

Ordinary fractionation methods such as distillation, precipitation, extraction, distribution, and adsorption techniques separate by both chemical structure and molecular weight. The only specific separation technique is GPC which segregates by molecular size. The fact that proper GPC conditions are sometimes difficult to meet in practice and that other properties may have an effect on the fractionation is only of secondary importance. At least, *there is* a principle that allows us to separate by one property, molecular size, alone. Once we have achieved a separation by size, we can then proceed to fractionate by chemical nature, selecting from the other techniques the one best suited to a specific material.

It is not necessary to do the size separation by GPC first and then follow it by structural separation. The order of techniques will be determined by several considerations: the range of molecular weights, the types and varieties of chemical structure, and the capacity of the selected separation systems. In complex organic mixtures there are various kinds of interactions that may interfere with separations. Frequently it may be necessary to start with GPC and thereafter sub-fractionate by alumina chromatography; in other mixtures, the reverse sequence may yield better separation. Indeed, the reverse sequence will generally be preferred because an alumina column can separate much greater quantities of material than a GPC column of equal size. In certain applications we may start with a general separation, follow with GPC, and then use alumina or silica gel chromatography.

Proper fractionation is very important; however, it is only part of the task. The other part is an adequate analysis of the fractions. Among the most useful analytical methods for heavy petroleum fractions are elemental analysis, NMR, IR spectroscopy, and molecular weight and density determination. Mass spectroscopy, though a powerful tool for volatile samples, is difficult to interpret on samples containing involatile or unstable components.

In this paper we wish to report a preparative fractionation of an asphalt and a novel evaluation of analytical data (*1*) obtained on some of its fractions.

Experimental

Three hundred grams of a Venezuelan asphalt was dissolved in 1.5 liters of benzene containing 10% methanol and fractionated on a GPC column 10 ft long and 6 in. in diameter. This is a high loading ratio

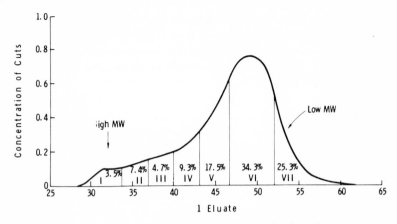

FIG. 1. Elution curve of Venezuelan asphalt obtained on large column.

of 430 mg sample/100 ml column volume (1). The column was packed with a polystyrene gel of 100,000 molecular weight exclusion limit.* A flow rate of 15 liters/hr was maintained. Thirty cuts were collected and reconstituted in such a way as to yield the 7 fractions shown in Fig. 1. After evaporation under a nitrogen stream on a hot plate and later under vacuum at 50°C, the fractions were extracted with 30 times their volume of n-pentane to separate the soluble "maltenes" from the insoluble "asphaltenes."

Ten-gram amounts of the maltenes were further fractionated on 3-ft × 1-in. columns packed with Fisher alumina, A-540, which had been deactivated with 10% water. The samples were eluted with n-pentane followed by cyclohexane, benzene, and, finally, benzene plus 20% methanol. The resulting 30–40 fractions were reconstituted by criteria based on IR, UV, and fluorescence spectra to render an oil and 3 distinct resin fractions. In some minor cases, cuts were left out if they could not be assigned to either the previous or the next fraction because they contained too much from both.

The total fractionation is presented schematically in Table 1, which also includes the number-average molecular weights of pertinent cuts. The molecular weights were obtained by vapor pressure osmometry as described earlier (2).

Selected fractions were subjected to elemental analysis, NMR, IR, molecular weight determination, etc. Heteroatom content, C/H ratio,

* Column design and gel preparation were carried out by Dr. O. L. Harle.

TABLE 1

Fractionation Scheme with Per Cent of Asphalt Sample and Molecular Weights

	Extraction chromatography on alumina plus 10% water				
GPC fraction	n-Pentane			Cyclo-hexane, Resin 2	Benzene plus 10% MeOH, Resin 3
	Asphaltene[a]	Oil	Resin 1		
I	3.5% 25,000	—	—	—	—
II	4.4% 20,000	—	—	—	—
III	4.1% 15,000	←————————————0.6%[b]————————————→			
IV	5.5% 8,000	1.0% 5,500	0.6% 2,500	0.4% —	0.6% 2,000
V	1.0% 5,000	5.7% 740	3.6% 950	1.0% 750	1.6% 800
VI	2.6% 1,900	←———————————13.6%[b]———————————→ ←———————————700———————————→			
VII	2.1% 960	15.0% 550	0.6% 550	1.1% 560	1.4% 650

[a] In each case, the first entry is the wt% of the fraction relative to the total amount of asphalt; the second entry is the number-average molecular weight of that fraction.

[b] These maltene fractions were not subfractionated.

and per cent of functional or structural groups can be plotted versus molecular weight for the different groups, such as oils, resins, and asphaltenes. The plots are not always very enlightening, as demonstrated by the examples given in Figs. 2 and 3. A more detailed structural breakdown is desirable because it provides additional information and also removes much of the scatter of the primary data.

PRINCIPLES OF INTEGRATED STRUCTURAL ANALYSIS

We developed a scheme (3) which combines the results from NMR, IR, elemental analysis, molecular weight, and density determinations to calculate the percentages of aromatic, naphthenic, and paraffinic carbon; number of ring systems per molecule, their size, compactness, composition, and substitution; and the number, length, and branching of paraffinic chains.

The analysis is limited by three approximations:

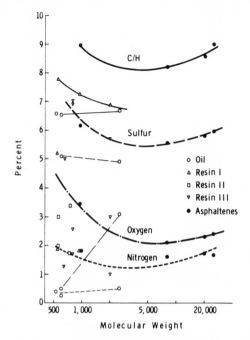

FIG. 2. Per cent O, N, S, and C/H ratio in asphalt fractions as a function of molecular weight.

(a) The calculated results are based on an averaging concept both for fused ring structures within a molecule and for molecules within a sample.

(b) Functionality of heteroatoms is approximated from IR data. Heteroatoms must be replaced by appropriate hydrocarbon groups, and proper adjustments must be made in carbon and hydrogen content as well as in molecular volume. The latter correction is based on relations given by van Krevelen (4).

(c) The model considers only six-membered rings. We cannot rigorously account for five-membered ring systems. The error introduced by their presence in petroleum fractions is generally quite small and can be further reduced by use of a high estimate for a compactness factor (to be defined later).

Despite these approximations, the results obtainable through "integrated structural analysis" are much more detailed than previously thought feasible. Our scheme was made possible by the intro-

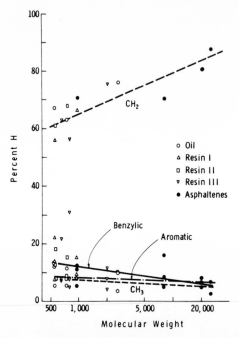

FIG. 3. Per cent H in CH₃, CH₂, aromatic, and benzylic configuration as measured by NMR in asphalt fractions as function of molecular weight.

duction of an improved set of relations between chemical structure and density (5), a complete analytical compilation of relations between structure variables, and by the use of "floating parameters."

The new density relations are based on 455 hydrocarbons spanning a molecular weight range from 70–619.

The inclusion of four "floating parameters" was necessitated by the fact that the number of structural variables considered exceeded the number of independent structural relations by 4. The floating parameters themselves are structural variables whose values must be estimated. They were selected in such a way that their numeric values fall into narrow ranges and that inaccurate assignments do not significantly affect the structural results.

An important consideration in our scheme is the number, size, and shape of fused ring systems. The definition of a "compactness factor" relating linearly actual, maximal, and minimal numbers of peripheral

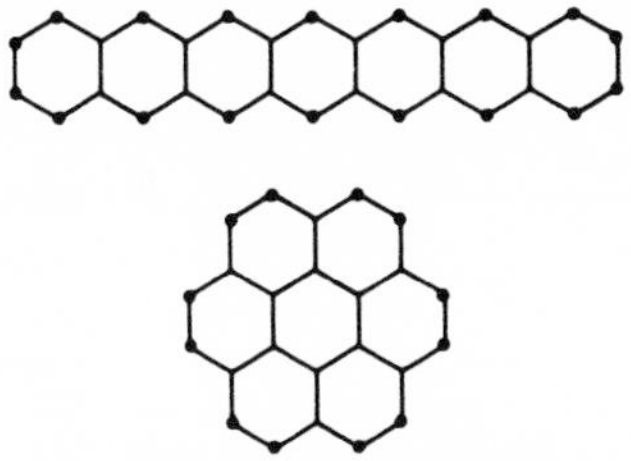

18 Peripheral Carbon Atoms,
$C_p \cdot C_p$ (Max.); $\bullet \cdot 0$

12 Peripheral Carbon Atoms,
$C_p \cdot C_p$ (Min.); $\bullet \cdot 1$

Compactness Factor $\bullet \cdot \dfrac{C_p \text{ (Max.)} - C_p}{C_p \text{ (Max.)} - C_p \text{ (Min.)}}$

FIG. 4. Illustration of the ring compactness factor which holds for any six-membered ring system: aromatic, naphthenic, and mixed.

ring carbons greatly simplified the analysis. Figure 4 illustrates this relation and also shows that open (catacondensed) ring systems have more peripheral ring carbons than compact (pericondensed) ones.

The large number of structure relations were reduced to 3 simultaneous equations in 3 unknowns which can be numerically solved by digital computer using the Newton-Raphson method (5).

Our scheme was tested on a model molecule. Calculated values of structural variables were very close to the actual values provided by the structure. The test also revealed a high sensitivity of the analysis to the accuracy of experimental data. On the other hand, the results were not much affected by improved estimates in the floating parameters.

RESULTS

We applied our method to three fractions of the Venzuelan asphalt, viz., VII Oil, VII Asphaltene, and II Asphaltene. According to the data listed in Table 2, the high and the low molecular weight asphaltenes are distinctly different in their structure. The high molecular weight asphaltenes have larger ring systems with higher compactness in their aromatic and lower compactness in their naphthenic parts, and they have a greater proportion of paraffinic chains. The asphalt oil is different from the asphaltenes only in degree, not in basic nature. It is quite aromatic with smaller, but still considerable, fused ring systems having an average of five rings. Predictably, it is much more naphthenic and has twice as many paraffinic chains per size, though not longer ones than the asphaltenes.

These data present only a brief example of the kind of information

TABLE 2

Comparison of Selected Structural Parameters of an Oil and Two
Asphaltene Fractions

Fraction	Ring systems per molecule	Aromatic rings	Naph- thenic rings	Aromatic carbons	Benzylic carbons	Naph- thenic carbons	Paraf- finic carbons
		Per ring system		Per 100 carbon atoms			
VII O	1.09	2.0	3.1	24.4	9.8	25.4	40.4
VII A	0.78	10.2	6.8	48.1	12.5	11.1	28.3
II A	25	8.0	4.4	37.7	5.2	12.9	44.2
Range (%)	±5	±5	±13	±1.2	±6	±7	±2

that can be obtained by our integrated structural analysis. More
details and various tests to show the accuracy and the limits of the
method are to be published elsewhere (3).

This paper demonstrates how GPC in combination with other
fractionation methods can yield distinct and large enough fractions
for extensive analysis. A novel "integrated structural analysis" applied
to such fractions affords new insight into the structural details of the
heavy ends of petroleum and similar materials.

REFERENCES

1. K. H. Altgelt, Separ. Sci., 5, 777 (1970).
2. K. H. Altgelt, Amer. Chem. Soc. Div. Petrol. Chem. Preprints, 13(3), 37 (September, 1968).
3. E. Hirsch and K. H. Altgelt, Anal. Chem., 42, 1330 (1970).
4. D. W. van Krevelen, Coal, Elsevier, New York, 1961, pp. 316 and 322.
5. E. Hirsch, Anal. Chem., 42, 1326 (1970).
6. H. Margenau and G. M. Murphy, The Mathematics of Physics and Chemistry, Van Nostrand, New York, 1943, p. 475f.

Author Index

Numbers in parentheses are reference numbers and indicate that an author's work is referred to although his name is not cited in the text. Numbers in italics show the page on which the complete reference is listed.

A

Abed, N. K., 354(28), *358*

Abramowitz, M., 255(18), *290*

Ackers, G. K., 120(2), *132*, 138(20), *143*, 195, *201*, 340, 348(9), *350*, 354, 355(33), 356(54), 357(58, 66), *358*, *359*

Adams, E. T., Jr., 340(6), *350*

Adams, H. E., 157(1), *164*

Adler, A. D., 344(13, 14), *350*, 353(25), 356(25), *358*

Aggarwal, S. L., 380(9), *390*

Agrawal, B. B. L., 353(19), *358*

Aitken, A. C., 255, *290*

Albaugh, E. W., 197, *201*, 550(3), *560*, 562(5), *567*, 582, *591*, 594(11), *610*

Alberty, R. A., 502(16), *527*

Aldhouse, S. T. E., 77, *80*, 89(16), 90, *102*, 244(9), *289*

Alexander, W. J., 429(3), 431(3), 433 (8), *453*, 469, *480*

Alfrey, T., 466(4), *480*

Alliet, D. F., 500(13), *527*

Altgelt, K. H., 5(3), *11*, 49, 63, 67(70a), *70*, *72*, 105(1), 112(1), 115(1), *117*, 166(6), *178*, 193(3, 5), 194(5–8), 198(3, 6), 199(5), *201*, 343(12), *350*, 352, 354(30), *357*, *358*, 550(8–10), *560*, 562(1, 2), *567*, 580, *591*, 594(6, 7), *610*, 614(3), *622*, 624(1), 625(1, 2), 626(3), 630(3), *630*

American Society for Testing and Materials, 470, *480*

Anderson, D. M., 354(36), 355(36), *358*

Anderson, L. L., 582(10), 585(10), *591*

Andrews, P., 352(9), 353(24, 27), 355 (24, 27), *357*, *358*

Aris, R., 53, *70*

Armonas, J. E., 76, *80*, 89, *102*, 244(9), 285, *289*

Arrington, M. C., 68, *72*, 131(29), *133*, 137(16), *143*, 166(5), *178*

ArRo Laboratories, 248(16), *290*, 500 (14), *527*

Awad, E. S., 354(28), *358*

B

Bailey, S., 429(5), *453*, 469(17), *480*

Balke, S. T., 83, 98, 99, 100, *101*, 136, *143*, 149(3), *153*, 244, 253(11), 258, 261, 276(11), *290*

Barnes, D. A., 410(2), 412(2), *415*, 392 (4), *405*

Barrall, E. M., II, 29(12), 36, 37, 64, *71*, 166(3), 167(3), 170(3), *178*

Barthel, P., 482(2), *491*

Barton, D. E., 256(24), *290*, 294, *326*

Bartunek, R., 483(5), *491*

Beau, R., 58(38), 64(38, 51, 52), 66(38), 67(38), *71*, 115(9), *117*, 503, *527*

Beecher, J. F., 380(9), *390*

Bender, P., 502(16), *527*

Benoit, H., 6(4), 8, *11*, *12*, 131, *132*, 137, 139(11), *143*, 199(16), *201*, 277, *290*, 496, 503

Berger, H. L., 136(1), *143*

Berger, K. C., 225(2), 230(2), *233*, 241 (5), *241*

Bergland, G. D., 297(10), 300(10), 316 (10), *326*

Bernard, R. A., 59, *71*

Bertoniere, N. R., 456(2), 458(7), *464*

Bethune, J. L., 357(59), *359*

Biesenberger, J. A., 200(17), *201*

Subject Index

A

Alkali crumb aging, 485–490

Anionic polymerization, stereospecific block copolymers, 378

Aspect ratio, 61

Asphalt, fractionation by GPC, 623–630

Asphaltenes, 625 ff.
from steam stimulated oil wells, 611–622

B

Band broadening, *see also* Peak broadening; Spreading, instrumental correction for, 81–101

Band spreading, in GPC of micellar solutions, 334–335

Belt detector, 33

Bio-Beads, 5, 15

Bio-Gel A, 22–23

Bio-Gel P, 22, 352

Bio-Glas, 19, 20, 64
of broad pore size distribution, 165–177

Biopolymers, macromolecular association in, 339–340

Block copolymers, α-methyl styrene–butadiene, 385–386
stereospecific, 377–390
styrene–butadiene, 379–382, 384–385, 387–390, 397, 401–402
vinylbiphenyl–isoprene, 386–387
vinyl toluene–butadiene, 382–384

Broadening, apparent and real distribution, with PMMA samples, 235–241

Butadiene–α-methyl styrene copolymers, block copolymers, stereospecific, 385–386

Butadiene–styrene copolymers, block copolymers, stereospecific, 379–382, 384–385, 387–390

composition, 391–405

graft copolymers, 403–405

Butadiene–vinyl toluene copolymers, block copolymers, stereospecific, 382–384

C

Calibration, 6, 135–143
curves based on polystyrene in THF, 493 ff.
curves for polymers in 2,2,2-trifluoroethanol, 493–527
of instrumental spreading, 149–152
linear curve, for instrument spreading correction, 243–289
nonlinear curve, for instrument spreading correction, 291–326
for polybutadiene (low-mw) with dual detectors, 407–415

Cellulose, alkali crumb aging, 485–490
chain length distribution by GPC, 429 ff.
fractionation, 429 ff., 467–469
steeping, 482–485

Cellulose acetate, chain length distribution by GPC, 429, 449–452

Cellulose nitrate, in viscose process, GPC study, 465 ff.

Chain length, extended, as calibration parameter, 136–137

Colorimeter, in sample detection, 34–35

Compactness factor, 628–629

Computer programs, for data treatment of GPC of high polymers, 152
on-line data recording, 536–547
viscose process analysis, 474–476